An Introduction to Metamorphic Pe

Second Edition

This second edition is fully updated to include new developments in the study of metamorphism as well as enhanced features to facilitate course teaching. It integrates a systematic account of the mineralogical changes accompanying metamorphism of the major rock types with discussion of the conditions and settings in which they formed. The use of textures to understand metamorphic history and links to rock deformation are also explored. Specific chapters are devoted to rates and timescales of metamorphism and to the tectonic settings in which metamorphic belts develop. These provide a strong connection to other parts of the geology curriculum. Key thermodynamic and chemical concepts are introduced through examples which demonstrate their application and relevance. Richly illustrated in colour and featuring end-of-chapter and online exercises, this textbook is a comprehensive introduction to metamorphic rocks and processes for undergraduate students of petrology, and provides a solid basis for more advanced study and research.

Bruce Yardley is Emeritus Professor at the University of Leeds. He previously taught at the Universities of Manchester and East Anglia and has spent sabbaticals at Otago, ETH Zurich and Wisconsin – Madison. He has worked on many aspects of metamorphic petrology and crustal fluid processes, and has served as Chair of the Metamorphic Studies Group (1991–3), Science Secretary of the Geological Society of London (2002–6), President of the European Association of Geochemistry (2005–6) and also President of the Mineralogical Society of Great Britain and Ireland (2019–20). He has held a Harkness Fellowship at the University of Washington (1974–6) and a Humboldt Prize at GFZ Potsdam (2009–11).

Clare Warren is a Senior Lecturer and metamorphic geologist and geochronologist at the Open University, UK, and has worked extensively on metamorphic rocks that form in subduction and continental collision zones. She has published a number of widely-cited papers, and has served as Treasurer and Secretary of the UK Metamorphic Studies Group and on the Mineralogical Society of Great Britain and Ireland Council. In 2020 she was the first recipient of the UK Metamorphic Studies Group's Barrow Award.

An Introduction to Metamorphic Petrology

SECOND EDITION

Bruce Yardley

University of Leeds

Clare Warren

The Open University, Milton Keynes

CAMBRIDGE
UNIVERSITY PRESS

CAMBRIDGE
UNIVERSITY PRESS

University Printing House, Cambridge CB2 8BS, United Kingdom

One Liberty Plaza, 20th Floor, New York, NY 10006, USA

477 Williamstown Road, Port Melbourne, VIC 3207, Australia

314–321, 3rd Floor, Plot 3, Splendor Forum, Jasola District Centre, New Delhi – 110025, India

79 Anson Road, #06–04/06, Singapore 079906

Cambridge University Press is part of the University of Cambridge.

It furthers the University's mission by disseminating knowledge in the pursuit of education, learning, and research at the highest international levels of excellence.

www.cambridge.org
Information on this title: www.cambridge.org/9781108471558
DOI: 10.1017/9781108659550

First published 2021

Printed in Singapore by Markono Print Media Pte Ltd

A catalogue record for this publication is available from the British Library.

ISBN 978-1-108-47155-8 Hardback
ISBN 978-1-108-45648-7 Paperback

Contents

Preface

This book is a completely new edition of *An Introduction to Metamorphic Petrology*, originally published by Longman in 1989. It is designed as a core textbook for second- and third-year undergraduate metamorphic petrology courses, and to support more-advanced teaching. Our aim is to provide the background knowledge and understanding of metamorphic rocks needed by a professional geologist who will not become a petrology specialist, and to give a thorough foundation in the basics of metamorphic petrology for future researchers in the field. We have assumed a basic knowledge of chemistry, physics, maths and mineralogy, and some familiarity with the petrological microscope. Where more-detailed knowledge is necessary, this is provided in the text or in text boxes. There are worked examples for some of the quantitative parts of the course.

The book is designed to be worked through from start to finish, with many of the later chapters building on material presented earlier. Chapters 1–3 provide background to metamorphism and the underlying theory of metamorphic mineral assemblages. Chapters 4–6 describe the metamorphism of the three main protolith types, and Chapters 7 and 8 describe the textures of metamorphic rocks and what they tell us about processes. Chapters 9 and 10 tie metamorphic petrology to the underlying tectonic processes that cause metamorphism and include applications of geochronology to metamorphic rocks. These sets of chapters could be tackled independently if required.

For this edition we have extended the sections dealing with determining the conditions of metamorphism and links between metamorphism and deformation, and added a major new section on dating of metamorphic minerals. Links between metamorphism and tectonics are completely updated. Since the text touches on many interdisciplinary topics, we have given Further Readings at the end of each chapter. The text has been rewritten throughout, making use of many new field examples, and the generous decision by Cambridge University Press to produce the book in colour has allowed us to include many photographs and produce colour figures. We have also provided questions for students at the end of each chapter, and some of these can be readily adapted to match specific materials that the instructor has been using in practical classes. Supplementary material is also available at the website [https://www.cambridge.org/IMP2e].

Bruce Yardley
Clare Warren

Acknowledgements

This book reflects an approach to metamorphic rocks that has been influenced by many mentors, colleagues and students. We owe a debt to everyone who has helped us hone our skills in teaching and research, pointed us in new directions and shown us the errors of our ways. In particular, a number of friends and colleagues have supplied us with their original photographs or published images. Our thanks to Barbara Kunz, Catherine Mottram, Alex Copley, Joe Cann, Geoff Lloyd, Neil Manktelow, Olivier Beyssac, Lucy Campbell, Becky Jamieson, Richard Taylor, Pedro Castiñeras, Jared Butler and Dave Prior. Also thanks to the many colleagues who graciously shared photos and figures that didn't eventually make it into the text.

We are grateful for perceptive and helpful discussions and comments along the way from anonymous reviewers and from Dave Pattison, Barbara Kunz, Thomas Müller, Becky Jamieson and Catherine Mottram. The project was only possible because of the support, understanding and patience of Susan Francis, Melissa Shivers and the rest of the team at Cambridge University Press, who made it possible to deliver the sort of book we wanted.

Last but not least, Clare owes thanks to Felix for putting up with so many lost weekends, while Bruce apologises profusely to Nick for putting her through this book writing nonsense again.

1 The Concept of Metamorphism

A **metamorphic rock** is one that has been changed from its original igneous or sedimentary form: it has grown new minerals in response to new physical or chemical conditions. A wide variety of processes can cause changes to the mineralogical composition of rocks, including heating, burial, deformation, fluid infiltration or shocks caused by meteorites hitting the Earth's surface. Most of these processes, and thus the formation of the vast majority of metamorphic rocks on Earth, take place near tectonic plate margins. As a result, metamorphic rocks provide us with a record of the ambient crustal conditions as rocks get buried, deformed, transformed into new varieties and then transported back up to the surface by a combination of tectonic and surface processes.

Many geologists tend to think that metamorphic rocks should contain large and attractive crystals. However, more objectively, a metamorphic rock can be defined as *any rock in which new mineral grains have grown in response to changed external conditions, so that it now either contains minerals which were not stable in the original sedimentary or igneous environment in which it first formed, or its original minerals have* **recrystallised** *to develop new textures.*

In practice, changes that take place during sediment diagenesis and those associated with ore deposit formation are conventionally excluded from the scope of metamorphism, even though some of these processes are, by this definition, metamorphic processes. The boundaries of the discipline of metamorphic geology are therefore to some extent arbitrary, and coloured by individual geologist's experience and focus.

The scientific understanding of metamorphic processes – and the rates and timescales involved – advanced more slowly than our understanding of igneous and sedimentary processes, perhaps because we cannot usually see metamorphism taking place in 'real time' (Box 1.1). Instead, we are reduced to inferring how and where metamorphism occurs from the rocks now exposed at the surface. However, metamorphic processes are extremely important for providing insight into how the Earth behaves, because most rocks in the Earth's crust have experienced metamorphism of some form or other.

Recent developments in experimental, analytical and computational techniques mean that we can gain a good idea of both the depth and temperature at which a particular metamorphic rock formed, and how long the **metamorphic cycle** of heating, burial, transformation, deformation and transport back to the surface took. In favourable circumstances, we can now determine when different minerals started to crystallise, both absolutely and relative to episodes of deformation, and whether the metamorphism was episodic or relatively continuous through time.

This chapter provides an overview of what metamorphic rocks are, illustrated by some examples of metamorphic rocks in the field, and discusses simple metamorphic changes. The chapter also outlines the factors that control metamorphism and introduces some of the basic terminology that underpins the science. By the end of this chapter you should be able to distinguish a metamorphic rock from a sedimentary or igneous rock, describe some of the mineralogy and textures common in metamorphic rocks and have a basic understanding of the types of tectonic settings in which metamorphic rocks are formed. All of these subjects will be returned to in more detail in subsequent chapters.

BOX 1.1 The development of modern understanding of metamorphism

Modern ideas about how metamorphic rocks form can be traced back at least as far as James Hutton's *Theory of the Earth* published in 1795. Hutton recognised that some of the rocks that formed the Scottish Highlands had originally been sediments and had been changed by the action of heat deep in the Earth. In the early 1800s a distinction was recognised between **contact metamorphism**, where new metamorphic minerals were found in rocks around an igneous intrusion, but generally grew without associated deformation, and **regional metamorphism**, where metamorphic minerals were present in rocks that were also affected by pervasive deformation over large areas (typically tens or hundreds of thousands of square kilometres). In both cases, studies documented variations in the mineral assemblages of metamorphic rocks which were likely to have evolved from the same starting material.

In the latter part of the nineteenth century, three main hypotheses about the cause(s) of metamorphism arose. In Britain, scientists focussed on Hutton's suggestion that *heat* was critical for forcing metamorphic changes. This idea clearly influenced George Barrow's early systematic studies of metamorphism across a region in the Scottish Highlands (Section 1.2.1); he linked the metamorphic changes to the heat from granite intrusions. In Switzerland and Germany, however, scientists more strongly emphasised the role of *pressure and deformation* over that of heat in causing regional metamorphism. A third, predominantly French, line of thought emphasised the role of *fluids*, which were considered to cause significant changes in rock chemistry during metamorphism. We are indebted to French scientists for pioneering the study of trapped bubbles of metamorphic fluids (**fluid inclusions**) in the minerals of metamorphic rocks (Chapter 2, Box 2.2).

Our present understanding of the conditions under which metamorphism takes place can be traced back to the beginning of the twentieth century, and in particular to the work of Victor Goldschmidt. He applied the then new ideas of thermodynamics to the calculation of the conditions at which certain minerals would form by reaction from others (for example, wollastonite forming from reaction between calcite and quartz). At around the same time, the Finnish geologist Pentti Eskola was studying the metamorphic changes evident in south-west Finland. He also applied the principles of chemical equilibrium to the interpretation of the different **mineral assemblages** or groups of minerals that occur together in different metamorphic rocks that he observed (discussed in Chapter 2). Comparison of his results with those of Goldschmidt led him to identify metamorphic mineral assemblages characteristic of particular pressure and temperature regimes, and to correlate assemblages formed under similar conditions in rocks formed from different original **protoliths**.

Eskola's ideas, put forward in the 1920s, were only slowly accepted. It took until the 1950s for his ideas to gain widespread acceptance, after experiments were carried out to determine the pressure and temperature conditions at which different mineral assemblages are stable. They showed that different types of metamorphic rocks could represent different temperature–depth relationships.

In the early 1960s the Japanese geologist Akiho Miyashiro recognised that metamorphic belts of a similar age, but metamorphosed under very different pressure conditions, often occurred next to each other, and coined the term **paired metamorphic belts** to describe this. The Plate Tectonics Revolution of the late 1960s provided a process-oriented framework that explained how these different associations could all be formed. Today, information from metamorphic petrology is central to understanding the tectonic settings in which ancient metamorphic rocks formed (Chapter 10).

Since the 1960s, many metamorphic minerals and mineral assemblages have been synthesised under carefully controlled experimental conditions. Their thermodynamic properties provide additional insights into the conditions of their formation, and huge databases of mineral thermodynamic parameters now underpin modern pressure–temperature calculations (Chapter 3). The work of assembling the data and creating the software to apply it to understanding metamorphic conditions was a major focus in the last part of the twentieth century, involving groups in many countries. Thanks also to vastly improved analytical methods, different software packages now make estimating the pressures and temperatures at which many metamorphic assemblages grew something that can be done without expert knowledge of the underlying thermodynamics (Chapters 2 and 3).

Over the past few decades, the focus of metamorphic geology has progressed from determining the pressure–temperature conditions of formation of metamorphic rocks to the investigation of the rates, timescales, durations and processes of metamorphism. This information defines the origins of the terrane in which the rocks occur in a way which was inconceivable at the time the first edition of this book was published. Major developments in constraining metamorphic processes and conditions have arisen from the development of new technologies for chemical analysis. High-precision techniques for elemental and isotopic analysis of ever-smaller volumes of rocks and minerals have constrained the behaviour of fluids during metamorphism (Chapters 2, 4–6), the relationship between tectonics, deformation and metamorphism (Chapters 8 and 10) and the timescales and rates of metamorphism (Chapter 9).

1.1   Metamorphic Rocks

Metamorphic rocks exhibit new textures and/or new minerals but many also retain some characteristics of their original **protolith**, the original, un-metamorphosed rock, such as its bulk chemical composition or features such as bedding. Examples of rocks that retain some obvious protolith features but have also developed new metamorphic minerals and textures are shown in Figure 1.1. The fact that some of these rocks have been metamorphosed may be difficult to tell at a first glance: the field context of the sample is often crucial. For example, the cross-laminations of the siltstones in Figure 1.1a are still clearly preserved although the original

Figure 1.1 Examples of metamorphic rocks that retain evidence for their original nature. (a) Cross-bedded metasiltstone with coarse staurolite-bearing interbeds of metapelite, Maine, USA. (b) Metamorphosed impure limestone with bivalve fossils, Lukmanier Pass, Switzerland. (c) Metamorphosed basaltic pillow lavas, Southland, New Zealand. (d) Deformed and metamorphosed pillow lavas, Wadi Huulw, Oman.

interbedded clay-rich layers are now pelitic schists with large, metamorphic crystals of staurolite. The marble in Figure 1.1b contains clear cross-sections of bivalve (gryphaea) fossils. This rock could in theory pass as a limestone, however other rocks in the near vicinity have clearly grown new metamorphic minerals. The metabasaltic pillows in Figure 1.1c have retained the distinctive shape that demonstrates eruption of lava into water, but the original near-black minerals have been completely transformed into new blue-green varieties. The original pillow shapes in Figure 1.1d have been flattened during extensive deformation and the rock contains new metamorphic minerals that have a distinctive blue-purple colour.

Although it is not normally possible to see metamorphic changes taking place in the same way that we can watch some igneous or sedimentary rocks being formed on human timescales, there are certain types of relatively rapid metamorphic changes occurring near the surface today. For example, in high-temperature geothermal fields, such as those exploited for power in Iceland, Italy, New Zealand and elsewhere, volcanic glass and minerals that originally formed in high-temperature igneous environments are being actively converted to clays, chlorite, zeolites, epidote and other minerals that are more stable in the cooler, wetter environment of the geothermal system. These metamorphic changes are taking place at depths of only tens to a few hundred metres and at rates that affect the economic exploitation of these systems (Chapter 5, Section 5.6).

1.2   What Do Metamorphic Rocks Look Like?

Before tackling some of the rather abstract concepts of how and why metamorphism takes place, it is worth having some sort of appreciation of what they actually look like in the field. Much modern metamorphic petrology is based on data collected in the laboratory. However, these data are little value unless they can be related back to the textures and field relations of natural rocks. Field studies provide the critical information that underpins the whole subject.

Metamorphic rocks may be found across regions spanning tens or even hundreds of square kilometres or may be very much more localised. Their mineral assemblages may be similar over large areas or may vary over short distances. In many cases, metamorphic rocks show clear evidence of their original rock type, but in detail there will be significant changes. Metamorphosed sedimentary rocks (**metasediments**) have very little porosity and the constituent grains no longer reflect the original sedimentary particles. Metasediments commonly have quite distinct physical properties. Metamorphosed igneous rocks may also retain the gross form of their precursors. However, instead of being made up of interlocking grains crystallised from a melt, they contain metamorphic minerals which may replace specific original crystals or demonstrate complete recrystallisation of the rock. Irrespective of the original rock type, if metamorphism was accompanied by deformation then the metamorphic minerals may be aligned in a tectonic fabric.

1.2.1   The South-East Highlands of Scotland

One of the classic examples of regional metamorphism is in the south-east Highlands of Scotland (Figure 1.2). In the late 1800s, George Barrow showed that there were systematic mineralogical changes in rocks of similar composition across this region and argued that they reflected differences in the temperature of metamorphism.

The rocks of the area were originally Neoproterozoic to Cambrian sediments with basaltic lavas and minor intrusions in parts of the succession. Original quartz-rich sandstones show relatively little mineralogical change across the area, but the **metapelitic rocks** (metamorphosed pelites, or muddy sediments originally composed predominantly of clay minerals) display new metamorphic minerals. All the rocks have fabrics of aligned minerals linked to deformation that took place during the

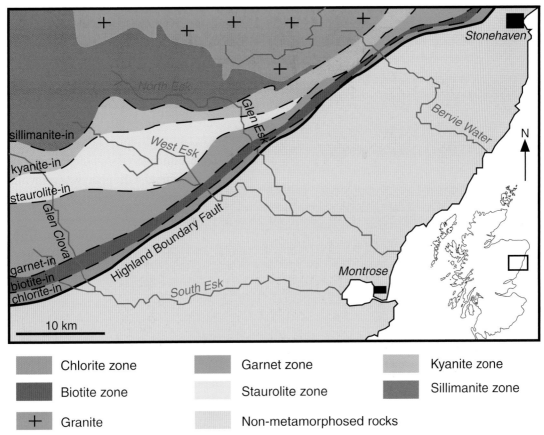

Figure 1.2 Simplified metamorphic map of index mineral zones in the Scottish Highlands, based on Tilley (1925). The Highland Boundary Fault separates older metamorphic rocks to the north and west from younger, unmetamorphosed sediments to the south and east.

metamorphism. The new minerals in the metapelites define a series of **metamorphic zones** (Figure 1.2), each with a characteristic mineralogy which reflects the conditions at which metamorphism took place.

The best-known section through the originally-mapped zones is in the valley of the River North Esk, known as Glen Esk. Immediately north of the Highland Boundary Fault, the exposed rocks are fine-grained and strongly foliated, and contain few mineral grains coarse enough to identify in the field. Under the microscope, however, fine-grained metamorphic *chlorite* can be seen, intergrown with white mica. Quartz and albite (sodic feldspar) are also present together with minerals such as pyrite and tourmaline, which are sometimes coarse enough to see in outcrop.

The next mappable metamorphic zone is marked by the appearance of visible grains of *biotite*. Not all of the rock types in the region are of the correct composition for biotite to grow, but in the metapelites, grains of biotite and muscovite are just large enough to be visible under the hand lens. These mica grains create a 'shimmer' on the surfaces along which the rock splits.

Subsequent metamorphic zones exposed up the valley are defined by the first appearance in metapelites of the metamorphic minerals garnet, staurolite, kyanite and then sillimanite. Each grew in turn in response to increasingly higher temperatures.

The appearance of such **index minerals** often coincides with an increase in grain size, and the index minerals form **porphyroblasts**, or grains that are often much coarser than the matrix grains around them (an example of garnet porphyroblasts is shown in Figure 1.3a). Many porphyroblasts enclose **inclusions** of matrix minerals or minerals that were previously in the matrix at the time that mineral was growing. Porphyroblasts with a very high density of inclusions are termed **poikiloblasts** (Figure 1.3b).

Metamorphic geologists use the relative term **grade**, or **metamorphic grade**, to refer to the **peak** (or maximum) conditions of temperature, or more rarely peak pressure, at which a rock formed. A high-grade rock, for example, has been metamorphosed at higher peak temperatures (and usually higher peak pressures) than a low-grade rock.

Each index mineral persists to higher grades than the zone that it characterises. Whether or not a particular index mineral has developed in any rock depends on the rock composition *and* the metamorphic conditions. For example, biotite and garnet are present in a wide range of schists but staurolite only appears in metapelites whose composition is specifically rich in aluminium and poor in calcium. Each rock contains many other minerals, for example muscovite, plagioclase and quartz, which co-exist with the index minerals (Figure 1.3). These other minerals persist

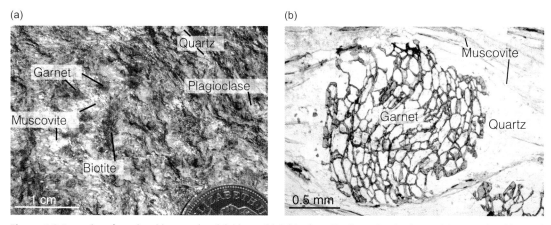

Figure 1.3 Examples of porphyroblasts and poikiloblasts. (a) Schist from Tyndrum, Scotland containing porphyroblasts of pink garnet, white crystals of plagioclase feldspar, greyish quartz, brown biotite and silvery muscovite. (b) Thin-section photomicrograph of a garnet poikiloblast in a muscovite schist (As Sifah, Oman). Note that this photomicrograph was taken in plane polarised light (PPL); unless otherwise specified, all photomicrographs in the book are taken in PPL.

across multiple zones and, despite not being useful as index minerals, they are still an important part of the mineral assemblage because they participate in the reactions by which the index minerals are created or destroyed. All of these concepts will be returned to in greater detail in Chapter 4.

The metamorphic zones in the south-east Highlands were originally defined by plotting the location of the different index minerals on a map and drawing lines through the first appearance of each index mineral in the direction of increasing grade. Index minerals may grow at somewhat different temperatures in layers of slightly different composition, and so the 'first appearance' method of drawing the boundaries served to smooth out the effects of random variation in rock composition. It means, for example, that some metasediments in the garnet zone will not in fact contain garnet and will be indistinguishable from biotite zone rocks (Figure 1.4). The zone boundaries are called **isograds**, meaning lines of constant metamorphic grade. Because isograds based on mineral appearance cannot be drawn accurately in areas where rock compositions vary, it is now customary to try to define isograds more rigorously by basing them on full mineral assemblages, rather than on the appearance of individual index materials.

It is generally assumed that higher-grade metamorphic rocks formerly contained the mineral assemblages of lower-grade zones, and that they progressively recrystallised and/or changed their mineralogy as metamorphism proceeded, rather than being converted directly from, say, unmetamorphosed sediment to high-grade schist. Evidence for this concept of **progressive metamorphism** includes the

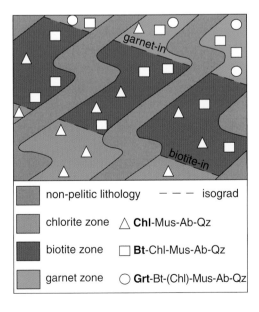

Figure 1.4 Schematic map illustrating the distribution of some index minerals and their associated typical mineral assemblages in a regionally metamorphosed metasedimentary sequence. Also shown are the isograds for the first appearance of the index minerals drawn on the basis of the assemblage information. Mineral abbreviations are detailed in Appendix 1.

preservation of minerals typical of lower-grade zones as **inclusions** inside larger grains of other minerals. These inclusions can persist into higher-grade zones where they would not otherwise be present. Similarly, zoned minerals, whose composition changed as they grew in response to changing pressure and temperature, preserve evidence of a history of metamorphic recrystallisation. These topics are covered in more detail in subsequent chapters.

1.2.2   Bugaboo Aureole, British Columbia, Canada

Another common type of metamorphism is developed in the **country rocks** around igneous intrusions in response to the heat given off as the magma crystallises and cools. In this situation, a **metamorphic** or **contact aureole** develops outwards from the source of heat.

In British Columbia, Canada, a granite–granodiorite magma intruded a Neoproterozoic turbidite sequence during the Late Cretaceous. The country rocks were already metamorphosed to chlorite-grade before the intrusion caused further, more-localised, metamorphism. Figure 1.5 shows the metamorphic zones and isograds that developed in the ~1 km wide metamorphic aureole around the intrusion.

You will notice that different metamorphic minerals formed in the aureole around the Bugaboo intrusion from those formed during regional metamorphism in the

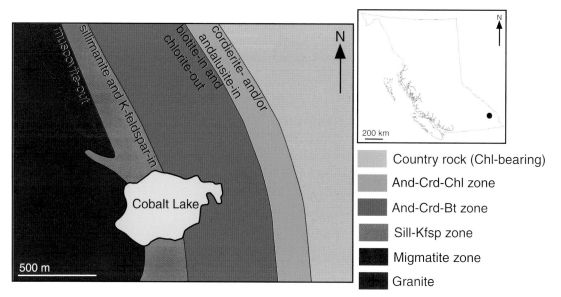

Figure 1.5 Simplified geological map of the metamorphic zones surrounding the Bugaboo batholith in British Columbia, Canada (shown in the inset). The colours of the zones are related to temperature and match those in Figure 1.2. Modified from Pattison & DeBuhr (2015).

south-east Highlands of Scotland. Cordierite and andalusite occur in the aureole, whereas garnet, staurolite and kyanite are absent. This suggests either that the physical conditions of metamorphism were different in the two areas or that the bulk rock chemistry of the protolith was different, or both. Different minerals provide us with different clues.

Multiple zones are present in both areas, suggesting the rocks in both formed over a range of temperatures. Since sillimanite is present in the high-grade rocks in both areas, it is reasonable to assume that the temperature ranges of metamorphism overlap. The mineralogical differences could, however, reflect different depths – and hence pressures – of metamorphism. We can investigate this by looking at the different Al_2SiO_5 **polymorphs** (minerals with the same chemistry but different crystal structure) that are found in the two regions. In the Scottish example, kyanite is succeeded at higher grade by sillimanite, whereas in the Canadian example, it is andalusite that is succeeded by sillimanite. The density of kyanite is 3.6 g/cm^3 whereas that of andalusite is only 3.15 g/cm^3. The effect of increased pressure, in general, is to favour the formation of denser minerals (further explored in Chapter 2, Section 2.3). Since both andalusite and kyanite have the same chemical composition and appear to form over a similar temperature range, we can infer that the contact metamorphism around the Bugaboo intrusion took place at lower pressure (and hence nearer the surface) than the regional metamorphism in the Scottish Highlands. Differences in mineral assemblages formed in sediments during their metamorphism at different pressures are explored further in Chapter 4.

1.3   Metamorphic Changes

The two examples just outlined show us that rocks undergo two principal types of readily observable change during metamorphism:

- **mineral changes,** which involve replacement of original minerals by new metamorphic minerals due to chemical reaction, and
- **textural changes,** which involve recrystallisation of minerals to produce new textures such as the alignment of platy minerals in slate or the progressive coarsening of limestone to marble.

These types of changes may occur together or take place more or less independently, according to the cause of metamorphism and the type of rock involved. They can occur during **prograde metamorphism,** as the rock is heated, or during **retrograde metamorphism,** as the rock is cooling (and generally also being transported back to the surface).

It is important to note here that two words are sometimes used interchangeably for the process of bringing rocks back up to the surface: uplift and exhumation. However, the two have different meanings: **uplift** involves the movement of rocks upwards compared to a fixed datum, e.g. the centre of the Earth or mean sea level. Uplift may happen during crustal thickening and mountain building but will not change the pressure acting on rocks within the uplifted crust. **Exhumation** on the other hand describes a number of different processes by which rocks are brought from depth towards the Earth's surface; this involves removal of overlying material and results in a decrease in pressure. Uplift accompanied by erosion is therefore also exhumation. This topic is covered further in Chapter 9.

One important feature of all metamorphic changes is that they mostly take place in the solid state. In other words, the rock is never completely disaggregated, as would be the case if extensive melting occurred. **Partial melting** (also known as **anatexis**) occurs in many metamorphic rocks at high temperatures, and **migmatites** – rocks that show evidence for partial melting – are considered metamorphic rocks (Chapter 4, Section 4.3). Features such as original compositional layering of sediments are retained through metamorphism (Figure 1.1a), even if distorted and disrupted by deformation. Whereas melts develop only at very high metamorphic grades, small amounts of other fluids are nearly always present in the pore spaces of rocks during metamorphic mineral growth; these fluids play a very important role in facilitating metamorphic changes (Section 1.6.3, Chapters 2, 7 and 8).

Many studies have shown that metamorphic rocks have very similar chemical compositions to their sedimentary or igneous protoliths, except for the removal or addition of volatile species (usually dominated by water, H_2O). This assumption of **isochemical** metamorphism, or unchanging chemical composition during metamorphism, is implicit in much of the development of our understanding of mineral reactions in metamorphic rocks.

Sometimes the chemical composition of a rock *does* change more significantly during recrystallisation. The process of chemical change during metamorphism is known as **metasomatism** and the resulting metamorphic products are known as **metasomatic rocks**. Examples of these are discussed in Chapters 5 and 6.

Rocks are made up of minerals, and the range of minerals present is a reflection of rock composition. Most common metamorphic minerals are composed of the **major elements** (those that have concentrations >1% in the Earth's crust): Si, Al, Ti, Fe, Mg, Ca, Na, K, C and H. These bond with O to form silicate and oxide crystal structures. Common rock-forming minerals such as garnet or feldspars can also take in specific **minor elements**, which have crustal concentrations between 1% and 0.1%, for example Mn, and **trace elements** with crustal concentrations of <0.1%, for example Rb, Sr, U, Th or Sm.

Most metamorphic rocks also contain small amounts of **accessory minerals**, which contain minor and trace elements not readily incorporated into the crystal structures of the main rock-forming minerals. Instead, these elements form minerals in which they dominate, for example apatite and zircon are accessory minerals that host P and Zr respectively. The substitution of elements into mineral structures may be very useful to geologists: for example the substitution of Rb for K in micas allows them to be dated by Rb–Sr geochronology (Chapter 9, Section 9.2) and the temperature-dependent substitution of Ti for Si in biotite or Zr in zircon allows crystallisation temperatures to be estimated (Chapter 3, Section 3.3.8).

1.4　Factors That Control Metamorphism

As we have already seen, metamorphism takes place when a rock is subjected to a new physical or chemical environment in which its existing mineral assemblage is no longer the most stable. Changes in the **physical environment** could mean a change in the temperature or pressure to which the rock is subjected (Box 1.2). A new **chemical environment** may mean the infiltration of a fluid that reacts with the rock, or simply interactions between two adjacent rock types which are no longer compatible under the conditions of metamorphism.

The degree to which a rock's mineral assemblage changes as the chemical and physical environment vary depends on the chemical composition of the protolith. A rock in which a wide range of the major elements are abundant, such as an original pelitic sediment or a lava flow, will likely undergo more mineralogical changes than a rock that is chemically simple (such as original limestone or quartz sand). This concept will be explored further in Chapters 2 and 3.

1.4.1　Temperature

Temperature is a measure of how hot something is (Box 1.2). Heat flows from bodies at higher temperatures to bodies at lower temperatures until the temperature difference is eliminated. The Earth has always slowly lost heat, but measurements of the amount of heat escaping from the Earth's surface in different places today show significant variability. Leaving aside the consequences of shallow groundwater flow, these variations reflect three major contributions to surface heat flow:

(a) the amount of heat flowing into the base of the crust from the mantle (higher in areas of mantle upwelling such as plumes);

(b) heat generated by radioactive decay within the crust (greater for continental crust than oceanic crust because of higher concentrations of radioactive elements such as uranium and thorium); and

(c) heat brought into the crust by rising bodies of magma.

BOX 1.2 Temperature and pressure units

The Celsius temperature scale, °C, is normally used for descriptive purposes in geosciences, but it is important to use the absolute or Kelvin temperature scale (on which water boils at 373.15 K, and freezes at 273.15 K), when dealing with thermodynamic properties of minerals or performing thermodynamic calculations. For all practical purposes, T K $= T$ °C $+ 273$.

The unit of pressure traditionally used in geology is the bar or kilobar (kbar). The SI unit of pressure, the pascal (in units of $kg/m.s^2$), is now becoming increasingly common. Fortunately, the conversion is simple: 1 bar $= 10^5$ Pa, 1 kbar = 100 MPa (megapascal or million Pa) or 0.1 GPa (gigapascal or billion Pa). In this book, we use the SI unit system. Pressure–temperature diagrams are generally labelled with pressure units of 100s MPa or GPa for easy comparison with older diagrams that have their axes labelled in kbar.

In addition, rapid exhumation of deeply-buried rocks in young mountain chains such as the Himalaya means that hot rocks are brought to the surface quickly, without much loss of heat. Crustal stretching in areas of rifting produces a similar effect by bringing hotter lower crustal and mantle rocks nearer to the surface. Finally, surface temperatures can also be affected by absorption of heat by thick accumulations of cold sediment.

Surface heat flow is a reflection of how rapidly temperature increases with depth: the higher the surface heat flow, the hotter the rocks at the same depth below the surface. The local rate of change of temperature with depth at any point in the Earth is known as the **geothermal gradient**, and, due to the variations in heat input discussed above, crustal geothermal gradients can vary considerably at different places. They are usually in the range 10–30 °C/km but extremes of 5>60 °C/km do occur. The range gets smaller as depths increase.

Because of the range of factors which contribute to heat flow, crustal geothermal gradients vary somewhat with depth. In general they tend to be steeper in the upper part of the continental crust than in the lower part because of the heat generated by radioactivity in crustal rocks. Below the lithosphere, convection causes a rather uniform asthenospheric geotherm over considerable depths. A **geotherm** shows the way in which temperature varies with depth below a particular part of the Earth's surface and is made up of elements with different geothermal gradients. The thermal structure of the major tectonic settings discussed in this book is shown in Figure 1.6.

The varying geothermal gradients that characterise different tectonic settings help us to interpret the settings in which different metamorphic rocks have formed. This concept is further explored in Chapter 10.

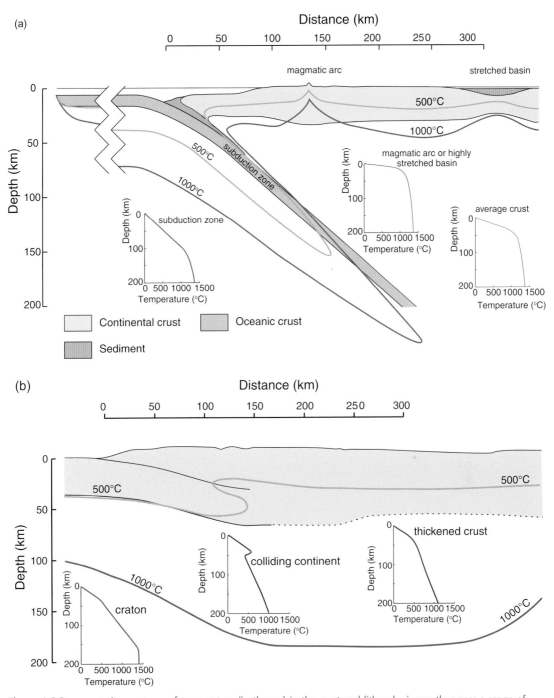

Figure 1.6 Representative contours of temperature (isotherms) in the crust and lithospheric mantle across a range of tectonic settings. Inset diagrams are representative geotherms. (a) Mid-ocean ridge, subduction zone, magmatic arc and a stretched basin. Compiled from England & Katz (2010); Richards (2011); van Keken *et al.* (2008). (b) Continental collision zone, based on models for the Himalaya, from Priestley *et al.* (2008); Wang *et al.* (2013). Thanks also to Greg Houseman.

1.4.2 Pressure

Pressure is a measure of the force per unit area to which a rock is subjected (Box 1.2). To a first approximation, metamorphic petrologists assume that it acts equally in all directions. Under this assumption, the pressure at any point in the crust is simply due to the weight of overlying rocks and is known as the **lithostatic pressure**, P_l:

$$P_l = \rho g h, \qquad\qquad\qquad [1.1]$$

where ρ is the mean density of overlying rock, h is depth and g is the acceleration due to gravity (9.8 m/s^2). An example calculation is provided in Box 1.3.

BOX 1.3 **Example calculation: Linking pressure and depth**

A metamorphic mineral assemblage crystallises at a pressure of 900 MPa. Assuming an upper continental crustal density of 2550 kg/m^3 and that the total pressure experienced by the rock was solely due to lithostatic pressure, at what depth did this assemblage form to the nearest kilometre?

Remember that 1 MPa = 10^6 Pa, and that the units of Pa are kg/m.s^2. Remember to always check that you are using the same units throughout the calculation (e.g. kg or g, m or km, a or Ma) to avoid mistakes.

From Equation [1.1], $P_l = \rho g h$.
Therefore $h = P_l / \rho g$;
$h = 900\,000\,000/(2550 \cdot 9.8)$
$h = 36\,014\,$m
$h = 36\,$km.

The multiple forces acting on any surface in a rock mass can be referred to in general as **stress**. There are two types of stress. **Normal stress**, represented by σ, acts perpendicular to the surface (e.g. the faces of the cube in Figure 1.7). **Shear stress**, τ, is a distorting force that arises if the normal stresses acting in different directions are unequal.

The state of stress in a rock mass is defined by the principal normal stresses, commonly called σ_A, σ_B and σ_C (Figure 1.7). These are mutually perpendicular and comprise the greatest compressive stress, the least compressive stress, and the intermediate compressive stress. In the Earth, one of the principal normal stresses is almost always vertical and corresponds to the lithostatic pressure defined in

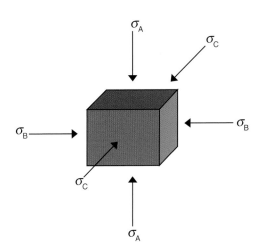

Figure 1.7 Schematic diagram showing the normal stresses acting on a rock. To a first approximation, lithostatic pressure $P_l = \sigma_A = \sigma_B = \sigma_C$.

Equation [1.1]. Where rocks are weak, the differences between the normal stresses are small and the mean stress

$$\sigma_{\mathrm{mean}} = (\sigma_A + \sigma_B + \sigma_C)/3 \qquad\qquad [1.2]$$

is very close to lithostatic pressure. The difference between the normal stress in any direction and the mean stress is known as the **deviatoric stress**.

Many metamorphic rocks are deformed – an observation that suggests the rock experienced non-uniform stresses at some point in its history. Changes in the shape of a rock mass in response to stress differences are known as **strain** and are described in detail in structural geology textbooks. Deformation requires deviatoric stress and can take place irrespective of whether the mean stress is large or small. High pressure does not itself cause distortion: there are many examples of rocks that have been metamorphosed at very high pressures without being obviously distorted or deformed. For example, Figure 1.8 shows a metamorphosed gabbro that was not deformed despite being buried to depths of >80 km during the Alpine orogeny and undergoing metamorphic reactions. The magmatic textures of the original plagioclase and augite are still preserved, though the original minerals have been **pseudomorphed**, or replaced, by new metamorphic minerals. The pyroxene is now omphacite, while the plagioclase has been replaced by intergrown albite, kyanite and zoisite.

As an illustration (feel free to try this experiment at home), hold a rectangular block of (soft) ice cream briefly at the bottom of a bucket of cold water. It will not be distorted, because the increased pressure due to the weight of overlying water acts equally in all directions. Next, place the ice cream block on a bench and place the bucket of water on top of it; the ice cream will be steadily

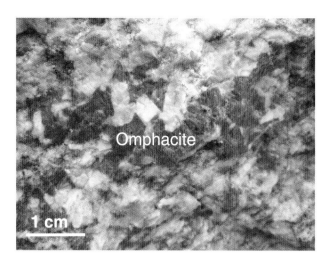

Omphacite

1 cm

Figure 1.8 A metamorphosed gabbro from the Saas Valley, Switzerland, retains the original ophitic igneous texture but the minerals have been transformed. Locality courtesy of Kurt Bucher.

squashed flat. In this case the pressure acting on the sides of the block is just atmospheric pressure, but the pressure acting vertically is atmospheric pressure plus the pressure due to the weight of the bucket of water. Since the vertical stress is greater than the lateral stresses, and ice cream is quite weak, the ice cream deforms.

Rock deformation is evidence that, at some stage, the normal stresses acting on a rock mass in different directions were unequal, i.e. the rock was subjected to deviatoric stress. The differences in normal stresses that can be sustained by any rock are limited by its strength. Experimental studies have shown that under most prograde metamorphic conditions, rocks are fairly weak and can only sustain normal stress differences of a few MPa. This stress difference is usually very small compared to lithostatic pressures of (typically) several hundreds of MPa. This is why metamorphic geologists generally assume that pressure in metamorphism is equivalent to lithostatic pressure so that the pressures at which mineral assemblages are found to be stable can be equated to depth of metamorphism. As a rough guide, the pressure exerted by a column of rock 10 km high is in the range 260–320 MPa according to rock density (and thus composition). The deformation resulting from deviatoric stress plays a major role in developing the textural characteristics of metamorphic rocks (discussed further in Chapter 8), but rarely influences the stable mineral assemblage except in cases where the deformation catalyses reactions or permits movement of fluids.

1.4.3 Fluids

Fluids in pores and grain boundaries play an integral part in metamorphism. Hydrous minerals, such as micas and amphiboles, are present in many metamorphic rocks formed at high temperatures, and so H_2O must have been present in the rock during their formation. Furthermore, since volatiles such as H_2O or CO_2 are often released from minerals as metamorphic reactions proceed, they must be present as

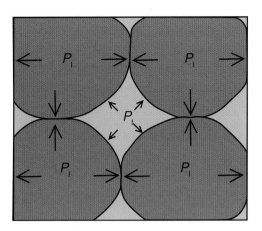

Figure 1.9 Interaction between the pressure of fluid in rock pores, P_f, and lithostatic pressure, P_l, for a rock in which $\sigma_A = \sigma_B = \sigma_C$.

fluid in the rock, even if this fluid is largely expelled. Samples of the metamorphic fluids that were present during metamorphism are sometimes preserved as fluid inclusions in mineral grains and are discussed further in Chapter 2, Box 2.2.

Fluid pressure is the pressure exerted by fluid present in pore spaces and along grain boundaries (Figure 1.9), and is a very important component of the total pressure that rocks experience. The lithostatic pressure, P_l, holds the grains or minerals in a rock together. The pore fluid pressure, P_f, acts to force the grains apart. If the effective pressure, P_e (which is $P_l - P_f$) becomes negative, then the fluid will tend to push the grains apart, increasing porosity, and the rock will eventually crack.

Within a few kilometres of the surface, the fluid pressure is usually close to the weight of an overlying column of water, and is known as the **hydrostatic pressure**. This is much lower than lithostatic pressure because water is much less dense than rock. In deep basins, however, sediments become buried sufficiently deeply that fluid can no longer be squeezed out and pass to the surface along pores or cracks. As a result, elevated fluid pressures, which may approach lithostatic pressure, are sometimes generated and are known from deep hydrocarbon basins. These elevated fluid pressures can only arise if the host rock is itself impermeable, or if it is beneath an impermeable cap. As rocks are heated and buried further and become meta-morphosed, the continued compaction, the release of mineralogically-bound water during metamorphic reactions, and the thermal expansion of existing fluid, can all drive up fluid pressures so that they approach lithostatic pressure. It is also likely that recrystallisation makes the metamorphic rock itself impermeable. As a result, we normally assume that fluid pressure during prograde metamorphism (i.e. while reactions are taking place that release volatiles) is approximately equal to lithostatic pressure: $P_f = P_l$. If fluid pressure exceeds the lithostatic pressure by more than the tensile strength of the rock (which is usually small), the rock is likely to crack by

hydraulic fracturing allowing fluid to escape along cracks during the process; some veins may mark the site of such fractures. In marked contrast, as metamorphic rocks cool, the fluid pressure is likely to drop to very low values because the reactions that take place during cooling absorb water rather than releasing it (Chapter 8).

As fluids are so important during metamorphism, understanding something of their behaviour at metamorphic pressures and temperatures is essential. At atmospheric pressure, liquid water undergoes a phase change to a gas (steam) by boiling at 100 °C, and this change involves a large volume increase. The boiling point of water is raised by increased pressure, as in a domestic pressure cooker, and the relationship between the pressure and temperature of boiling water defines the boiling curve (red line on Figure 1.10). The density difference between boiling water and steam gradually decreases along the boiling curve until the molar volume of the liquid and the molar volume of the gas are the same. This condition defines the end of the boiling curve and is known as the **critical point**. For water, it is located at approximately 22 MPa and 375 °C (red dot on Figure 1.10). At pressures greater than 22 MPa, known as the **critical pressure**, water cannot be boiled however hot it becomes. Instead, water that is being heated will expand steadily (and therefore also decrease in density), but not boil. Similarly, at temperatures greater than 375 °C

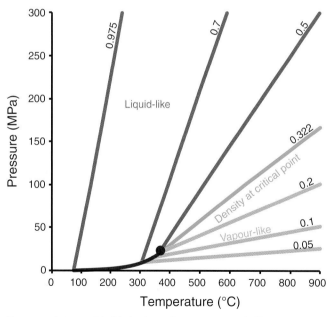

Figure 1.10 Supercritical behaviour of pure water at shallow metamorphic pressure and temperature conditions. The boiling curve and critical point are in red, with the blue, green and yellow lines representative lines of equal volume (isochores). The density of water in g/cm^3 is labelled on each isochore. The green isochore is for the density of water at the critical point, with the blue lines showing 'liquid-like' water and the yellow lines showing 'vapour-like' water. The data are from Haar (1984).

(the **critical temperature**), steam will compact as pressure is increased but will not condense to a separate liquid phase, no matter how much it is compressed. Water at pressures or temperatures greater than the critical point is referred to as a **supercritical fluid** rather than a liquid or a gas. For most metamorphic conditions, supercritical water is similar to liquid water at the surface in terms of its density and related chemical properties (such as its ability to dissociate salts). However, hot supercritical water is less viscous than liquid water: it would slop out of a bucket very easily if you could carry it around under metamorphic conditions.

To illustrate the significance of water density, consider the effect of intruding magma into water-bearing sedimentary rocks. At shallow depths, where the fluid pressure is less than 22 MPa, any water in the rock will boil. The significant change in volume (by a factor of 5 or more) as it turns into a gas may even result in the rock fracturing and bursting apart. Even if the pores and cracks remain filled with steam, the actual mass of water remaining in the rock will be so small that it will effectively have been 'boiled dry'. Water will not be available to facilitate metamorphic reactions.

In contrast, at greater depths, where the fluid pressure exceeds 22 MPa, heating the water to near-magmatic temperatures will not cause boiling. A significant mass of supercritical water may remain in the pore space of the rock even at very high temperatures. This water is available to promote recrystallisation and the growth of metamorphic minerals.

1.5   Types and Settings of Metamorphism

Metamorphism of the Earth's crust generally takes place in zones of active tectonics, such as along plate boundaries. Each different tectonic setting develops distinctive pressure and temperature conditions and these metamorphic conditions in turn produce diagnostic types of metamorphic rock. We will return to the link between tectonics and metamorphism in much more detail in Chapter 10, but there are a few broad terms that refer to different types of metamorphism that are useful to know before delving into more detail.

1.5.1   Regional Metamorphism

Regional metamorphism, which was introduced earlier in this chapter with the Scottish Highlands example, is the overall term used to classify the large areas of metamorphosed rocks that are characteristic of many mountain belts, especially those formed during continent–continent or arc–continent collision such as the Caledonides, Appalachians and Himalayas. Regional metamorphism is almost invariably accompanied by deformation and folding, and regional metamorphic rocks typically have planar textures or **fabrics** that result from the deformation.

For example, mudstones develop a cleaved slaty fabric as they are compressed, and later develop schistosity due to the planar alignment of larger platy minerals (micas) in the rock with further metamorphism. Metamorphic grade can vary from very low (rocks that retain many original sedimentary or igneous features) to very high (rocks which have experienced melting). The source of heat that leads to regional metamorphism is widespread and diffuse, and may include heat derived from radioactivity within the crust and heat from the mantle. Sometimes, there is a general association with igneous intrusions that may have contributed to the overall regional rise in temperature. Typical rock types formed during regional metamorphism include **slates, phyllites, schists** and **gneisses** (defined in detail in Section 1.6). Regional metamorphism also encompasses two more specialised types of metamorphism.

Burial metamorphism is a form of low-grade regional metamorphism which affects thick volcano-sedimentary successions and is characterised by the growth of hydrous minerals (Chapter 5, Section 5.3) without any evidence of pervasive deformation (such as cleavage). The distinctive mineral assemblages of burial metamorphism are developed from volcanic glass and other reactive high-temperature igneous material. Under the same conditions, siliciclastic sediments may undergo recrystallisation of their original sedimentary minerals, but these changes are usually classed as diagenesis.

High-pressure metamorphism is a specific type of regional metamorphism with distinctive mineral assemblages that are stable only at high pressures but relatively low temperatures. These rocks contain unusual variants of common minerals such as the sodic pyroxene omphacite and the sodic amphibole glaucophane. In extreme cases, metamorphic diamond is known, implying burial to about 120 km depth (Chapter 5, Section 5.5).

1.5.2   Contact Metamorphism

Contact metamorphism, which we introduced earlier in this chapter with the Canadian example, is the metamorphism that results from an increase in temperature in the country rocks that immediately surround an igneous intrusion. The rocks thus affected form a metamorphic aureole, whose metamorphic grade decreases away from the contact with the intrusion. Typically, aureole rocks are not deformed during metamorphic recrystallisation, and form tough rocks with randomly oriented interlocking crystals called **hornfels** (Section 1.6). The country rocks may have previously undergone regional metamorphism, for example contact aureoles around granites may develop in slates which previously experienced low-grade regional metamorphism. These **polymetamorphic** rocks may retain vestiges of the earlier regional planar fabrics, for example slaty cleavage. In cases where some deformation did accompany igneous intrusion, the contact metamorphic rocks may be schistose and look very similar in hand specimen to regional metamorphic rocks. The products of contact metamorphism are very varied; those most commonly encountered formed around granite plutons in the upper or middle crust. However, in certain local settings, such as where volcanic rocks bake sediment very close to the surface, distinctive and unusual rocks and minerals may result.

1.5.3 Dynamic or Cataclastic Metamorphism

Dynamic metamorphism takes place along fault planes or shear zones as a result of the intense deformation of rock in the immediate zone of movement. Often the mechanical deformation is accompanied by recrystallisation or by growth of new minerals due to movement of fluid into the zone of deformation. The rocks formed during this type of metamorphism are typically fine-grained and highly-deformed **mylonites**, but, in the most extreme cases, melting can take place leading to the formation of **pseudotachylytes** (Section 1.6.3). Both are discussed further in Chapter 8.

1.5.4 Hydrothermal Metamorphism

Hydrothermal metamorphism is the result of the circulation of hot water through a body of rock along fissures and cracks and involves metasomatism as an integral part of the process. This sort of metamorphism is often associated with igneous activity because steep temperature gradients such as those present around shallow intrusions are required to drive fluid convection. Hydrothermal activity is important in geothermal fields and is responsible for the formation of associated mineral deposits. Contact metamorphism may also develop in the same general area.

   The most widespread type of hydrothermal metamorphism is **sea-floor metamorphism**, which takes place at mid-ocean ridges (Chapter 5). Basaltic rocks dredged from the ocean floor are commonly altered: for example, plagioclase has been replaced by albite, and olivine and glass have been replaced by minerals such as chlorite and epidote. Active convection cells that circulate seawater through young oceanic crust have altered the basalt chemical composition.

1.5.5 Shock Metamorphism

Shock (or impact) metamorphism is associated with meteorite impacts and is the result of near-instantaneous passage of shock waves through the underlying rock. Unlike other types of metamorphism, it is unrelated to terrestrial tectonics. The metamorphic changes range through formation of conical fracture patterns (known as shatter cones) in rocks, the development of deformation bands within crystals and the formation of high-pressure minerals including coesite and stishovite (high pressure forms of quartz). As the stress relaxes after the shock wave has passed, brief rises in temperature can produce extremely heterogeneous melts or even vaporise the rock completely. Impact craters are quite rapidly eroded, buried or destroyed by plate tectonic processes on Earth, but shocked minerals in the ash-fall sediments can provide evidence of such events in the geological past. In less tectonically-active parts of the solar system, such as the Moon, impact metamorphism is the dominant geological process.

1.6   Naming Metamorphic Rocks

The purpose of a rock name should be to clearly describe and characterise it, particularly in the field, and to convey useful information about it. What constitutes useful information depends on your point of view and the rock being described, so depending on the point being made, it can be acceptable to use different names for the same rock. For example, the nature of the protolith is important when mapping stratigraphy in metamorphic rocks. In other studies where the objective is to distinguish rocks that were metamorphosed under different conditions, the index mineral is more useful. The rock texture, e.g. slate or schist, is normally an essential part of the name, to convey the character of the rock. Each of the features that are important for metamorphic rock names are discussed in more detail below. A suggested universal naming system provides a consistent terminology and allows comparison of rocks irrespective of the focus of the study (Fettes & Desmons 2007).

1.6.1   Protolith Names

The prefix 'meta' may be used in front of any sedimentary or igneous rock name to denote its metamorphic equivalent. Some metamorphic rocks have specific names, such as marble – which is used instead of meta-limestone. Protolith names may be used as nouns with additional compositional qualification, e.g. diopside marble, and/or as adjectives qualifying a textural term e.g. metapelitic schist. Some of the common names, and their adjectival forms, are outlined in Table 1.1.

1.6.2   Index Mineral Names

Some essentially monomineralic rocks are named for their dominant mineral, e.g. quartzite, marble, serpentinite or amphibolite (Section 1.6.4). The names of any significant metamorphic minerals may be used as qualifiers in metamorphic rock names, e.g. garnet–mica schist or forsterite marble. A number of other names referring to particular mineral associations are described under 'Specific Names' below (Section 1.6.4).

Table 1.1 Common metamorphic rock types and their original materials

Original material	Metamorphic rock type (noun/adjective)
Sediment	Metasediment/metasedimentary
	Paragneiss at high grade
Argillaceous or clay-rich sediment (pelite)	Metapelite/metapelitic
Arenaceous or sandy sediment (psammite)	Metapsammite/metapsammitic
Quartz sand	Quartzite/quartzitic
Marl	Calc-silicate/metacalcareous
Limestone or dolostone	Marble
Basalt or gabbro	Metabasite or metagabbro/metabasaltic or metagabbroic
	Mafic is also used
Granite	Metagranite/metagranitic
	Orthogneiss where highly deformed

1.6.3 Rock Texture Names

Rock textures are important for naming metamorphic rocks, and indicate whether or not oriented fabric elements are present to dominate the rock's appearance, and the scale on which they are developed. They will be explored in more detail in Chapters 7 and 8. Textural names are usually nouns, qualified by adjectives indicating the parent material or the present mineralogy (e.g. pelitic hornfels; garnet schist). In many regionally metamorphosed rocks, sheet silicates, and sometimes other minerals, develop a **preferred orientation** aligned perpendicular to the maximum compression direction, giving rise to a **planar fabric** or **foliation**. The names used for planar fabrics, outlined below, depend on the grain size and general appearance of the rock and the boundaries between them are gradational. In cases of doubt, it is worth remembering that the majority of metamorphic rocks can be described as schist! The following paragraphs and the associated photographs illustrate the main textural rock types found in areas of regional metamorphism (from low to high grades) followed by the textural terms associated with metamorphism in the presence and absence of intense deformation.

Slate is fine-grained foliated rock, formed from clay-rich sediments, that cleaves along its foliation planes due to the alignment of very fine **phyllosilicate** (sheet silicate) grains. The rock has a dull appearance on fresh surfaces (Figure 1.11a). The individual aligned grains are too small to be seen with the naked eye but can be imaged using Scanning Electron Microscopy (SEM) techniques (Figure 1.11b). The SEM image shows the strong alignment of most of the phyllosilicate grains, and the larger, interspersed, more tabular quartz grains.

Phyllite is similar to slate but the slightly coarser phyllosilicate grains are sometimes discernible in hand specimen and give a silky or shimmery appearance to

(a)

(b)

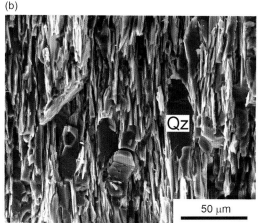

(c)

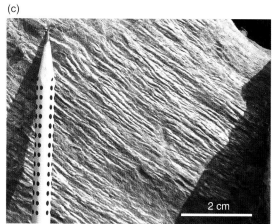

Figure 1.11 (a) Slate with thin sandy interbeds, Abereiddy Bay, Wales. (b) SEM image of a fractured slate surface with aligned phyllosilicates surrounding tabular quartz grains (QZ), Easdale, Scotland. (Photo: Geoff Lloyd.) (c) Phyllite, Valais, Switzerland. (Photo: Barbara Kunz.)

cleaved surfaces (Figure 1.11c). The cleavage surfaces are commonly less perfectly planar than in slates. Phyllites also form from fine-grained sediments such as muds.

Schist is characterised by parallel alignment of moderately coarse grains, usually clearly visible with the naked eye (Figure 1.3). This type of fabric is known as **schistosity**, and where deformation is fairly intense it may be developed by other minerals, such as hornblende, as well as by phyllosilicates. Mica schists form from clay-rich sedimentary protoliths similar to those of most slates, but other schistose rocks can be derived from a range of sedimentary and igneous precursors.

Gneiss is a term for coarse (grain size of several millimetres), feldspar-rich rocks which commonly have a planar fabric that is defined by compositional layering rather than mineral alignment. This is termed **gneissic layering** and typically layers richer in quartz and feldspar alternate with layers richer in garnet, mica, pyroxene or amphibole. This layering is generally unrelated to original layering and forms when different minerals segregate during deformation at high temperatures (Figure 1.12 a–c). Gneiss may form from metasedimentary or meta-igneous

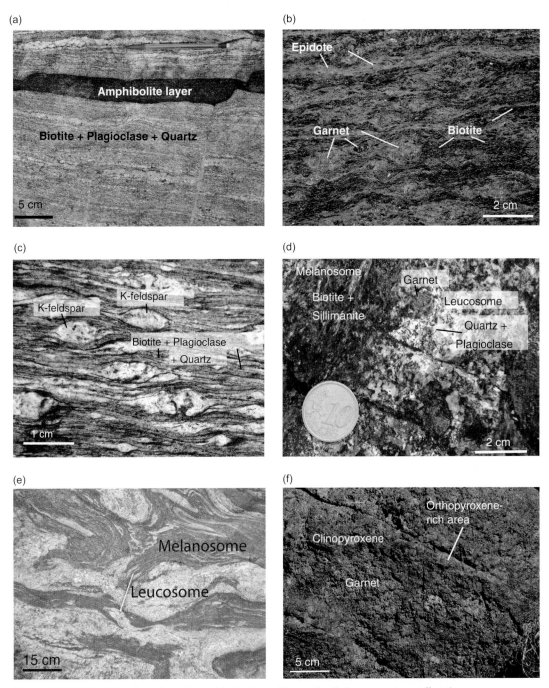

Figure 1.12 (a) Granitic orthogneiss with amphibolite layer, Bhutan. Small shear zones have offset the amphibolite layer and quartz veins. (b) Paragneiss, Ontario, Canada. (c) Augen gneiss, Val D'Ossola, Italy. (Photo: Barbara Kunz.) (d) Migmatitic metapelite with adjacent leucosome and melanosome (restite), Connemara, Ireland. (e) Migmatite with extensive development of leucosomes. Pervasive ductile deformation has disrupted both the leucosomes and the melanosomes. (Photo: Barbara Kunz.) (f) Mafic granulite, Ben Strome, Scotland, containing garnet, clinopyroxene and orthopyroxene. (Photo: Richard Taylor.)

protoliths, and it is often difficult to determine the original protolith in the field. In general, meta-igneous gneiss (**orthogneiss**; Figure 1.12a), tends to be dominated by quartz and feldspar, whereas metasedimentary gneiss (**paragneiss;** Figure 1.12b), commonly contains garnet and may have appreciable biotite. Augen gneiss (Figure 1.12c) is a distinctive type of gneiss that forms during intense ductile deformation. Feldspars in the original gneiss are more resistant to deformation than weaker quartz and mica, so form distinctive 'eye'-shaped lenses in the rock.

Migmatite contains darker, schist-like parts and paler veins or irregular patches with an igneous appearance rich in quartz and feldspar; in fact the name itself means 'mixed rock' (Figure 1.12d and e). Migmatites have an extensive and complex terminology all of their own, but the most useful terms for a field description are the **leucosome** (the lightest coloured parts) and the **melanosome** (the darkest part), usually visible between or around the leucosomes. The proportions of leucosome and melanosome may vary, depending on how much of the rock has melted and/or how much of the melt has mobilised. The melanosome is also known as the 'restite', i.e. the minerals that have not melted. Migmatites most commonly form from metapelitic protoliths (discussed in more depth in Chapter 4, Section 4.3 and Box 4.3) but metabasic migmatites may form in very high-grade terranes.

Granulite is a type of gneiss characterised by a lack of mica and is associated with very high-grade metamorphism. Typical minerals include feldspar, sillimanite, pyroxene, amphibole and garnet (Figure 1.12f). Granulites may form from both metasedimentary and meta-igneous protoliths, and will be covered further in Chapters 4 and 5.

Hornfels is a tough rock made of random fabric of interlocking grains, formed during high-temperature contact metamorphism in the absence of deformation (Figure 1.13). Hornfelses commonly contain porphyroblasts or poikiloblasts, and their formation will be discussed further in Chapter 4.

Figure 1.13 Hornfels from Skiddaw, UK, showing large elongate grains of andalusite weathering proud of the surface, and round holes where grains of cordierite have weathered away. The matrix contains biotite, quartz and feldspar but they are rather too fine-grained to identify in the field.

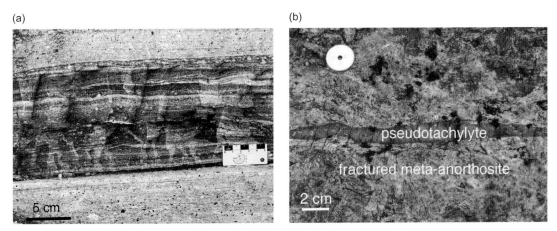

Figure 1.14 Examples of highly deformed rocks. (a) Mylonitic fabric developed in highly strained Bedár orthogneiss, Spain. Both the paler and darker layers are highly deformed; the darker layers are finer grained, and it is this which makes them look darker. (Photo: Pedro Castiñeras.) (b) Pseudotachylyte (black vein infill) developed in a fractured meta-anorthosite (granulite), Lofoten, Norway. (Photo: Lucy Campbell.)

Mylonite is a fine-grained rock produced in zones of intense ductile deformation (Figure 1.14a). As the rock deforms, pre-existing grains deform and recrystallise as smaller grains. The overall appearance of the rock in outcrop is commonly flaggy, even though individual hand specimens may show no marked mineral alignment because of the thorough recrystallisation. Mylonites will be discussed in more detail in Chapter 8.

Pseudotachylyte is a black glassy rock, found in seams in high-grade rocks (Figure 1.14b). It gets its name from its apparent similarity to 'tachylyte', which is a black volcanic glass formed by cooling of basaltic magma. Pseudotachylyte forms during extreme dynamic metamorphism, when rocks are melted by frictional heating and will be discussed further in Chapter 8.

1.6.4   Specific Names

A few common metamorphic rock types have their own specific names.

Quartzite is a metamorphic rock formed predominantly of quartz (Figure 1.15a). It is very resistant to weathering and commonly develops a smooth surface. Lichens and other surface plants grow poorly on quartzite, so it is often easy to spot in the field, even in vegetated regions.

Marble is a metamorphic rock formed predominantly of carbonates, most commonly calcite (Figure 1.15b). It tends to weather with a rough surface texture, and may develop karstic features identical to those that form in massive limestones.

Serpentinite is a green, black or reddish rock composed predominantly of serpentine (Figure 1.15c). It forms by hydration of igneous or metamorphic peridotites (olivine-rich ultrabasic rocks).

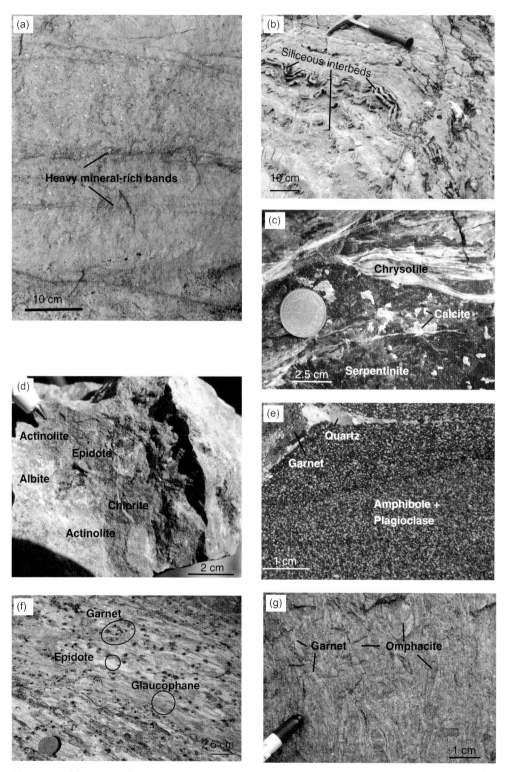

Figure 1.15 (a) Quartzite from NW Scotland showing relic sedimentary layering, with coarse quartz in some bands. (b) Calcite marble with thin siliceous interbeds picking out small-scale folds, NW Namibia.

Greenschist is a green metabasite, usually composed predominantly of chlorite, epidote and actinolite, whose different shades of green provide the rationale behind the name (Figure 1.15d). Albite is also usually present.

Amphibolite is an essentially bimineralic dark-green to black rock, usually foliated, composed of hornblende and plagioclase, with minor amounts of quartz, biotite, epidote, chlorite or garnet also possible (Figure 1.15e). Most amphibolites form from basic igneous protoliths but some metamorphosed calcareous sediments may also be dominated by plagioclase and amphibole.

Blueschist is a dark, lilac-blue-grey foliated metabasite. It owes its colour to the presence of abundant sodic amphibole, typically glaucophane but is seldom truly 'blue' unless quite coarse (Figure 1.15f).

Eclogite is a metabasite composed of garnet and omphacitic clinopyroxene which forms under metamorphism at very high pressures (Figure 1.15g). Under these pressures plagioclase feldspar is completely absent. Other common minerals in eclogites include quartz, kyanite, micas, amphiboles, and epidote group minerals.

SUMMARY

This chapter has outlined the major types of metamorphic rocks, the factors that can cause rocks to experience metamorphism, and the types of geological setting in which metamorphism takes place. This information provides the basis for the chapters that follow.

Metamorphism affects a very diverse range of original rocks and takes place in many different tectonic settings. The study of metamorphic rocks can provide considerable insight into the nature of the deep crust and the processes that take place there. Metamorphic rocks therefore provide information that is complementary and supplementary to modern geophysical 'snapshots' through different tectonic settings.

During metamorphism, new minerals crystallise, and pre-existing minerals may change shape in response to changing external conditions. Three main factors control changes in rock mineralogy: pressure, temperature and fluids. A major assumption in studies of most metamorphic rocks is that their chemical composition

Figure 1.15 (*cont.*) (c) Serpentinite, Zermatt, Switzerland. The outcrop surface has small white calcite crystals on it that formed during weathering. (d) Greenschist, central Iran. Epidote, actinolite and chlorite exhibit distinct shades of green. (e) Coarse garnet-bearing amphibolite, Ontario, Canada. (f) Blueschist, Ile de Groix, France. Garnet porphyroblasts are set in a matrix dominated by glaucophane and epidote. (Photo: Barbara Kunz.) (g) Eclogite, Stadlandet, Norway. At this scale, the rock appears bimineralic but it also contains quartz and rutile.

has remained unchanged during metamorphism, except for loss of volatiles. However, there are metamorphic rocks that have experienced significant chemical change or metasomatism.

Metamorphic rocks may be named in a number of different ways according to the essential points that are important at the time of naming. These essential points may include the nature of the protolith, any index minerals present and/or the rock texture.

EXERCISES

1. Name four common index minerals found in metapelites, and three common non-index minerals of metapelites. What makes a mineral a good index mineral?

2. What is a polymorph? Name three minerals mentioned in this chapter that are polymorphs of each other and explain how two of them are useful for determining relative pressures of metamorphism.

3. What is the difference between a porphyroblast and a poikiloblast?

4. Name three factors that control metamorphic changes in rocks.

5. Which of the following rock names describe a metasediment and which describe a meta-igneous rock: quartzite, greenschist, eclogite, paragneiss, orthogneiss, marble, slate, phyllite, amphibolite?

6. Put the following rocks in order of metamorphic grade: gneiss, slate, schist, phyllite. Describe the main textural and mineralogical features that define each.

7. A sequence of rocks containing layers of mudstone, quartz sandstone, limestone and basalt is metamorphosed during a regional metamorphic event. How would you name the metamorphic rocks in the field if you don't have any information about the degree of metamorphism the rocks have experienced?

8. A metamorphic rock experienced pressures of 1.4 GPa. Assuming a continental crustal density of 2550 kg/m^3, and assuming that metamorphic pressure was equal to lithostatic pressure, at what depth was it metamorphosed? Use Equation [1.1] and take a value for g of 9.8 m/s^2. Give your answer to the nearest kilometre (km).

9. If basaltic magma with a temperature of 1200 °C is extruded onto the ocean floor at a mid-ocean ridge at 3.5 km below the surface of the ocean, will the sea water boil? (Use Figure 1.10 to help you; you may ignore the differences between sea water and fresh water).

10. Figure 1.8 illustrates a rock that has been metamorphosed to very high pressures without being deformed. Explain how this is possible.

11. Describe the first-order observations that might help you determine whether a basaltic igneous rock has been metamorphosed under regional, contact or hydrothermal conditions?

FURTHER READING

Evans, B.W. (Ed.)(2007). *Metamorphic Petrology*. Landmark Papers No. 3. Mineralogical Society of Great Britain and Northern Ireland.

Fettes, D. & Desmons, J. (Eds.) (2007). *Metamorphic Rocks: A Classification and Glossary of Terms. Recommendations of the International Union of Geological Sciences. Subcommission on the Systematics of Metamorphic Rocks*. Cambridge University Press, 244 pages.

Fossen, H. (2016). *Structural Geology*, 2nd Edition. Cambridge University Press, 524 pages.

Fowler, C. M. R. (2005). *The Solid Earth: an Introduction to Global Geophysics*, 2nd Edition. Cambridge University Press, 685 pages.

Ingebritsen, S., Sanford, W. & Neuzil, C. (2006). *Groundwater in Geologic Processes*. Cambridge University Press, 536 pages.

Sanders, I. (2018). *Introducing Metamorphism*. Dunedin Press, 147 pages.

Whitney, D. L. & Evans, B. W. (2010). Abbreviations for names of rock-forming minerals. *American Mineralogist*, **95**, 185–7.

Yardley, B. W. D., MacKenzie, W. S. & Guilford, C. (1990). *Atlas of Metamorphic Rocks and their Textures*. Longman, Harlow, 120 pages.

2 Chemical Equilibrium in Metamorphism

The fundamental assumption underlying most studies of metamorphic rocks, including the examples described in Chapter 1, is that their mineral assemblages reflect the physical conditions of pressure and temperature which prevailed at the time when they grew. This assumption is grounded in numerous field studies that have shown that the mineral assemblages in any particular rock type vary systematically and predictably across an area. Additionally, rocks of the same composition always appear to develop the same mineral assemblage when they are subject to the same metamorphic conditions. These observations are the basis for applying the theory of chemical equilibrium to metamorphic rocks. This chapter outlines the information that can be gleaned by treating rocks as equilibrium chemical systems along with the limitations and pitfalls of so doing.

The geochemical approach outlined in this chapter allows us to:

- identify which metamorphic rocks in an area are likely to have experienced similar metamorphic conditions,

- approximate the temperature and pressure (a proxy for the depth of burial) at which metamorphism took place, and
- show how these conditions have varied between outcrops of similar composition with different mineral assemblages.

By the end of this chapter you should understand what the phase rule is, how the phase rule is applied to rocks to determine the number of phases that could have equilibrated together, and how it makes complex systems understandable. You should be able to interpret *P–T* and compositional phase diagrams and use them to constrain the *P–T* conditions of a particular mineral assemblage.

2.1   Equilibrium

Let's consider what happens in a lump of rock undergoing metamorphism from the perspective of the atoms it contains. Our arbitrary lump of rock is a **system** in chemical terms. The atoms make up crystalline grains of different minerals and probably also an intergranular fluid. Each of these constituents is known as a **phase**. Phases make up the system and may be solid or fluid. Each phase is physically distinct and can be separated using physical methods, at least in principle. For example, plagioclase and quartz are common phases in pelitic schists. The plagioclase will have a composition intermediate between anorthite (Ca-rich) and albite (Na-rich). Anorthite and albite do not count as separate phases because the plagioclase grains cannot be separated into albite particles and anorthite particles, however finely they are ground. The plagioclase grains are **solid solutions**, and anorthite and albite, the extreme Ca-rich and Na-rich compositions, are known as **end-members**. Many phases have variable compositions because they are solutions.

Our system can be subjected to some specific conditions of pressure and temperature, and if we maintain these conditions unchanged for a sufficiently long time, the atoms in the system will group themselves into the most stable configuration possible. The system is then at **equilibrium**. This configuration may involve a number of solid or fluid phases depending on the atoms available and the conditions to which the system is subjected. The solid phases present when the system has come to equilibrium amount to a **stable mineral assemblage**. Even at equilibrium, atoms are constantly in motion and may move between phases. However, there is no overall change in the amount or composition of each phase present over a period of time. In other words, when a system is at equilibrium, nothing happens at the observable scale!

Suppose the temperature (or pressure) of the system is now slowly changed. The co-existing phases will eventually cease to be in equilibrium with one another;

some may grow at the expense of others, and new phases may appear. Such changes in rocks are known as **metamorphic reactions**; these lead to the production of a new metamorphic assemblage that is at equilibrium under the new conditions.

All chemical systems with the same chemical composition (i.e. the same types of atoms present in the same proportions), that are subjected to the same conditions, will develop the same assemblage of phases if equilibrium is attained. Regional studies, such as those outlined in Chapter 1, suggest that all specimens of similar bulk composition collected within the same metamorphic zone will have the same mineral assemblage and this is the evidence that the assumption of chemical equilibrium during metamorphism is reasonable. Likewise, it follows that where mineral assemblages vary between rocks of the same composition, the rocks must have experienced different metamorphic conditions and can usefully be divided into metamorphic zones.

2.2   The Phase Rule

Most metamorphic rocks are chemically quite complex and are made up of a relatively large number of minerals. To understand how they formed we first need to know whether these minerals all co-existed at equilibrium. For example some mineral phases are only present because reactions did not go to completion and so are out of equilibrium with others. If we can predict from theory how many mineral species can co-exist stably at equilibrium in a particular rock, then we can judge whether the mineral assemblage in the rock is likely to be an equilibrium association, or may contain minerals formed at different times under different conditions that never equilibrated with each other.

Predicting the number of phases a rock could develop at equilibrium can be done using the **phase rule**, which was first applied to metamorphic rocks by geochemist V. M. Goldschmidt early in the twentieth century. This classic law of physical chemistry is particularly useful in metamorphic petrology, because we often deal with rocks that undergo mineralogical changes while their composition remains the same.

The phase rule can be demonstrated by imagining a glass of water (at atmospheric pressure) as a simple chemical system. We can convert the water into ice or steam by changing its temperature. Ice, liquid water and steam are all physically separable phases, however since they have the same chemical formula we can make them all out of a single chemical **component**, i.e. H_2O. Components are the chemical constituents needed to make the phases we wish to consider in our system; from the point of view of the phase rule it is *the smallest number of chemical constituents needed to define the compositions of all* *the phases* that is significant. For example, a system that contains only andalusite and kyanite has only one component, Al_2SiO_5, whereas a system that contains andalusite, corundum and quartz must have two components, Al_2O_3 and SiO_2, to make all the phases.

Returning to our glass of water, we have a one-component system (H_2O), in which we wish to consider the relationships between three phases at different temperatures and pressures. The question "what is the temperature of a glass of water?" seems nonsensical because liquid water is stable over a range of temperatures. On the other hand the question "what is the temperature of water coexisting with steam (i.e. boiling)?" has an obvious answer, 100 °C. To be precise it is of course necessary to know the exact atmospheric pressure in the laboratory where the water is being boiled, but provided this information is available, the temperature of the boiling water can be obtained from standard tables without recourse to a thermometer.

The conclusions that we can draw from the glass of water, for the behaviour of a one-component system are as follows.

1. One phase can occur alone over a range of temperatures (T) and pressures (P). This one-phase, one-component system is said to have two **degrees of freedom** because it is possible, within limits, to vary its temperature and its pressure *independently* without changing the number or nature of the phases present.
2. Two phases can only co-exist in equilibrium in the one-component system at a single temperature for any given pressure. As long as the two phases co-exist, the system has only one degree of freedom because any change in the pressure at which they co-exist leads to a related change in temperature and vice versa.

These conclusions are summarised for the general case by **the phase rule**, which can be expressed as:

$$F = C - P + 2 \qquad [2.1]$$

where F is the number of independent degrees of freedom for the system, C is the number of chemical components and P is the number of phases. Note that C is always the minimum number of chemical ingredients that are needed to make all the phases. For most minerals, this means that we use the oxides of elements as components, rather than the elements themselves, because O is not able to be added or removed independently.

One consequence of the phase rule is that, in very general terms, mineral assemblages with a large number of phases have only a small number of degrees of freedom. As a result, the range of conditions under which the assemblage grew is relatively precisely constrained, and it is more easily possible to determine them. This is a good guide when deciding which rocks to select for estimating conditions of metamorphism.

2.2.1   Compositional Variation

Although pressure and temperature are very important variables in metamorphism, solid solution means that the composition of many common metamorphic minerals is also variable. While it is often convenient to imagine the degrees of freedom of a mineral assemblage as representing the variables P and T, there is no fundamental reason why this should be so; variable compositional parameters may be more appropriate. However, no matter how many chemical substitutions are possible within the minerals of a particular rock, the number of degrees of freedom calculated from the phase rule still gives the number of **independent variables**, or variables that can change *independently*.

As an example, consider a rock containing white mica + quartz + kyanite + alkali feldspar + aqueous fluid, in which the mica and feldspar are solid solutions intermediate between the Na and K end-members. The phases

mica	$(Na, K)Al_3Si_3O_{10}(OH)_2$
quartz	SiO_2
kyanite	Al_2SiO_5
feldspar	$(Na, K)AlSi_3O_8$
fluid	H_2O

can all be made from five oxide components: Na_2O, K_2O, H_2O, SiO_2, Al_2O_3. From the phase rule, since $C = P = 5$, the assemblage has two degrees of freedom. However, there are four possible variables: P, T, feldspar composition and mica composition. This means that only two variables can be changed independently; the two remaining variables become **dependent variables** and take on values dictated by those of the independent variables. Hence all mixtures of these five phases that we might choose to make and *separately subject to the same pressure and temperature* (i.e. taking P and T as independent variables) will end up having mica and feldspar of the same composition when equilibrium has been attained.

Suppose a neighbouring rock does not contain kyanite. It therefore consists of only four phases, and the number of independent degrees of freedom is increased to three ($F = 5 - 4 + 2$). At any specified value of P and T, one degree of freedom remains, but both mica and feldspar compositions are variable. This means that the Na/K ratios of mica and feldspar are related, and do not vary independently of one another. For our given value of P and T, only one composition of feldspar can co-exist with any specified mica composition in this simple rock, and vice versa.

2.2.2   Application of the Phase Rule to Natural Rock Systems

It is often quite difficult to make the jump from the idealised and simplified systems studied by experimentalists or discussed in textbooks to the natural associations of real minerals found in metamorphic rocks. Chemical analysis of a typical pelitic schist, for example, would show that it contains significant amounts of the major elements, represented by the oxides SiO_2, TiO_2, Al_2O_2, Fe_2O_3, FeO, MgO, MnO, CaO, Na_2O, K_2O and H_2O. Smaller amounts of P, S, B, F, Sr, Ba, Zr and traces of other elements will also be present. Suppose that the phases in the rock are muscovite, biotite, garnet, chlorite, plagioclase, quartz, tourmaline, ilmenite, pyrrhotite, apatite and zircon. This system has at least 18 components but only 12 phases (assuming a pore fluid was present during metamorphism), hence there are at least eight degrees of freedom. Does this mean that the assemblage can exist over such a wide range of conditions that it is unlikely to be a useful indicator of conditions of metamorphism, or can some of the minor constituents be ignored?

Accessory minerals such as tourmaline, apatite, zircon and the sulphides each concentrate a particular minor or trace element that does not readily enter any of the other minerals present, e.g. B in tourmaline and Zr in zircon. As a result, these minerals are not usually involved in the major metamorphic reactions because there is no other phase for the minor or trace element to enter. Only under specific conditions will they participate in metamorphic reactions, and then they may become very valuable guides for unravelling metamorphic history (Chapter 9). For most of the rock's metamorphic history, however, they are inert, and sit passively in the rock surrounded by other phases that can react together. It is therefore reasonable to discount both these accessory phases and their distinctive minor element components when applying the phase rule.

Other minor element components do not form their own minerals but occur in small amounts in solid solution in the common rock-forming minerals. Examples include Sr and Ba in feldspars or Mn in Fe–Mg minerals. The concentration of, for example, Sr in feldspar, can vary independently of pressure and temperature depending on how much feldspar and how much Sr are present in the rock, and this variation can be considered as one of the 'excess' independent degrees of freedom. Unless unusually high levels are present, such variation in minor element concentration has little effect on the stability of the major silicate phases and, as a first approximation, can also be ignored.

Overall, the chemical components which are important for determining whether a mineral assemblage may have formed at equilibrium are therefore those which are major constituents of more than one of the phases. On this basis, applying the phase rule gives a good idea of whether the minerals present in a rock are likely to constitute an equilibrium assemblage.

2.3 Metamorphic Phase Diagrams

As long as mineral assemblages follow the rules of chemical equilibrium, it is possible to construct diagrams relating them to the conditions in which they formed or the rock compositions in which they grew. Some of these **phase diagrams** are based on experimental data that constrain the stability of different phases under different conditions. Others are simply geometrical arrangements showing how the same rock composition can be made up of different minerals according to the conditions.

2.3.1 *P–T* Diagrams

One of the commonest ways of representing the stability fields of different meta-morphic mineral assemblages is on a plot of pressure against temperature (called a **P–T diagram**), because such a diagram allows us easily to visualise where in the Earth a particular mineral assemblage might form.

It follows from the phase rule that if a rock has no more components than there are phases present, the mineral assemblage will be stable over a distinct region of the *P-T* diagram because it has at least two independent degrees of freedom. The assemblage is therefore said to occupy a **divariant field**. Figure 2.1 shows the divariant fields in which andalusite, sillimanite and kyanite are stable. Since these phases all have the same composition, they make up a one-component system; all can be made from the component Al_2SiO_5. Any two of them can co-exist in equilibrium only at a single temperature for each pressure and the system then has one degree of freedom. This is the case along the boundaries between divariant fields, and these boundaries are known as **univariant curves**. In Figure 2.1 the univariant curves intersect at a single point at which all three phases can occur in equilibrium. From the phase rule, a system with $C + 2$ phases has no degrees of freedom and can therefore be stable only at a unique pressure and temperature

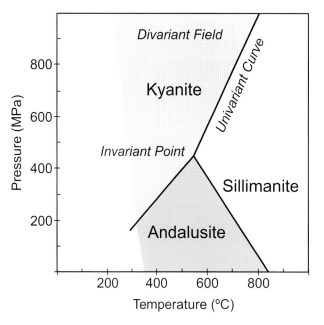

Figure 2.1 *P–T* diagram illustrating the divariant fields in which andalusite, kyanite and sillimanite are stable, with the intervening univariant curves and the invariant point (or aluminium silicate triple point). The boundaries are from Pattison (1992) and Spear & Pattison (2017).

known as an **invariant point**. The invariant point in this system is commonly referred to as the **aluminium silicate triple point**.

Many of the laws that govern metamorphic phase diagrams are purely geometrical. They follow from the phase rule without any further knowledge of geology or chemistry beyond the compositions of the minerals concerned; a detailed account is given in Appendix 2.

2.3.2 Compositional Phase Diagrams

The value of the P–T diagram is that it shows the actual range of physical conditions under which specific assemblages are stable relative to others. However, we often want to be able to determine whether different samples have different mineral assemblages because they were metamorphosed under different conditions or because they differ significantly in their chemical composition. **Compositional phase diagrams** allow us to do this, even where we cannot estimate the actual temperatures and pressures at which the rocks formed.

Compositional phase diagrams are constructed for a specific value of P and T to show the possible mineral compositions and mineral assemblages that may be at equilibrium under the chosen conditions. Like P–T diagrams, they have a basis in the phase rule. Provided the P–T conditions selected lie within a divariant field on a P–T diagram and do not coincide exactly with a univariant curve, the number of phases present in any rock will be no more than the number of components in the system.

Compositional phase diagrams are useful because a wide range of information can be shown on them, and they are easy to construct. All that is needed is knowledge of the composition of each mineral and the assemblages in which they occur.

Figure 2.2 shows a representative system made up of the three components CaO, SiO_2 and Al_2O_3, and is drawn for a single P–T condition. Points represent minerals of fixed composition. **Tie-lines** connect phases that co-exist; for example, in Figure 2.2, anorthite can co-exist with grossular, Al-silicate and quartz.

On these diagrams, three-phase assemblages are represented as triangles. At the P–T condition for which the diagram is drawn, each assemblage can only develop in rocks whose bulk composition lies within that particular coloured triangle. In Figure 2.2 for example, the mineral assemblage developed in a rock of bulk composition X, will be corundum + grossular + Al-silicate, whereas a rock of composition Y at the same conditions will be made up of Al-silicate + anorthite + quartz. These assemblages are known as **compatible assemblages** because the triangles that represent them do not overlap. Note that rock compositions plot in the same place on the diagram irrespective of the P–T conditions for which it is drawn. This makes it possible to predict the mineral assemblages that will develop as P–T conditions change.

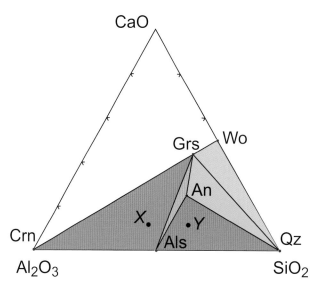

Figure 2.2 Compositional diagram showing the compositions and stable relationships between a number of phases in the system CaO–SiO$_2$–Al$_2$O$_3$, for some arbitrary conditions at moderate pressure and high temperature. Mineral abbreviations are listed in Appendix 1. Only phases directly connected by tie-lines can co-exist and possible divariant assemblages are coloured in different shades of green. Unless lying exactly on a tie-line, any bulk composition in this system will crystallise as a three-phase mixture according to which triangular field it lies within. X and Y are bulk compositions discussed in the text.

Minerals not directly connected by tie-lines cannot stably co-exist; for example, Figure 2.2 shows that anorthite cannot co-exist with corundum or wollastonite at the *P–T* conditions for which the diagram was drawn.

A major limitation in using this type of diagram is that the bulk composition of the rock has to be reduced to three components (corresponding to the apexes of the triangle). This is generally fine for simple systems (see Section 2.5, below), but more complex systems require additional assumptions or simplifications, some examples of which will appear in Section 2.6 and in Chapters 4–6. Often these assumptions or simplifications are unsatisfactory, sometimes they are untenable.

Despite these limitations, compositional phase diagrams are generally useful. For example, in an area where multiple metamorphic zones are present, a series of compositional diagrams can be plotted for successive *P–T* conditions (from either field or experimental evidence). The series helps to illustrate the changes that we might expect in mineral assemblages as conditions change and enables us to deduce the reactions which have taken place. This is illustrated for a real example in Section 2.5.

2.4   Metamorphic Reactions and the Phase Rule

Reactions that take place during metamorphism involve a number of phases and the reacting assemblage can be described using the phase rule. Different reactions may involve the appearance of new minerals, changes in the abundance of existing minerals and/or changes in the mineral compositions.

2.4.1 Discontinuous Reactions

The simplest reactions take place at a specific temperature and pressure, for example the boiling or freezing of pure water. An assemblage consisting of both reactants and products has only one degree of freedom, and is in equilibrium only along a univariant curve. Simple reactions like this are known as **discontinuous** or **univariant reactions**. Where some of the phases in a rock system are solutions, this simple relationship no longer applies because there are extra components but no increase in the number of phases. Natural examples of truly discontinuous reactions are rather rare, because few metamorphic minerals occur as pure end-members. However, the dehydration of serpentine to olivine + talc (discussed below in Section 2.5) is a reasonably good approximation.

2.4.2 Continuous Reactions

When the phases involved in a reaction include solutions with a variable composition, it is possible for reactions to take place under different conditions according to their composition. A simple example is the boiling of salty water, which takes place at progressively higher temperatures with increasing salt concentration (Figure 2.3). Since the steam given off when the salt solution boils is essentially pure H_2O, the concentration of salt in the remaining liquid increases progressively, along with temperature, as boiling proceeds. Applying the phase rule, the system has two components (NaCl, H_2O) but even when reactants and products co-exist there are only two phases (liquid, steam). A system with salt solution reacting to steam therefore has two degrees of freedom even while the reaction is taking place. The curve showing the composition of boiling salt solution as a function of temperature in Figure 2.3 *looks* like a univariant curve, but only because the figure has been drawn for a fixed pressure. This means that one of the possible degrees of freedom

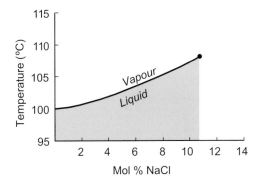

Figure 2.3 Boiling curve for salt solutions in the system NaCl–H_2O, showing the variation in boiling temperature (i.e. liquid–vapour equilibrium temperature) with salt content (expressed as mole % NaCl in the liquid phase). The pressure is fixed at 1 atmosphere and the point at which the boiling curve ends corresponds to the onset of crystallisation of solid halite. Data are from Haas (1976).

(pressure) has been fixed. The figure shows that the two remaining variables, temperature and composition of the boiling solution, vary dependently at constant pressure. By specifying two parameters (e.g. P and T or P and solution composition) the condition of the system is completely specified in accordance with the phase rule.

Reactions of this type that take place progressively so that reactants and products can co-exist over a divariant field (e.g. a fixed interval of pressure and temperature) are known as **continuous reactions**. They invariably involve at least one phase of variable composition and lead to a change in the composition as well as in the abundance of different phases as the reaction proceeds. Continuous reactions are widespread in metamorphism and various examples are discussed in Chapters 4–6.

2.4.3   Cation-Exchange Reactions

A third type of reaction involves only a change in phase composition, with no growth or dissolution of the phases themselves. These reactions are known as **cation-exchange** reactions and take place in rocks with at least two minerals which exhibit the same type of solid solution. The most widespread examples are among minerals that are rich in Fe and Mg (also known as **ferromagnesian minerals**). For example, as the temperature of a garnet–biotite schist changes, Fe^{2+} and Mg^{2+} are exchanged between garnet and biotite. The exchange can be written in a schematic fashion as:

$$Fe\text{-garnet} + Mg\text{-biotite} = Mg\text{-garnet} + Fe\text{-biotite} \qquad [2.2]$$

although of course there are only single equilibrium compositions of garnet and biotite present in the rock at any one time. Many other types of solid solution behave similarly if they are represented in two or more co-existing minerals.

Note that here we have written out a balanced equation using '=' to show the equivalence of the two sides. This is because the equation is intended to show that they are in equilibrium. Where an equation represents a reaction and one side is growing at the expense of the other, we will use the symbol '→'.

Cation-exchange reactions take place progressively over a range of temperatures, and involve the exchange of cations between the co-existing minerals so that the relative proportion of the cations within each mineral changes with temperature, even though their relative proportion in the rock as a whole remains constant and the number of moles of each mineral does not change (Box 2.1).

Although different minerals may show the same type of ionic substitution, it is normally the case that different mineral structures show different preferences for the alternative cations. As a result, cations are said to be **partitioned**

BOX 2.1 Cation-exchange reactions in basic terms

The essence of cation-exchange reactions is that metal cations switch between minerals so that, while the composition of each mineral changes, the amounts remain the same. Suppose two cars are parked together, one with a higher roof than the other. Four tall people and four short people approach and get into the cars at random, four in each. Two cars and eight people then drive off and before long the tall people in the car with the low roof realise that they would be more comfortable in the other car. So when they stop for a break, the tall people in the low-roofed car swap seats with the short people in the high-roofed car. After the break, two cars carrying eight people drive off again, but the people have been redistributed. People swapping between car seats is analogous to cations swapping between minerals and in both cases the exchange has happened so that there is a better fit of the moveable object (person or cation) with the framework it has to fit into (car or mineral structure). Overall, the number of cars and the number of people are the same before and after the exchange. In rock terms, this means that the number of phases and the overall number of the exchanging cations also remain constant.

---

between the minerals. Partitioning results in the **fractionation** of elements or isotopes into some phases relative to others. Ions that substitute for one another in a mineral structure to form a solid solution must be of broadly similar size and usually carry the same charge. However, detailed differences in the size and shape of the sites available in each mineral mean that different minerals will show a preference for one cation relative to others which can substitute into the same structural site. Hence, if there are several different ferromagnesian minerals present at equilibrium in a rock, the ratio of $Mg^{2+}/(Mg^{2+} + Fe^{2+})$, (commonly known as the mole fraction of the Mg end-member, or X_{Mg}) of each mineral will always be different.

Note that the X_{Mg} value and other compositional ratios are calculated using the molecular proportions of the elements (or their oxides) in an analysis, not the weight percentages. Hence for a mineral which is a solid solution between Fe and Mg end-members, $X_{Mg} = n_{Mg^{2+}}/(n_{Mg^{2+}} + n_{Fe^{2+}})$, where n denotes the number of atoms in the formula.

Garnet, for example, shows a strong preference for Fe over Mg, and so has a lower value of X_{Mg} than other, co-existing, Fe–Mg silicates such as staurolite, biotite or pyroxene. The ratio X_{Mg} for a specific mineral is referred to using the standard mineral abbreviation (Appendix 1), so in the case of garnet it becomes X_{Mg}^{Grt}. Many petrologic and experimental studies have determined the relative values of X_{Mg} for different co-existing minerals. In the case of minerals common in pelitic schists, it is $X_{Mg}^{Grt} < X_{Mg}^{St} < X_{Mg}^{Bt} < X_{Mg}^{Chl} < X_{Mg}^{Crd}$.

The ratio by which Fe and Mg are partitioned between a pair of ferromagnesian minerals is defined by the **distribution coefficient** (K_D). For a general case of Fe $\rightleftharpoons$ Mg exchange between phases A and B this is given by:

$$K_D = \frac{(Mg/Fe)^A}{(Mg/Fe)^B} = K_{D_{Fe-Mg}^{A-B}} \qquad [2.3]$$

For example, in the case of exchange between garnet and biotite, garnet conventionally corresponds to phase A in Equation [2.3], while biotite is phase B. K_D in Equation [2.3] is then the same as the equilibrium constant for the cation-exchange reaction written in Equation [2.2]. For a suite of garnet–biotite rocks metamorphosed at the same P–T conditions, the value of K_D is the same irrespective of whether the minerals are overall rich in Fe or overall rich in Mg.

Cation-exchange reactions occur between ferromagnesian minerals because the precise degree to which Fe^{2+} is partitioned into a particular mineral relative to Mg^{2+} is dependent on temperature, with smaller differences in X_{Mg} between any two co-existing phases at higher temperatures. As a result, K_D tends towards a value of 1 with increasing temperature.

While the X_{Mg} value of the individual minerals changes with temperature, the X_{Mg} value of the bulk rock remains constant, just as in the example of swapping car passengers in Box 2.1. In that example, the two cars always contained a total of four tall passengers and four short ones. Overall, the X_{Mg} values of the individual ferromagnesian minerals must reflect the composition of the rock. In Mg-rich garnet–biotite schists, both the garnet and the biotite will have relatively high X_{Mg} values compared to an Fe-rich garnet–biotite schist in which both minerals will have low values of X_{Mg}.

2.5 Application of Chemical Equilibrium to Natural Rocks: an Example

Before attempting to use the principles of chemical equilibrium to interpret meta-morphic rocks in general, it is worth demonstrating the validity of the approach in a simple system. The example below describes the metamorphism of rocks whose chemistry can be closely approximated by a three-component chemical system, enabling the use of simple compositional phase diagrams. This simplification means that none of the minerals are treated as solid solutions and so only discontinuous (univariant) reactions are considered.

Ultrabasic rocks are composed predominantly of MgO, SiO_2 and H_2O, although CaO, Al_2O_3 and some FeO are also usually present in minor amounts. Bodies of dunite (only olivine) with wehrlite (olivine + clinopyroxene), lherzolite (olivine + clinopyroxene + orthopyroxene) and related rock types are common in orogenic belts and have often been extensively hydrated on the sea floor or during subduction to become serpentinite bodies. Subsequent higher-temperature metamorphism may cause breakdown of serpentine and even lead to regeneration of the original igneous minerals which grow as new meta-morphic grains with distinctive texture.

In the Italian Alps, heat provided by the intrusion of the Bergell tonalite into the serpentinite body exposed in Val Malenco caused a contact metamorphic aureole to form in the serpentinite (Trommsdorff & Connolly 1996; Trommsdorff & Evans 1972). The map of the metamorphic aureole, including the sample locations on which the isograds are based, is shown in Figure 2.4. Away from the intrusion, the serpentinite body mainly contains antigorite (serpentine) and relic igneous forsterite (olivine) and diopside (clinopyroxene) with minor chlorite and magnetite. This regional assemblage is Zone A. Three contact metamorphic zones (Zones B–D), each characterised by a particular mineral association, surround the intrusion:

> Zone A: antigorite + forsterite + diopside (regional assemblage)
> Zone B: antigorite + forsterite+ tremolite
> Zone C: talc + forsterite + tremolite
> Zone D: anthophyllite + forsterite +tremolite

To a good approximation, the metamorphic reactions that take place at the isograds between Zones B and C and between Zones C and D involve only phases that can be made from the components MgO, SiO_2 and H_2O (the other minerals are effectively inert). The compositions of the reacting phases, antigorite, olivine, talc, anthophyllite, enstatite and the H_2O fluid assumed to have been given off by the dehydration reactions, can be plotted on a triangular diagram with MgO, SiO_2 and H_2O as the corner compositions, as shown in Figure 2.5a.

The mineral assemblage found in Zone B is represented in Figure 2.5b by the triangle of tie-lines that connect the phases present in the rock: antigorite, forsterite and H_2O. Although only antigorite and forsterite are present in Zone B today, it is assumed that an aqueous fluid phase was also present during metamorphism. Since this assemblage has three components and three phases it follows from the phase rule that it has two degrees of freedom, and can thus exist over a range of temperature and pressure. Note that tremolite cannot be shown because it contains an additional component, CaO; this is returned to in the following section.

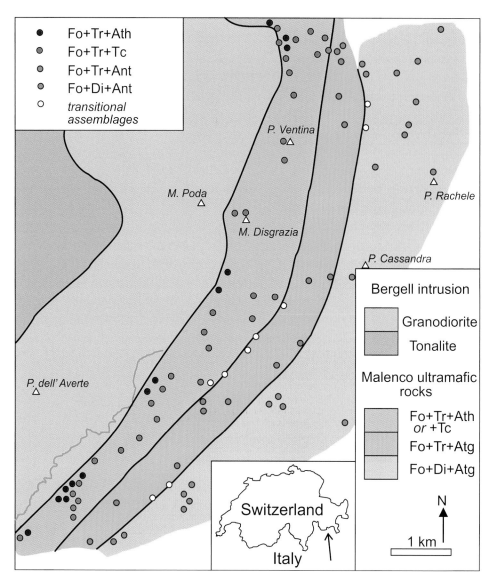

Figure 2.4 Map of the distribution of mineral assemblages in metamorphosed ultrabasic rocks in the aureole of the Bergell tonalite, Italy. Sampled assemblages are shown as coloured spots and define the zones shaded in the same colour. Note that the highest-grade anthophyllite zone is not shown separately on the map, although the anthophyllite-bearing assemblages are identified. Modified from Trommsdorff & Connolly (1996).

The exercise is repeated for the Zone C assemblage in Figure 2.5c. The difference from Zone B is that antigorite is no longer present and any rock with a composition that lay within the antigorite–forsterite–H$_2$O triangle in Zone B will now lie within the forsterite–talc–H$_2$O triangle in Zone C. The fact that these two different mineral assemblages can develop in rocks of the same composition confirms that they come from different metamorphic zones and do not just arise because rocks have different

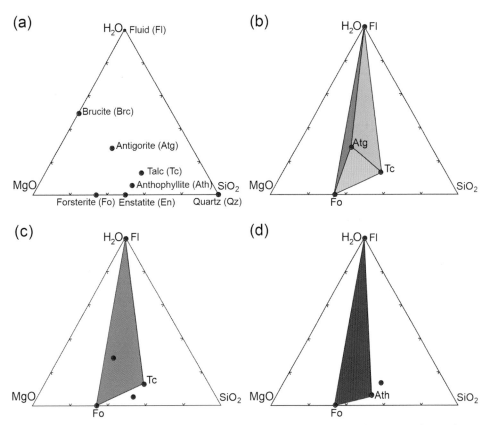

Figure 2.5 MgO–H$_2$O–SiO$_2$ triangular diagrams showing: (a) the compositions of the major phases of metamorphosed ultrabasic rocks; (b)–(d) phase compatibilities in successive zones of the Bergell aureole (Figure 2.4). Ca-bearing minerals are not shown. Further information about the *P–T* conditions under which these assemblages are stable is presented in Chapter 3. Note that this diagram has been constructed using the molecular proportions of the components. For example enstatite (MgSiO$_3$) plots at the midpoint of the MgO–SiO$_2$ side of the triangle because it contains 1 mole of MgO and 1 mole of SiO$_2$.

compositions. The composition of antigorite itself lies within the talc–forsterite–H$_2$O triangle, meaning that antigorite cannot be stable under the temperature (and pressure) conditions experienced by Zone C. The position of the minerals on the triangular diagram tells us that a rock with the chemical composition of pure antigorite would recrystallise to talc + forsterite + H$_2$O under Zone C conditions. This tells us that the reaction that takes place in passing from Zone B to Zone C is:

$$\text{antigorite} \rightarrow \text{forsterite} + \text{talc} + H_2O \qquad [2.4]$$

which can be balanced as:

$5Mg_3Si_2O_5(OH)_4 \rightarrow 6Mg_2SiO_4 + Mg_3Si_4O_{10}(OH)_2 + 9H_2O$. This reaction represents the ultimate high-temperature stability limit of serpentine and so this type of reaction is known as a '**terminal**' reaction.

A comparison of the assemblage of Zone D (Figure 2.5d) with that of Zone C (Figure 2.5c), shows a different type of transformation has occurred. Talc does not lie within the anthophyllite–forsterite–H_2O triangle and cannot co-exist with forsterite under the conditions that Zone D experienced, because talc and forsterite lie on opposite sides of a new anthophyllite–H_2O tie-line. Phases can only co-exist over a range of P and T conditions if they are connected by a tie-line on a diagram of this type. The reaction that takes place between Zones C and D results in the replacement of the talc–forsterite tie-line, present in Zone C assemblages, with an anthophyllite – H_2O tie-line (found in Zone D). Hence the reaction must be:

$$\text{forsterite} + \text{talc} \rightarrow \text{anthophyllite} + H_2O \qquad [2.5]$$

or $4Mg_2SiO_4 + 9Mg_3Si_4O_{10}(OH)_2 \rightarrow 5Mg_7Si_8O_{22}(OH)_2 + 4H_2O$.

A reaction of this type, which does not represent the ultimate limit of stability of any of the phases involved, consists of a 'tie-line flip' on the phase diagram. The reason why talc is not found in Zone D in these particular rocks is that there was always more olivine available initially than was required to react with the amount of talc present, and so reaction [2.5] continued until all the talc was used up. Talc can survive under Zone D conditions in rocks that are richer in silica than serpentinite.

The rocks of the Bergell aureole provide a demonstration of a progressive sequence of contact metamorphic zones, each characterised by a divariant assemblage, with univariant assemblages occurring near the zone boundaries. The principle of interpreting metamorphic rocks in terms of chemical equilibrium seems to be well founded in this case. The compositional phase diagrams represent the mineral assemblages found in the field and allow us to determine whether different rocks could have formed under the same conditions or must be from different zones. They also allow us to deduce the reactions that took place as metamorphic grade increased.

This example is particularly straightforward because solid-solution effects are small; to a first approximation additional components are lacking, and so the reactions are discontinuous. Some further examples are provided at the end of this chapter. The plotting of assemblages and reactions involving solid-solution minerals will be demonstrated in Chapter 4.

2.6   Phase Diagrams for Too Many Components: the Projection

Representing the reactions that take place in the Bergell aureole by the system $MgO – SiO_2 – H_2O$ is a simplification because other components and phases are also present. For example, one isograd involves the calcic phases diopside and tremolite. This introduces the chief problem with metamorphic phase diagrams: how to represent systems of more than three components on a flat piece of paper. We could represent four components as a 3D tetrahedron, but this gets messy and awkward to interpret. The problem can often be overcome by using a **projection,** which shows only a particular subset of the total mineral assemblage. In this example, all the assemblages shown in Figure 2.5b–d contain H_2O as a phase. We could therefore represent them all by just showing the pair of minerals that co-exist with H_2O in each zone, plotting their compositions in terms of their proportions of MgO to SiO_2. This is done by drawing a line from the H_2O apex of the $H_2O – MgO – SiO_2$ triangle through the composition of the mineral until it intersects the $SiO_2 – MgO$ side at a point corresponding to the **projected composition** of that mineral. The procedure is shown for talc in Figure 2.6a. Using this technique, we can represent all mineral assemblages *that include* H_2O on a one-dimensional $MgO – SiO_2$ diagram, and this can be used to deduce the reactions in exactly the same way as the triangular diagrams of Figure 2.5, by assuming that H_2O is always available to balance

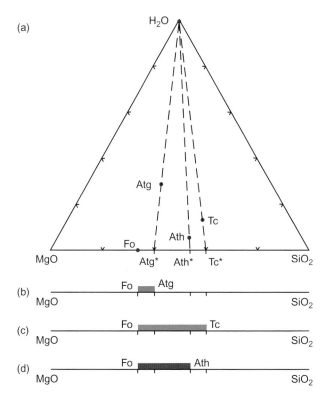

Figure 2.6 Projections of Mg–silicate mineral compositions. (a) Illustration of the projection of phase compositions in the system $MgO – H_2O – SiO_2$ onto the $Mg – SiO_2$ join. Projected compositions denoted by *. (b)–(d) Mineral assemblages (all also include an aqueous fluid phase) from the Bergell aureole (Figure 2.4), shown projected onto an $MgO – SiO_2$ binary diagram from H_2O.

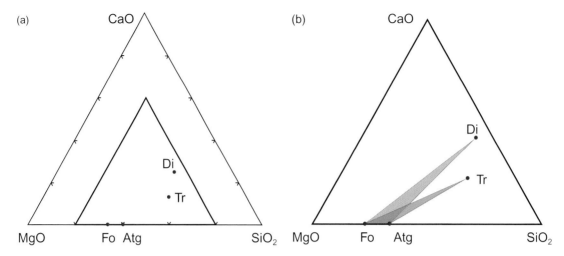

Figure 2.7 Use of a projection to investigate assemblages with Ca–Mg silicates. (a) Diopside and tremolite projected from H_2O onto the $CaO–MgO–SiO_2$ triangle. (b) Phase compatibilities in Zones A and B of the Bergell aureole respectively, shown within the portion of the $CaO–MgO–SiO_2$ triangle outlined in (a).

reactions as necessary. The $MgO–SiO_2$ projections for Zones B, C and D in the Bergell aureole are shown in Figure 2.6b–d.

Having reduced the three-component $MgO–SiO_2–H_2O$ system to one line introduces the possibility of adding a fourth component to construct a triangular diagram. To illustrate the reaction from Zone A to Zone B, CaO is an essential component, and in Figure 2.7a all the participating phases are shown on a $CaO–MgO–SiO_2$ diagram, projected from H_2O. Again, the projection means that only assemblages containing H_2O as an additional phase can be represented, and the phases are plotted according to the relative proportions of CaO, MgO and SiO_2 in their formulae, neglecting any H_2O. When it comes to working out reactions from the changes in tie-lines between zones, water will always be available to appear as a phase in the reaction so that it balances. The assemblages of Zones A and B are shown in Figure 2.7b. The tremolite–forsterite tie-line characteristic of Zone B crosses the diopside–antigorite tie-line representative of Zone A assemblages, and from this tie-line flip you should be able to deduce the reaction (this exercise is set at the end of the chapter). Do you need to include water as a separate phase to balance the hydrous minerals? Some projections used in subsequent chapters are more difficult to plot than this, but the underlying principle is still the same.

2.7   The Influence of Fluids on Metamorphic Reactions

Fluids play an important role in metamorphic reactions. Except at the highest temperatures, most mineral reactions proceed by a process of **solution-**

reprecipitation, i.e. reactant minerals dissolve into a water-bearing pore fluid at the same time that product minerals grow from it nearby (Chapter 7, Section 7.5.1). The pressure of the fluid in rock pores directly influences the temperature at which many reactions take place and has an important impact on how rocks deform. High fluid pressure can encourage rocks to fracture (Chapter 1, Section 1.4.3), but water also facilitates ductile deformation. Under most prograde metamorphic conditions, rocks deform by pressure solution in the presence of water, while water also facilitates intracrystalline deformation. This is discussed further in Chapter 8.

Many metamorphic reactions release fluid, most commonly H_2O and CO_2. These reactions are known as **devolatilisation reactions** because they release volatiles, and the conditions under which they occur depend not only on pressure and temperature, but also on the fluid pressure. The temperature needed for a dehydration reaction to occur increases as the pressure of H_2O in the pore space increases. This is an example of what is sometimes termed Le Chatelier's Principle, and similar effects are discussed further in Chapter 6.

Devolatilisation reactions generate more pore fluid, which may result in a local increase in fluid pressure and flow of fluid to regions of lower fluid pressure. The upper limit to which the fluid pressure can rise is controlled by the overall lithostatic pressure, P_l, and the tensile strength of the rock, beyond which the rock will be cracked by hydraulic fracturing, thus preventing any further rise in fluid pressure (Chapter 1, Section 1.4.3). Regionally metamorphosed rocks often contain veins that are likely to mark sites where fractures have formed and fluid has escaped, but many metamorphic rocks have dehydrated without veins developing; for example, veins are rare in thermal aureoles. It seems likely that often the rising fluid pressure inflates pores and push grains apart, so that the newly-generated fluid can escape as fluid pressure rises before fractures can develop.

Conversely, loss of fluid causes fluid pressure to drop, so that the lithostatic pressure can cause pores to collapse. Thus prograde metamorphism probably involves a balance between reactions, which tend to increase porosity, and fluid loss which tends to decrease porosity. Because of these mechanical limits to fluid pressure, it is commonly assumed that metamorphism takes place with $P_f = P_l$ and this is also the condition used in many experimental studies. If this is the case, fluid is retained in rocks during metamorphism despite being at very high pressure, which means that rocks undergoing prograde metamorphism must be very impermeable, and it must normally be difficult for fluid to flow through them.

Fluid is almost always present during progressive metamorphism of sedimentary rocks. The fluid present in pores initially is the sedimentary formation water, typically a brine. This evolves chemically as more fluid is released from minerals by metamorphic reactions, and cation exchange takes place with minerals. Nevertheless, it is likely that most metamorphic fluids are dominated by water as the solvent, with variable amounts of chloride salts in solution. NaCl predominates, but

KCl, CaCl$_2$ and FeCl$_2$ can all be important. The concentration of silica in meta-morphic fluids is usually controlled by quartz solubility, as we shall see in the following section. Since fluid makes up only a very small part of the rock mass and interacts with the rock at elevated temperatures over extended periods of time, it is reasonable to assume that metamorphic pore fluid is a saturated solution of the rocks that host it. The dissolved load of many metal ions in pore fluid is dictated by the concentration of chloride, the most abundant anion, but the relative proportions of different cations changes with P and T through exchange between fluid and minerals, much like cation-exchange reactions between minerals. Some important rock-forming components, including silica and alumina, dissolve as weak acids and bases rather than metal cations; their concentrations in solution mainly depend on T and P rather than the chlorinity of the pore fluid, although concentrations of dissolved gases are also important.

Fluid inclusions in minerals are one of the most important sources of information about the compositions of the fluid phase present during metamorphism. Box 2.2 provides a brief introduction.

BOX 2.2 Fluid inclusions

Much of what we know about metamorphic fluids comes from the study of fluid inclusions in minerals: bubbles of fluid trapped when a host mineral grew, or trapped subsequently as a result of the host mineral cracking and resealing. Some examples of fluid inclusions are shown in Figure 2.8. Fluid inclusions provide information about how fluid composition varies with metamorphic grade on a regional scale, and play an important role in developing our understanding of how fluids affect the evolution of both high- and low-grade metamorphic rocks (Touret 2001). Typically, they appear as small bubbles within minerals, commonly between 1 and 20 microns (μm) in diameter, but sometimes larger. They may have been trapped as the host crystal grew, or subsequently when the host has cracked and the crack resealed. Figure 2.8a illustrates a typical metamorphic vein quartz with planes of small fluid inclusion bubbles marking healed fractures and containing the fluid that was present when the host crystal cracked, rather than when it grew. The other examples in this figure are of unusually large inclusions. Fluid inclusions are usually trapped as a single, homogeneous fluid phase and the hole in the crystal which the fluid occupies does not change in volume significantly as the rock cools and is exhumed to the surface. However, the fluid inside the hole does shrink on cooling, lowering the internal pressure, and at temperatures below the critical temperature of the fluid (Chapter 1, Section 1.4.3), a vapour bubble splits off from the remaining liquid (Figure 2.8b). Other changes on cooling can

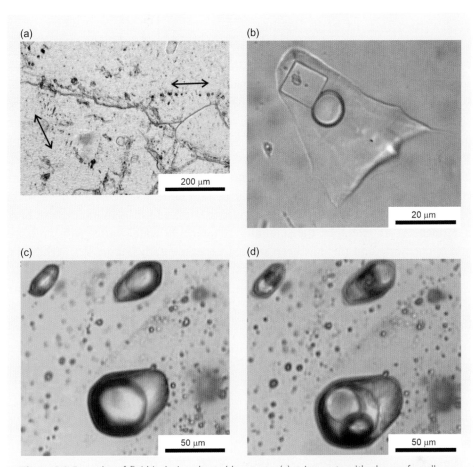

Figure 2.8 Examples of fluid inclusions hosted by quartz. (a) vein quartz with planes of small fluid inclusions decorating healed cracks. The arrows indicate two intersecting inclusion trails (b) brine fluid inclusion, with an approximately triangular shape, containing an approximately circular vapour bubble, strongly outlined because of the large density difference with the surrounding brine, and a square halite daughter crystal. (c) and (d) $CO_2 - H_2O$ fluid inclusions photographed at two different temperatures. In (c), photographed at +25.3 °C, the large inclusions contain two immiscible phases, water surrounding a bubble of liquid CO_2. In (d), the same field of view is shown at −58.3 °C, close to the CO_2 triple point. The CO_2 bubble now includes a dark bubble of CO_2 vapour and a rounded crystal of CO_2 ice in the progress of melting. The water-rich portion is also frozen and there appear to be clathrate crystals ($CO_2 - H_2O$ ice) close to the interface.

include the precipitation of dissolved solids as daughter crystals (Figure 2.8b) and the separation of dissolved gasses as immiscible fluids (Figure 2.8c and d). The major characteristics of fluid inclusions are investigated by **microthermometry**, i.e. observing the temperatures at which bubbles or solids within them appear or disappear using a microscope fitted with a heating–freezing stage to vary the temperature of the sample (Goldstein & Samson 2003). The pair of images in Figure 2.8c and d illustrate this.

2.8   The Concept of Buffering

When the phase rule is applied to a natural mineral assemblage it often transpires that the system has only a small number of independent degrees of freedom, even though there are a relatively large number of possible variables, e.g. pressure, temperature, fluid pressure and phase composition. In this case the independent variables and the mineral assemblage effectively fix the value of some of the other variables or **buffers** them at a specific value for a given pressure and temperature. Knowing that, for example, mineral trace element contents or fluid compositions are fixed by the mineral assemblage in a rock, provides an invaluable source of additional information to help understand how it formed. As an example, the solubility of quartz in water varies with pressure and temperature and this has been studied experimentally (Figure 2.9). While the concentration of silica in water could vary considerably if no rock was present, we can use Figure 2.9 to predict the concentration of silica in the metamorphic fluid for any quartz-bearing rock under any metamorphic conditions, as long as the fluid is predominantly water. This is because silica concentration in the fluid is buffered by the presence of quartz.

A simple but important example of buffering is the way in which an assemblage containing magnetite and hematite can control the concentration of available oxygen in the fluid phase at a constant value, irrespective of the actual amounts of magnetite and hematite present. The relevant equilibrium is:

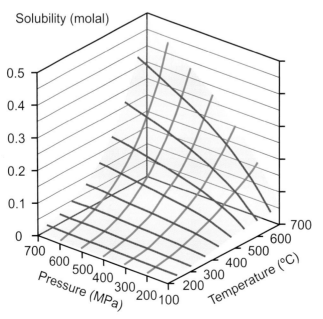

Figure 2.9 Solubility of silica in quartz-saturated supercritical water as a function of pressure and temperature. This is a 3D diagram with solubility as the vertical axis. Note that the pressure axis does not extend below 100 MPa. Simplified from Rimstidt (1997). The molal unit of solubility is the number of moles of the solute (here SiO_2) per kg of solvent (here H_2O).

$$\underset{\text{hematite}}{3Fe_2O_3} = \underset{\text{magnetite}}{2Fe_3O_4} + \underset{\text{in fluid}}{\frac{1}{2}O_2} \qquad [2.6]$$

The level of oxygen in the fluid phase is described in terms of its contribution to the total fluid pressure. Since oxygen makes up about 20% of the air at the surface today, the portion of atmospheric pressure, 1 bar, that is due to oxygen molecules, is about 0.2 bar or 0.02 MPa. This is known as the **partial pressure** of oxygen in air. For the purposes of chemical equilibrium studies, the thermodynamically effective pressure of a gas species in the fluid phase, or **fugacity** (f), is used instead of partial pressure. Like other measures of pressure, fugacity normally has units of pascals (Pa). Conditions with a relatively high f_{O_2} are referred to as **oxidising** whereas conditions with lower f_{O_2} are referred to as **reducing**. These terms are used in a strictly relative sense; even under the most oxidising conditions encountered in rocks, free oxygen molecules are effectively absent.

Applying the phase rule to the assemblage hematite + magnetite + $H_2O - O_2$ fluid, we have three phases and three components (e.g. $H_2O - FeO - O_2$). There are therefore two degrees of freedom in the system. If P and T are treated as independent variables, then f_{O_2} is a dependent variable with a fixed value at any specified P and T to which co-existing hematite and magnetite are subjected. Co-existing magnetite and hematite are said to buffer the fugacity of oxygen in the rock in which they occur; this is equivalent to saying that they buffer the chemical potential of oxygen.

Hematite is not particularly widespread in metamorphic rocks, and hematite–magnetite assemblages are relatively oxidising compared to many common combinations of minerals with Fe^{2+} and Fe^{3+}. Many assemblages in which magnetite, an Fe^{3+}-bearing phase, co-exists with quartz and an Fe^{2+}-bearing silicate mineral, such as olivine or almandine garnet, will also serve as an oxygen fugacity buffer, provided it is possible to write a reaction with magnetite + quartz on one side and the Fe-silicate + O_2 on the other. An example is the QFM (quartz–fayalite–magnetite) buffer, which is based on the equilibrium:

$$3SiO_2 + 2Fe_3O_4 = 3Fe_2SiO_4 + O_2 \qquad [2.7]$$

Values of f_{O_2} in metamorphism are typically extremely small, up to around 10^{-15} MPa under high-grade metamorphic conditions, and as low as 10^{-50} MPa at low grades. Firstly, this means that metamorphic rocks are readily oxidised if they come into contact with the atmosphere, or simply encounter surface-derived fluids. Secondly, the negligible levels of available oxygen in the crust mean that adjacent rocks often retain distinctive f_{O_2} values through metamorphism, unless extensive and focussed flows of fluid pass through them. In the absence of magnetite, the presence of Fe^{3+} in other minerals such as epidote or biotite can also provide a control on f_{O_2}

although, since the amount of Fe^{3+} substitution in such solid solutions can vary, they do not strictly buffer f_{O_2} to a unique value.

The idea that the composition of fluids (or solid solutions) can be controlled, or buffered, by the co-existing minerals follows directly from the phase rule and is fundamental to understanding metamorphic processes. Once understood, it is a very valuable tool in interpreting metamorphic rocks and we shall meet it at several places in the following chapters.

2.9 Practical Limitations to the Application of Chemical Equilibrium to Metamorphic Rocks

Very few rocks contain an assemblage of mineral grains which were all perfectly equilibrated with one another at a particular point in the metamorphic history and were then returned to the surface unchanged. So, does equilibrium exist during metamorphism? The best evidence comes from regional studies that demonstrate: (a) uniformity of mineral assemblage in rocks of similar chemical composition within well-defined areas; and (b) systematic variations in mineral assemblage for rocks of a particular chemical composition over a larger distance.

A somewhat different problem is presented by an individual rock or thin section. Almost every metamorphic rock experienced a prolonged history of recrystallisation, with probably one stage dominating the minerals present today. Usually, but by no means invariably, this will be the metamorphic episode that took place around the time of attainment of the highest (or peak) temperature experienced by the rock. However, the rock may retain minerals that were present during the prograde metamorphic history and have not broken down completely (relics), or grains that grew later than the main episode during partial recrystallisation, e.g. during retrograde metamorphism. How can these be distinguished?

In an ideal rock, each mineral will be found in contact with every other mineral within a single thin section, and the grains will show no evidence of breakdown. This ideal textural arrangement is approached in some carbonate-rich rocks, or in rocks metamorphosed at very high temperatures, but usually textures are more equivocal. The following criteria can provide useful pointers, but should not be followed slavishly.

1. *Is the rock obviously layered?* Compositional and mineralogical differences between layers may be observed even on the scale of a single thin section and minerals in different layers are not necessarily in equilibrium with one another (e.g. Figure 1.1a).

2. *Does a particular mineral occur in contact with the grain boundary network of the rock?* Most movement of material between reacting grains in a rock takes place along the grain boundaries rather than through mineral structure, except at the very highest temperatures. As a result, grains that become isolated from the continuous network of grain boundaries in the rock may be preserved as relics because other components are unable to react with them. There are two modes of occurrence of such relics: (a) as cores of zoned grains of a single mineral species, such as sodic cores of calcic amphiboles or the conspicuous coloured zones seen in tourmaline in many thin sections of schist (Figure 2.10a); or (b) as isolated **inclusions** within a different mineral species. For example, chloritoid or staurolite may occur as relic inclusions in garnet or plagioclase in pelitic schists from higher-grade zones than those where they are normally stable (Figure 2.10b, Chapter 7, Section 7.1). A cautionary note, however: a grain that appears in a two-dimensional thin section to be an inclusion may sometimes have been in connection with the rock matrix in the third dimension.

3. *Is there a strong association between a particular pair of minerals?* This is a difficult criterion to apply, but many retrograde minerals tend to replace (pseudomorph) a specific higher-temperature mineral, which helps distinguish them from the peak-metamorphic assemblage. For example chlorite often fringes and replaces garnet or biotite, and mats of fine-grained white micas (known as sericite) often replace Al_2SiO_5 minerals, staurolite or feldspar (Figure 2.10c). Fortunately, pseudomorphs of fine-grained retrograde alteration products such as **sericite** are usually quite conspicuous. Even so, interpreting the relationships between retrograde replacement minerals can be tricky, especially if they formed during low-grade metamorphism of igneous rocks, where each different igneous mineral may be replaced by a different metamorphic mineral. There are some cases where, even at high temperature, specific minerals tend to be associated with one another. For example, sillimanite often grows on biotite, even though it may be in equilibrium with other minerals in the rock (Figure 2.10d).

4. *Are there theoretical reasons for supposing that all minerals were not in equilibrium with one another?* This is a line of argument that must be applied only with great caution and requires a good understanding of the ranges of conditions over which particular minerals and mineral assemblages may be stable. However, it is sometimes apparent that, from phase rule considerations, a rock contains too many phases. A rock in which andalusite, kyanite and sillimanite are all present might just conceivably have been metamorphosed at the unique value of P and T at which all these three can co-exist, but it is much more probable that, as conditions of metamorphism changed, each polymorph formed at different times in its own stability field, but failed to break down completely as it became unstable.

(a)

(b)

(c)

(d)

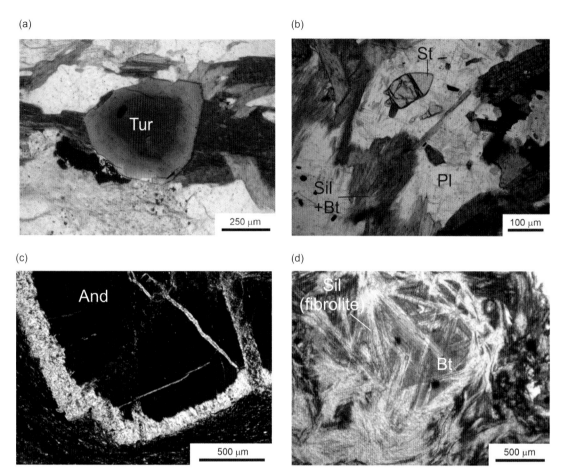

Figure 2.10 Photomicrographs of textures indicative of metamorphic history and process. (a) Compositionally zoned metamorphic tourmaline in staurolite schist. (b) Relic inclusion of staurolite in plagioclase from a garnet–biotite–sillimanite schist with fibrous sillimanite replacing biotite and penetrating the plagioclase. (c) Partial retrograde replacement of an andalusite porphyroblast by sericite. (d) Fibrous sillimanite (fibrolite) showing preferential growth on structurally-controlled orientations in biotite.

Even where an assemblage does not overtly contravene the phase rule criteria for equilibrium, it can sometimes be argued that one mineral in a rock would not have been stable at the same conditions as the others, although it is an argument that is easily abused. For example, chlorite can be stable up to grades equivalent to the sillimanite zone in mafic rocks of suitable composition, and is therefore not necessarily a retrograde phase in high-grade rocks. On the other hand, the association of chlorite + K-feldspar is stable only under very low-grade (chlorite zone) conditions and therefore chlorite in, say, a garnet-bearing schist that contains K-feldspar is almost certain to be of retrograde origin.

Where chemical analyses of minerals are available it is possible to use the partitioning of elements between mineral solid solutions as a criterion for equilibrium. Distribution coefficients (K_D) for the experimentally-constrained partitioning of Fe and Mg between pairs of ferromagnesian minerals are known for most metamorphic conditions. When very different results are obtained for a particular mineral pair in an individual sample it is likely to mean that those two minerals did not in fact co-exist in equilibrium.

5. *Textures indicative of reaction.* It has been pointed out earlier in this chapter that many metamorphic reactions take place over a range of conditions, and that within that range, products and reactants can co-exist in equilibrium even though one is tending to grow at the expense of the other. For this reason the fact that a particular mineral appears to be corroded and breaking down does not necessarily indicate that it was out of equilibrium with the other phases in the rock. Reactions can progress continuously over a range of conditions and examples of this will be provided in Chapters 4 and 5.

SUMMARY

This chapter has introduced the basic rationale behind many of our interpretations of metamorphic rocks, which are frequently based on the assumption that all or part of the mineral assemblage originally grew together at equilibrium under some specific physical conditions to which the rock was subjected. Changing conditions of pressure and temperature may lead to reaction and the production of new assemblages. While some reactions take place abruptly at a particular temperature and result in the disappearance of certain minerals and the growth of new ones, other reactions may take place over a range of conditions within which reactants and products co-exist but change progressively in composition and abundance.

Armed only with a knowledge of the general compositions of the minerals found in an area, it is possible to unravel the effects of rock composition from those of metamorphic conditions in determining mineral assemblages, and hence work out which rocks were likely metamorphosed under similar conditions and which reflect a different metamorphic environment. In this chapter we have pointed out a range of variables that can affect the stability of mineral assemblages. In addition to P and T, these include fluid pressure, P_f, as well as the chemical composition of both rocks and minerals. Fortunately, these variables do not all change independently. The phase rule makes it clear that, depending on the number of degrees of freedom of the assemblage, some of these parameters automatically take on pre-determined values if others are fixed, much as the boiling point of salt water at 1 atmosphere pressure is fixed if the salt content is known, and vice versa.

The closely related idea that the composition of a solution, whether solid or fluid, may be determined by the nature of the other phases that co-exist with it gives rise to the concept of buffering. The silica content of most metamorphic fluids is buffered by co-existence with quartz because they are usually saturated solutions of quartz, and this type of relationship will prove valuable in Chapter 3 as a means of estimating metamorphic temperatures. In a more abstract way, we can think of the oxygen content of the same fluid as being buffered if there are minerals present that contain Fe in different oxidation states, so that a reaction involving oxygen can be written between them.

Metamorphic rocks seldom recrystallise completely as metamorphic conditions change, and an understanding of metamorphic processes and the role of fluids allows us to understand and interpret rocks whose mineral assemblages record a history of how they have evolved through time. This is a theme which we will return to in Chapters 7 and 8.

EXERCISES

1. Why is it important to be able to identify assemblages of minerals which have formed at equilibrium with one another?

2. For each of the reactions listed below, determine whether they are continuous, discontinuous or cation-exchange reactions. Use the mineral formulae given in Appendix 1 if necessary.

   (a) jadeite + quartz = albite,
   (b) (Na,K) white mica + quartz = andalusite + (Na,K) alkali feldspar + water,
   (c) Fe-garnet + Mn-ilmenite = Mn-garnet + Fe-ilmenite.

3. Explain the terms *dependent variable* and *independent variable* and their relation to the phase rule.

4. A rock contains the following phases at the peak of metamorphism:
   quartz: SiO_2; muscovite: $KAl_3Si_3O_{10}(OH)_2$; K-feldspar: $KAlSi_3O_8$; sillimanite: Al_2SiO_5; water: H_2O.

   (a) What is the least number of chemical components needed to describe the composition of all of these phases? How many degrees of freedom does the assemblage have?

   (b) A more careful examination of the rock reveals that it also contains albite $(NaAlSi_3O_8)$. How many components are now needed?

   (c) Has the number of degrees of freedom changed?

5. Deduce the reaction between Zone A and Zone B in the Bergell aureole (Section 2.5) using the $CaO - MgO - SiO_2$ diagrams of Figure 2.7. First, write the reaction without attempting to balance it exactly, but adding water as appropriate. Next,

use the mineral formulae provided in Appendix 1 to balance the reaction. Why is it reasonable to assume that water was always present as a fluid phase in this aureole?

6. Anorthite (Ca-feldspar) is not stable at high pressures.

    (a) Sketch a version of Figure 2.2 to show the phase relations at high pressure, i.e. without anorthite present (you may assume all the other phases are still stable).

    (b) What will be the mineral assemblages of rock compositions *X* and *Y* under the high-pressure conditions? How do the high-pressure assemblages differ from those stable at the pressure conditions for which Figure 2.2 is constructed?

    (c) Deduce the reaction by which anorthite breaks down at high pressure.

FURTHER READING

Jamtveit, B. & Austrheim, H. (2010). Metamorphism: the role of fluids. *Elements*, **6**(3), 153–8.

Phillpotts, A. R. & Ague, J. H. (2009). *Principles of Igneous and Metamorphic Petrology*, 2nd Edition. Cambridge University Press, 667 pages.

Spear, F. S. (1993). Metamorphic phase equilibria and pressure-temperature-time paths. *Mineralogical Society of America, Monographs*, **1**, 799.

Yardley, B. W. D. & Bodnar, R. J. (2014). Fluids in the continental crust. *Geochemical Perspectives*, **3**(1), 127 pages.

3 The Pressure–Temperature Conditions of Metamorphism

Ever since metamorphic rocks were first recognised, geologists have sought to understand the conditions under which they formed and hence learn something about how rocks are buried and reworked inside the Earth. The recognition that many metamorphic mineral assemblages formed close to equilibrium, discussed in Chapter 2, opened up the possibility that metamorphic conditions could be determined once the conditions for equilibrium were quantified. In the course of the twentieth century, it gradually became possible to do this, and hence to determine the depths and temperatures at which specific rocks were formed. As we saw in Chapter 1, the variation of temperature with depth can be characteristic of particular tectonic settings and so P–T values for metamorphism have become essential for understanding the tectonic settings in which ancient metamorphic belts developed. The development of new experimental and analytical techniques and improved computer modelling approaches has meant that our ability to make accurate and precise estimates of the conditions at which particular metamorphic mineral assemblages can exist at equilibrium has improved rapidly. In this chapter we first introduce the underlying principles which enable us to make estimates of temperatures and pressures in metamorphism. We then introduce a number of approaches to estimating metamorphic conditions, ranging from classical qualitative methods (which are valuable during fieldwork) to comparison with experimental results and the application of thermodynamic databases. In Chapters 4–6 we present examples of how some of these approaches can be applied to a variety of rock types.

3.1 Metamorphic Reactions: the Thermodynamic Principles

The study of metamorphic reactions and the attainment of equilibrium is governed by the laws of thermodynamics, and a detailed account is beyond the scope of this book. However, some basic principles that are essential to understanding metamorphic rocks are outlined here in a qualitative way, and references for more rigorous study are given at the end of the chapter.

A fundamental concept is that any substance can be assigned a 'free energy' reflecting its energy content. At equilibrium, atoms are arranged into the substance or substances which result in the lowest free energy. There is more than one way of formulating a free energy, but in geology the **Gibbs free energy** (G) is generally used because the Gibbs free energy of a phase varies with its composition (if it is a solid or liquid solution) and with pressure and temperature, and these are the variables with which geologists are most concerned. G is an **extensive property** – the more material you have, the more free energy it contains. In contrast, an **intensive property** such as temperature has the same value irrespective of the size of the system. For extensive properties it is usual to normalise them to 1 mole of the substance, which in petrology is defined as the **formula weight** in grams. For example, from the atomic weights of Si and O, we obtain a formula weight for quartz (SiO_2) of 60.083. The **molar volume** (V) of quartz is then the volume of 60.083 g quartz, which is 22.69 cm^3 under standard conditions of 25 °C and 0.1 MPa.

The molar Gibbs free energy of any phase is a distinct quantity that can, in principle, be determined. However, we often deal with mineral solid solutions where it is the energy of a specific end-member in the solution that is of interest. For example, if the Gibbs free energy of albite in solution in plagioclase is greater than that of albite in solution in an adjacent orthoclase, then we might expect albite to move out of the higher-energy environment (the plagioclase crystal) into the lower-energy one (orthoclase). The measure used for the Gibbs free energy of the components of solutions is the **chemical potential** (μ), defined as:

$$\mu_i = \left(\frac{\partial G_p}{\partial n_i} \right)_{T,P,n_j} \qquad [3.1]$$

i.e. for an end-member or solution component i, its chemical potential μ_i is equal to the change in the Gibbs free energy of the phase in which it occurs (∂G_p) resulting from a change in the number of moles of i (n_i) when temperature, pressure and the number of moles of all other components (n_j) remain constant.

We can define the molar Gibbs free energy of a solution phase containing j components as the sum of contributions from each of those j components:

$$G_p = \left(\sum_{i=1}^{i=j} \mu_i n_i \right) / n_p \qquad [3.2]$$

where n_p is the number of moles of the phase made from n_i moles of each of the constituent components.

Hence the Gibbs free energy (G_S) of a system made up of m phases becomes:

$$G_s = \sum_{k=1}^{k=m} G_{p,k} . n_{p,k} \qquad [3.3]$$

where $G_{p,k}$ is the molar Gibbs free energy of the k th phase, of which $n_{p,k}$ moles are present.

Clearly, if the same atoms can be combined in a number of different ways involving different assemblages of phases present in different proportions, then the value of G_s will be different in each case. For example, at P–T conditions within one of the divariant fields shown on Figure 2.1, the Gibbs free energy of each of the possible forms of Al-silicate will be different.

At equilibrium, the Gibbs free energy of a system is at a minimum for the prevailing conditions of pressure and temperature, i.e. the stable combination of atoms is the one that yields the smallest value of G_s.

The reason why metamorphic reactions take place is that the value of G for every substance changes with variation in P or T, but for different minerals the magnitude of the change is different. As a result, the stable mineral assemblage at equilibrium under one set of conditions may no longer have the lowest possible free energy after P or T change.

As an example, consider a rock made up solely of andalusite. As conditions change, the same atoms can rearrange themselves into kyanite or sillimanite when these have a lower Gibbs free energy, since the chemical composition of each of these phases is the same, Al_2SiO_5. Andalusite has a lower density than kyanite (3.15 g/cm^3 versus 3.60 g/cm^3), and therefore high-pressure conditions are likely to stabilise kyanite as the denser phase. If temperature is kept constant, the variation in Gibbs free energy with pressure for any phase is given by:

$$\left(\frac{\partial G}{\partial P} \right)_T = V \qquad [3.4]$$

where V is the molar volume of the phase. Because neither mineral is very compressible, V remains relatively constant over large ranges of pressure. As a result, if G is plotted against P for any phase at a particular temperature, the result is a line that is nearly straight, with a slope of V. The notation V° is used for the molar volume measured under standard conditions (25 °C and 0.1 MPa) and to a good approximation the value of V° can be used to predict how Gibbs free energy

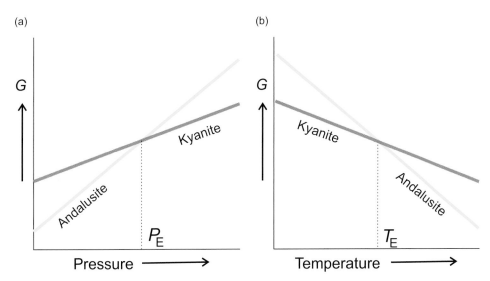

Figure 3.1 Schematic plots illustrating the variation in the molar Gibbs free energy (G) of andalusite and kyanite, (a) with respect to pressure at constant temperature, and (b) with respect to temperature at constant pressure. P_E and T_E are the pressure and temperature respectively at which the two phases can co-exist in equilibrium.

changes with pressure. This has been done schematically for andalusite (V° = 51.530 cm³) and kyanite (V° = 44.090 cm³) in Figure 3.1a, which is a plot of Gibbs free energy against pressure constructed for a specific, but arbitrary, temperature. The line representing kyanite has a shallower slope than that for andalusite because the molar volume is less, and hence the lines cross at a point corresponding to $P = P_E$, the pressure at which both phases have the same Gibbs free energy and are therefore equally stable and can co-exist in equilibrium. At pressures less than P_E, $G_{And} < G_{Ky}$ and so any kyanite will tend to react to form andalusite, while at pressures greater than P_E the reverse is true. The actual value of P_E depends on the temperature for which the plot is drawn.

There is a rather similar relationship between G and temperature when P is held constant. In this case it is a property called **entropy** (S), representing the degree of randomness in the atomic arrangement, which determines the way in which G changes with temperature. This is illustrated for andalusite and kyanite in Figure 3.1b, and here it is the phase with the larger value of S (i.e. andalusite) that becomes stable as T increases. Again, there is a unique temperature, T_E, at which kyanite and andalusite are equally stable for every pressure. If kyanite is heated above T_E it will tend to react to andalusite, while andalusite will tend to react to kyanite if cooled below T_E.

The relationships illustrated in Figure 3.1 are particularly simple because they concern only two phases of the same composition. However, in principle, there is no difference between comparing the Gibbs free energy of two phases of the same composition, and comparing the Gibbs free energy of two mineral assemblages, each comprising a number of phases, *provided that it is possible to write a balanced chemical reaction between the two assemblages.* Equilibrium is attained where the Gibbs free energy of both sides is the same. The equation $\Delta G = 0$ is perhaps *the* most important thermodynamic equation for metamorphic petrology. Where thermodynamic data are available for all the phases involved in two assemblages of equivalent composition, it is possible to calculate the *P–T* conditions for which $\Delta G = 0$, i.e. the equilibrium curve on a *P–T* diagram. Equally, if the equilibrium curve is determined experimentally and thermodynamic data for some of the phases are known independently, then thermodynamic data for other minerals can be extracted from the experimental results.

It is possible to apply the phase rule to rocks without any knowledge of the thermodynamic properties of the phases (e.g. Chapter 2). The predictions of the phase rule are consistent with the conclusions drawn from thermodynamic considerations. Andalusite and kyanite constitute a two-phase, one-component system and therefore if both phases co-exist, there is only one degree of freedom, as with boiling water. There is only a single *T* for any given *P* at which they can co-exist stably, and this will be T_E of Figure 3.1b.

A final thermodynamic property of fundamental significance for the way in which the Earth's crust behaves is **enthalpy** (*H*). The enthalpy of a phase reflects its heat content and so the reactants and products of a reaction may have different enthalpies in the same way as they may have different volumes. The enthalpy difference between reactants and products is:

$$\Delta H_r = H_{\text{products}} - H_{\text{reactants}}$$

and ΔH_r is known as the **heat of reaction**. If ΔH_r is positive, additional heat must be put into the system to enable the reaction to proceed even when the equilibrium temperature has been attained. Such reactions are known as **endothermic**: a good example is the boiling of water. Metamorphic reactions that liberate water from hydrous minerals (which usually take place during prograde metamorphism) are similarly endothermic. Other reactions, for example the hydration of peridotite to serpentinite or the breakdown of plagioclase as basalt is converted to eclogite, give off heat and are called **exothermic**.

3.2   Qualitative Relationships: Metamorphic Facies

Metamorphic zones were recognised in a number of different types of rocks long before the experimental equipment existed to reproduce the conditions under which their distinctive assemblages had formed (Chapter 1, Section 1.2.1). However, although zonal schemes have been devised for different regions based on a range of rock types including pelitic schists, metamorphosed basic and ultrabasic igneous rocks, calc-silicate rocks, marbles and even cherts, the precise correlation of zones based on different rock types can be difficult. To overcome this problem, the Finnish geologist Pentti Eskola devised a scheme of **metamorphic facies** (originally termed mineral facies), to capture associations of mineral assemblages and rock compositions that are characteristic of a distinctive range of metamorphic conditions. Because metamorphosed basic igneous rocks are widespread, and their distinctive mineral assemblages can be correlated with zones in other lithologies (except at very low grades), their mineral assemblages provide the basis for the facies scheme. The diagnostic mineral assemblages of the facies are stable over a wider range of P–T conditions than some mineral zones recognised in other rock types, but metabasites occur in almost all metamorphic belts. Facies based on mineral assemblages as a guide to metamorphic conditions might seem redundant in the modern world, where, as we shall shortly see, precise calculations of the temperature and pressure at which individual rocks have formed have become almost routine. However, these calculations are only possible once samples have been taken back to the laboratory and subjected to extensive analysis. In contrast, suites of rocks can usually be assigned to a metamorphic facies in the field, something which can be of great value to a geologist seeking to map structures that juxtapose rocks of different metamorphic grade, or engaging in general mapping or exploration.

The characteristic mineral assemblages of metabasic rocks from each of the facies are described in more detail in Chapter 5. Eskola deduced the relative pressures and temperatures at which rocks of the different facies had formed, based on two principles: (a) rocks formed at higher pressures will tend to contain denser minerals than rocks of the same composition metamorphosed at lower pressures; and (b) devolatilisation reactions almost invariably take place with increased temperature, so that progressively higher temperature rocks will have progressively lower volatile (H_2O and CO_2) contents.

The facies scheme has been revised numerous times since originally proposed. Our scheme (Figure 3.2) has fewer divisions at low temperatures and pressures than many others. For example some schemes include an additional 'epidote amphibolite facies' between the greenschist and amphibolite facies, which approximately corresponds to the garnet zone in metapelitic rocks (Chapter 4, Section 4.2). However, this gradational variation is difficult to identify in other rock types, and therefore we have not included it here. Likewise, we have not separately distinguished facies associated with contact metamorphism at low pressure, although the assemblages of low-pressure metapelites are clearly distinctive, as we saw in Chapter 1. Details of the assemblages typical of rocks in each facies are discussed further in Chapters 4, 5 and 6.

It is convenient to consider the facies in three groups: those of moderate pressure and temperature, those of high pressure and those of low temperature. Although we

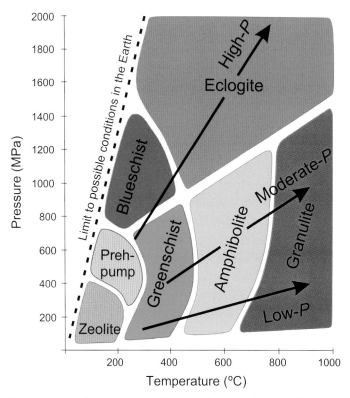

Figure 3.2 *P–T* diagram showing the approximate conditions under which mineral assemblages of different metamorphic facies are stable. The eclogite facies extends to higher pressures than shown on this plot, with coesite eclogites requiring pressures in excess of 2500 MPa (see Chapter 5, Section 5.5.3). The *P* and *T* values shown on the axes are the current best estimates and are subject to change, whereas the assemblages that assign a rock to a particular facies, will not. Black arrows show examples of the way in which *P* and *T* can vary together within an individual orogenic belt to define facies series and are the basis for the treatment of metamorphism in Chapters 4–6.

have not distinguished distinct low-pressure facies, there are distinctive assemblages of low-pressure metamorphism which will be discussed in Chapter 4, and distinctive tectonic associations which will be introduced in Chapter 10.

3.2.1   Groupings of Facies in Metamorphic Belts

We have already seen in Chapter 1 that sequences of metamorphic rocks in the field often span a range of grades, so several facies are commonly represented in any one metamorphic region. In practice, certain facies tend to occur together in the same metamorphic belt, giving rise to a **facies series** (Figure 3.2).

Facies of low-to-moderate pressure and moderate-to-high temperature (the greenschist, amphibolite and granulite facies) probably account for the bulk of metamorphic rocks exposed on Earth. The field examples described in Chapter 1 belong to the greenschist and amphibolite facies; granulite facies rocks form at still higher temperatures where they have often experienced melting and developed minerals also found in igneous rocks. Although it is not easy to identify distinct low- and moderate-pressure facies in the field, metapelites do show very different zonal schemes, as we saw in Chapter 1, and so low- and moderate-pressure facies series will be discussed separately in later chapters.

The blueschist and eclogite facies represent distinct, high-pressure conditions characterised by the production of relatively dense phases. The diagnostic metabasic rocks of these facies are blueschist and eclogite respectively (Chapter 1; described in more detail in Chapter 5). Rocks metamorphosed in the high-pressure facies commonly show evidence of overprinting under moderate-pressure facies conditions during their return to the surface.

The low-temperature (or low-grade) facies can be precursors to any facies series and commonly contain rocks that are too fine-grained for the minerals to be identified in hand sample. For example, slates are an example of metasedimentary low-grade rocks (Chapter 4). Some igneous rocks develop distinctive new minerals such as zeolites, although igneous features typically remain (Chapter 5). We divide low-grade rocks into the zeolite and prehnite–pumpellyite facies (Figure 3.2). Both facies may be developed regionally or in zones of hydrothermal alteration around volcanic centres. Some metamorphic belts, for example 'slate belts', record only low-grade facies assemblages, but elsewhere, low-grade facies may pass gradually into high-, moderate- or low-pressure facies.

3.3   Adding Numbers: Quantitative Estimates of Pressure and Temperature

When metamorphic zones and facies were first recognised, geologists had little idea of the absolute values of pressure and temperature to which most metamorphic rocks had been subjected. With the development of experimental techniques for the study of mineralogical phase equilibria at high pressures and temperatures in the 1950s and 1960s, the conditions of formation of many naturally occurring mineral assemblages could be quantified. Metamorphic geologists began to glean a reasonably good idea of the sorts of pressures and temperatures required for the formation of many common rock types. It is not practical to carry out experiments using all possible rock compositions at all possible conditions, and so the emphasis shifted to performing experiments with end-member mineral compositions whose results could be used to generate thermodynamic databases of mineral properties. More precise estimation of metamorphic conditions uses these data, coupled with mineral analyses, to calculate conditions for the specific rock samples that are being investigated. Mineral indicators of metamorphic *P* and *T* are known as **geobarometers** and **geothermometers** respectively, and are typically based on specific subsets of minerals in the rock under investigation.

In addition to looking at specific mineral geobarometers and geothermometers, it is also possible to solve for the conditions at which the full assemblage of minerals in a rock have co-existed, using a computer program that combines data from a thermodynamic database with mineral compositions (Section 3.3.9). An increasingly popular method of estimating *P–T* conditions of metamorphism is to calculate diagrams that predict the types and abundances of different minerals present in a specific rock composition across a range of *P–T* conditions (Section 3.3.10).

Most methods of determining *P–T* conditions are based on the thermodynamically-stable configuration of minerals in the rock, but there are successful approaches based on features which are not developed at equilibrium, as we shall see in Section 3.3.11.

A common theme across all these approaches is that the results they yield may be completely meaningless if the rock textures have not been carefully assessed and evaluated first (Chapter 2, Section 2.9). If any of the minerals whose presence is used to estimate metamorphic conditions were not in fact in equilibrium with the others, equilibrium calculations are pointless. *It is not possible to learn anything of value about metamorphic rocks and how they formed without first studying their mineralogy and textures.*

3.3.1   Petrogenetic Grids

The simplest way of estimating the *P–T* conditions for the formation of a particular mineral assemblage is to identify where that assemblage sits on a *P–T* diagram showing the univariant curves for the reactions that control the appearance and breakdown of the relevant minerals. This sort of diagram is called a **petrogenetic grid,** and it is normally constructed for a specific rock type. Experiments constrain the location of many univariant curves on these diagrams, providing limits to most of the common assemblages of metamorphism. Thermodynamic database compilations of these experimental data (Section 3.3.9) allow the construction of a wider range of grids, including reactions for which direct experimental data are not available; examples of these are provided in Chapters 4, 5 and 6. Many of the reactions that are plotted on a petrogenetic grid involve the release of volatiles, most commonly water. The temperatures at which such reactions take place depends on fluid pressure as well as total pressure (Chapter 1, Section 1.4.3).

The Bergell pluton in Switzerland, introduced in Chapter 2, Section 2.5, provides a good example. The petrogenetic grid for the isograd reactions is shown in Figure 3.3. It indicates that the temperatures attained in the different zones of the Bergell aureole ranged from below 450 °C in the outer aureole to around 650 °C near the contact. The highest temperature reaction in Figure 3.3, at over 600 °C, is reaction

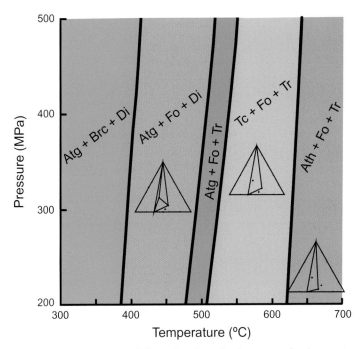

Figure 3.3 Petrogenetic grid illustrating the univariant curves for the reactions corresponding to the isograds in contact metamorphosed serpentinites from the Bergell aureole at Val Malenco, described in Chapter 2. The phase compatibility diagrams for the MgO–SiO$_2$–H$_2$O system from Figure 2.5 are included for reference. Based on Trommsdorff & Connolly (1996).

[2.5], which marks the replacement of talc by anthophyllite. Reaction [2.4], marking the disappearance of antigorite, takes place at a little over 500 °C and the reaction by which diopside is replaced by tremolite takes place at somewhat lower temperatures, in agreement with the isograds shown in Figure 2.4. Additionally, the mineral brucite, not mentioned in Chapter 2, is sometimes present outside the aureole and breaks down at temperatures of around 400 °C.

The majority of metamorphic minerals form solid solutions of variable composition, and their different end-members will have different stability limits. This means that experiments performed on a specific mineral composition may not accurately define the conditions of formation of the same assemblage in a rock in which the solid-solution mineral has a different composition. For example, the white micas found in pelitic schists metamorphosed under moderately high-grade conditions, such as in the staurolite zone, are solid solutions between paragonite (Na) and muscovite (K). At high grades, white mica breaks down to produce alkali feldspar, which is also a solid solution between Na and K end-members. The breakdown reactions for the two white mica end-members are:

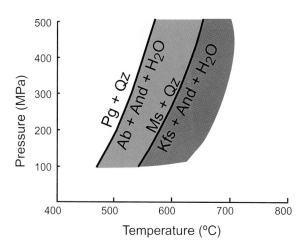

Figure 3.4 Petrogenetic grid showing the experimental univariant curves for the breakdown of the K and Na end-member white micas, muscovite and paragonite, to alkali feldspars (K-feldspar and albite respectively) and andalusite. Data from Chatterjee (1972) and Chatterjee & Johannes (1974).

$$NaAl_3Si_3O_{10}(OH)_2 + SiO_2 \rightarrow NaAlSi_3O_8 + Al_2SiO_5 + H_2O \qquad [3.5]$$

$$KAl_3Si_3O_{10}(OH)_2 + SiO_2 \rightarrow KAlSi_3O_8 + Al_2SiO_5 + H_2O. \qquad [3.6]$$

The experimentally-determined univariant curves for these reactions (with $P_{H_2O} = P$) are shown in Figure 3.4. Natural white micas, which do not normally have pure end-member compositions, will break down at intermediate conditions according to their actual composition. The difference between the P–T conditions of the two end-member curves is not particularly great in this case, and both the product and the reactant assemblages involve Na–K solid-solution minerals. The Na-content of the muscovite in appropriate rocks is usually small, and so it is reasonable to use the experimentally-determined muscovite breakdown reaction on a petrogenetic grid to indicate the approximate conditions under which muscovite–quartz assemblages react to K-feldspar–Al-silicate assemblages, irrespective of the presence of Na in the natural minerals.

There has been a recent tendency among metamorphism specialists to underestimate the value of petrogenetic grids in favour of P–T conditions calculated directly via computer programs and thermodynamic databases. However, like facies, petrogenetic grids are particularly useful for obtaining an immediate indication of the likely conditions under which different metamorphic rock units were formed, for example during fieldwork, since no analytical data are required. This information can then inform the development of a more-precise mapping and sampling strategy. Some metamorphic mineral assemblages are stable over such a restricted range of P–T conditions that their presence defines metamorphic conditions remarkably well. On the other hand, not all rocks have suitable compositions for the reactions shown on a petrogenetic grid to actually affect them. For example quartzites will mainly contain quartz and perhaps feldspar, regardless of the metamorphic conditions they have experienced.

3.3.2   Geothermometers and Geobarometers: the Principles

There are several different approaches to determining the P–T conditions of formation of a mineral assemblage once it has been established that the assemblage formed at equilibrium. The most widely-used ones are based on mineral reactions, which may be univariant, continuous or cation-exchange reactions. Obviously the most useful reactions for determining pressure of metamorphism will be those that take place at nearly the same pressure over a wide range of temperature, while a reaction with the opposite characteristics will make a good temperature indicator.

In Section 3.1 we pointed out that the variation in the Gibbs free energy of a mineral (or assemblage of minerals) with changing pressure depends on the molar volume of the mineral, whereas variation in Gibbs free energy with temperature depends on entropy. We can define the volume change of a reaction:

$$\Delta V_r = V_{products} - V_{reactants}$$

Similarly the entropy change is:

$$\Delta S_r = S_{products} - S_{reactants}$$

A reaction that makes a good pressure indicator will have a large ΔV_r but small ΔS_r. This means that variations in pressure will lead to large changes in the relative free energies of reactants and products, making reaction more likely. Fluctuations in temperature, on the other hand, will have relatively little effect. In contrast, reactions with a large ΔS_r but small ΔV_r are predominantly temperature-sensitive. Some reactions are almost exclusively sensitive to temperature, especially cation-exchange reactions, because the volume change accompanying such reactions is very small. Devolatilisation reactions are also predominantly temperature-sensitive under most crustal P–T conditions because of the large increase in entropy that accompanies release of volatiles from ordered mineral structures. Most pressure-sensitive reactions are, however, also somewhat sensitive to temperature and so good geobarometers are much rarer than good geothermometers.

The following sections provide an outline of the main approaches to determining the pressures and temperatures at which metamorphic rocks have formed. Further examples for specific types of rock are given in subsequent chapters. In all cases, estimating the P–T conditions at which metamorphic rocks have formed involves first a detailed textural investigation, as described in Chapter 2, Section 2.9, to establish which minerals are likely to have co-existed in equilibrium. It is then usually necessary to analyse the composition of the co-existing minerals, taking care to allow for chemical zoning if present. The resulting mineral analyses are then recalculated on a molar basis and can be incorporated into the appropriate equations or software, as discussed in the following sections.

3.3.3   Directly Calibrated Geothermometers and Geobarometers

When petrologists first began to estimate metamorphic conditions, they took the equilibrium conditions that had been determined experimentally for individual end-member reactions as their starting point, and tried to estimate how the natural mineral compositions might have shifted them. Each reaction was evaluated independently, and estimates based on different reactions did not always agree well. As more experimental data became available, the thermodynamic properties of minerals could be estimated from multiple sets of experiments to give an internally consistent dataset, and calculations based on different reactions then give consistent results. Simple experimental calibrations of individual reactions are less often used today, because thermodynamic datasets usually give more reliable results than individual sets of experiments. Nevertheless, because some important metamorphic minerals are complex solid solutions, there are some geothermometers and geobarometers that are based on experiments appropriate for specific rock types and conditions.

3.3.4   Markedly Continuous Reactions

Many important natural metamorphic reactions can be approximated by univariant curves on a petrogenetic grid, even where there is some solid solution in both products and reactants.

However, there are other common reactions in which product and reactant minerals may display different types of solid solution. For example the reaction:

$$\underset{\text{albite}}{\text{NaAlSi}_3\text{O}_8} \rightarrow \underset{\text{jadeite}}{\text{NaAlSi}_2\text{O}_6} + \underset{\text{quartz}}{\text{SiO}_2} \qquad [3.7]$$

found in some rocks metamorphosed at high pressures, is univariant in the pure end-member system. However, jadeite (a pyroxene) can undergo markedly different atomic substitutions from albite (a feldspar), e.g. towards diopside, $\text{CaMgSi}_2\text{O}_6$ (i.e. $(\text{Ca}, \text{Mg}) \leftrightarrow (\text{Na}, \text{Al})$) or aegirine, $\text{Na Fe}^{3+}\text{Si}_2\text{O}_6$ $(\text{Fe}^{3+} \leftrightarrow \text{Al})$. Neither of these types of substitution can occur in albite and in natural rocks containing Ca, Mg, Fe^{2+} and Fe^{3+}, the assemblage albite (pure) + quartz + Na-rich pyroxene solid solution is not uncommon. Reaction [3.7] is continuous in these rocks, and reactants and products can co-exist with several degrees of freedom over a large *P-T* interval.

The experimental breakdown of albite by reaction [3.7] (Holland 1980) results in a univariant curve for the end-member reaction, shown in Figure 3.5. The assemblage jadeite + quartz is restricted to very high pressures in the pure system. However, natural jadeite-rich pyroxenes can co-exist with albite and quartz in rocks formed at much lower pressures. Curves for the co-existence of a range of pyroxene compositions with albite and quartz (Figure 3.5) show that such pyroxenes are stable over a much wider range of pressures than pure jadeite. A simple rule of thumb is that impure phases will be stable over a wider variety of *P–T* conditions, and thus have larger stability fields than pure phases. For example, salt water has a

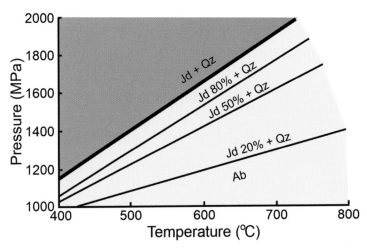

Figure 3.5 *P–T* diagram to illustrate the effect of solid solution on the equilibrium between albite, quartz and a jadeite-bearing clinopyroxene (omphacite). The equilibrium boundary in the pure jadeite–albite–quartz system, which is univariant, is represented by the upper curve. Successively lower curves are for the equilibrium between omphacites with successively lower percentages of jadeite, with albite and quartz. Note that albite remains stable throughout the area shaded yellow. Data from Holland (1980).

higher boiling point and lower freezing point than pure water because salt does not readily enter into either steam or ice.

A comparable equilibrium, which has been widely used as an indicator of metamorphic pressure, is based on the breakdown of anorthite with increasing pressure (Newton & Haselton 1981):

$$3CaAl_2Si_2O_8 = Ca_3Al_2Si_3O_{12} + 2Al_2SiO_5 + SiO_2 \qquad [3.8]$$
$$\underset{\text{anorthite}}{} \qquad \underset{\text{grossular}}{} \qquad \underset{\text{kyanite}}{} \qquad \underset{\text{quartz}}{}$$

Here, both the anorthite and grossular normally occur as dilute components of plagioclase and garnet solid solutions respectively. Many pelitic schists with kyanite or another Al-silicate phase also contain garnet and plagioclase, and so even though they cannot be thought of as rocks with a high calcium content, this equilibrium between Ca end-members is still applicable.

The *P–T* conditions of metamorphism are difficult to determine on a petrogenetic grid from markedly continuous reactions of this sort because the reactants and products may be stable together over a very wide *P–T* range, depending on mineral compositions. However, if the compositions of the natural phases are known, it is often possible to use thermodynamic calculations to estimate the *P–T* conditions of formation of a natural assemblage from the equilibrium conditions of the end-member reaction. To do this, the thermodynamically-effective concentration (or **activity**) of the end-member components present in the reaction must be estimated from the analysed chemical composition of the natural minerals. Box 3.1 outlines the principles by which these relationships between activity and composition are deduced.

BOX 3.1 Linking experimental to natural systems via activity

More often than not, natural solid-solution minerals have compositions that differ from those of the pure end-members used in experiments. To calculate P–T conditions for rocks from continuous equilibria involving solutions, it is therefore necessary to correct for these compositional differences. This is done by determining the activities of the end-member components of the natural mineral solid solution. **Activity**, as normally used in geology, is the thermodynamically effective concentration of the component in the solution. For example, in a pure H_2O fluid, the activity of water (a_{H_2O}) is 1, whereas if the fluid is a mixture of H_2O and CO_2, then $a_{H_2O} < 1$. Similarly, in a pure jadeite pyroxene, $a_{jd} = 1$. In an intermediate clinopyroxene, $a_{jd} < 1$, and the value will depend on the precise clinopyroxene composition. Once the activities of the end-member components of natural minerals are known, it becomes possible to calculate precisely how much the equilibrium curve will be displaced from its position in the pure system. As a rule of thumb, where a reaction involves components which occur in impure phases, the stability field of the side of the reaction with less pure phases will be extended relative to the side on which the phases are purer (activity is closer to 1). Thus the addition of salt extends the stability field of water relative to both ice and steam, since these contain less salt than co-existing water (e.g. Chapter 2, Figure 2.3).

The relationships between activity and composition are often complex, and not always well known or understood, especially at low to moderate temperatures. If natural minerals can have any composition between the end-member extremes, it is usually relatively straightforward to model how the activity of each end-member varies with the composition of the natural solid solution. Where there are gaps in the possible compositions because of immiscibility, there is often more uncertainty. For a number of important minerals, solid solutions are continuous at high metamorphic temperatures but may be discontinuous, with miscibility gaps, at low temperatures.

Formally, the activity of a mineral end-member is a dimensionless ratio whose value varies according to the way in which the ratio is defined. In metamorphic petrology, activity is defined relative to the pure mineral end-member at the metamorphic P and T (this is known as a standard state), so that the pure phase always has an activity of 1. In this way, the activity remains the same irrespective of the metamorphic conditions, which are, after all, the unknown quantity we want to determine. An example of how activity varies with composition for plagioclase solid solutions is shown in Figure 3.6 for a temperature of 700 °C; at lower temperatures, activity values do not follow

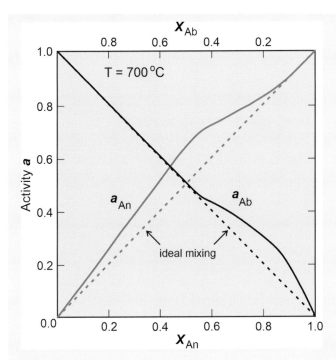

Figure 3.6 Relationship between the activities (a) of albite and anorthite end-members in plagioclase solid solutions, and their composition (expressed as the mole fraction of anorthite, X_{An}). Dashed lines are reference lines for ideal behaviour, solid lines fit the experimental results of Orville (1972). The diagram is valid for a temperature of 700 °C.

composition so closely. Thermodynamic databases (Section 3.3.9) used for calculations in metamorphic petrology now include many sophisticated models for activity–composition relationships in a very wide range of metamorphic minerals.

3.3.5   Cation-Exchange Reactions

Cation-exchange reactions (Chapter 2, Section 2.4.3) do not involve growth or breakdown of minerals, merely exchange of cations between phases. Therefore the volume change due to the ion exchange in one phase is usually almost exactly compensated by the volume change accompanying the reverse exchange in the other phase. As a result, ΔV_r is close to 0 and cation-exchange reactions are mainly dependent on temperature. The parameter K_D (Chapter 2, Equation [2.3]) describes the way in which Fe and Mg are distributed between garnet and biotite (or any other pair of Fe–Mg minerals), and it changes with temperature. There are thermodynamic reasons for predicting that a plot of $\ln K_D$ versus $1/T$ will yield a straight line, and so variation in K_D, if large enough, can make a useful geothermometer.

One of the best-known cation-exchange reaction geothermometers is the exchange of Fe^{2+} and Mg^{2+} between garnet and biotite (Equation [2.2]; Figure 3.7). The plot of $\ln K_D$ against $10^4/T$ provides a quantitative calibration of the temperature-sensitivity of the reaction. The simple linear fit to the data provides a way of calculating T for a natural rock once K_D has been calculated from mineral analyses, as outlined in Box 3.2.

BOX 3.2 Calculating garnet–biotite temperatures

The best fit through the data points in Figure 3.7 gives the following equation for ln K_D:

$$\ln K_D = (-2109/T) + 0.782, \qquad\qquad [3.9]$$

where T denotes temperature in Kelvin. We saw in Section 2.4.3 that the distribution coefficient for Fe–Mg exchange between garnet and biotite (reaction [2.2]) can be calculated from mineral analyses:

$$K_D = \frac{\left(Mg^{2+}/Fe^{2+}\right)^{Grt}}{\left(Mg^{2+}/Fe^{2+}\right)^{Bt}}.$$

Suppose a rock contains garnet for which Mg^{2+}/Fe^{2+} is determined to be 0.05, co-existing with biotite with $Mg^{2+}/Fe^{2+} = 0.30$. It follows that

$$K_D = 0.05/0.30$$

$$\ln K_D = -1.79.$$

Substituting into Equation [3.9] gives

$$-1.79 = (-2109/T) + 0.782.$$

We can then rearrange this to solve for T:

$$-1.79 - 0.782 = -2109/T = -2.574.$$

Hence $2.574T = 2109$ and so $T = 2109/2.574$, which is 819 K or 546 °C.

In detail, small volume changes associated with this reaction add a small source of uncertainty to the application of Equation [3.9] to natural rocks. The overall result is that temperatures calculated for rocks metamorphosed at mid to lower crustal depths are likely to be too low by a few tens of degrees (Spear 1993). The value of K_D is also often influenced by the presence of additional components that were absent from the experiments; for example Ca and Mn in garnet or Ti in biotite. Various refinements to the calibration of this thermometer have been

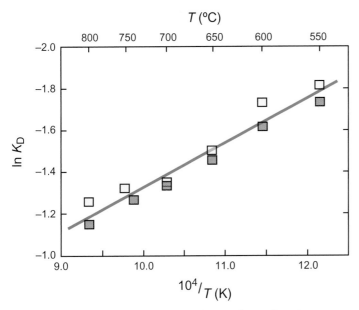

Figure 3.7 Experimental results for the exchange of Fe and Mg between garnet and biotite at a series of temperatures appropriate for medium- to high-grade metamorphism at 207 MPa. Yellow squares are results for experiments in which the initial biotite was the Mg-end-member, red squares are experiments with Fe-biotite initially. The green line is the best fit to the data points, see Box 3.2. From Ferry & Spear (1978).

proposed over the years to allow for such effects, but none is of undoubted universal applicability.

Cation-exchange thermometry has now largely been superseded for rocks whose assemblages have a large number of co-existing minerals, because for these the computational methods described below are likely to be more accurate. However, it is still a valuable tool for determining the conditions of formation of rocks with simple assemblages stable over a wide range of conditions, such as schists containing plagioclase + quartz + biotite + garnet (Wu 2015). Cation-exchange thermometers can also be used in combination with thermobarometers based on markedly continuous reactions (Section 3.3.4) to deduce the P–T conditions of metamorphism more accurately; an example is provided in Figure 3.9a.

3.3.6   Oxygen Isotope Thermometry

In much the same way that solid-solution minerals can exchange cations, many rock-forming minerals and fluids can also exchange oxygen atoms of different masses. Three isotopes of oxygen exist in nature, ^{16}O, ^{17}O and ^{18}O, and all are stable, i.e. they do not undergo radioactive decay. Different minerals show different preferences for oxygen atoms of different masses in their structure in much the same way that ferromagnesian minerals show different preferences for Fe relative to Mg. A fractionation factor, α, can be defined which is analogous to the K_D parameter used to monitor the progress of cation-exchange reactions:

$$\alpha_{A-B} = \frac{\left(^{18}O/^{16}O\right)_A}{\left(^{18}O/^{16}O\right)_B},$$

where the A and B suffixes denote different minerals.

Oxygen isotope fractionation factors vary with temperature in a comparable way to K_D for cation-exchange reactions (Section 3.3.5) with the differences between most co-existing minerals becoming smaller as temperatures increase. Therefore oxygen isotopic exchange has been most widely used as a geothermometer for low-temperature mineral assemblages. For example, in the Swiss Alps, oxygen isotope thermometry was used to provide temperature constraints on the formation of different structural generations of quartz–calcite veins in low-grade rocks (Kirschner *et al.* 1995).

3.3.7   Mutual Solid-Solution Geothermometry

Some mineral pairs exhibit limited mutual solid solution at low temperatures, but are completely miscible at high temperature. The phase diagram for this sort of system exhibits a solvus curve above which only one phase of variable composition is found, but below which two co-existing immiscible phases occur (Figure 3.8). The solvus curve indicates the amount of mutual solid solution that can occur at any particular temperature, and this will be at a maximum when both phases co-exist. Thus measurement of the composition of the co-existing

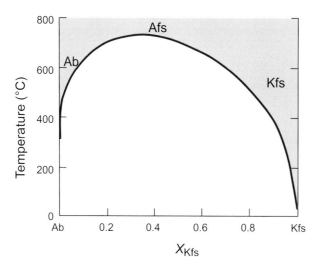

Figure 3.8 The alkali feldspar solvus, shown as a red line connecting the compositions of co-existing albite (Ab) and K-feldspar (Kfs) as a function of temperature. Possible single alkali feldspar (Afs) compositions lie within the pink field. This diagram does not attempt to show changes in alkali feldspar structure with temperature. Figure simplified from Brown & Parsons (1989).

minerals should define the temperature at which they crystallised (so long as they crystallised together of course). In contrast to oxygen isotope thermometers, most solvus geothermometers become more sensitive to temperature as temperature increases.

The example illustrated in Figure 3.8 is the alkali feldspar solvus between albite and K-feldspar. The increasing mutual solubility of these end-member alkali feldspars with increasing temperature is accompanied by structural changes that are also temperature-dependent. At very high metamorphic grades, any composition of alkali feldspar is possible, but under most metamorphic conditions albite and K-feldspar can co-exist. The solvus is asymmetrical and as a result the K-content of albite is very low (and is further suppressed if Ca is present). The Na-content of K-feldspar co-existing with albite is, however, a potential temperature indicator where the two feldspars have grown together. In rocks formed at very high temperatures, alkali feldspar commonly develops perthite textures during cooling and these are rather harder to interpret (Parsons & Lee 2009).

The extent of solid solution between calcite and dolomite likewise provides a useful geothermometer for some veins as well as marbles (Chapter 6). Under most conditions dolomite is rather pure, but co-existing calcite contains variable amounts of Mg, depending on the temperature at which they grew together (Mizuochi *et al.* 2010). As with the alkali feldspar solvus, recrystallisation during cooling can render the results invalid, but there are examples of it being applied successfully to high-grade marbles.

Another thermometer of this type is based on mutual solution between Ca-rich clinopyroxenes and orthopyroxene, and is applicable to very high-grade metamorphic rocks. As with alkali feldspar perthite, high-T (high-temperature) pyroxenes typically unmix during cooling and form intergrowths, but they are less prone to further chemical changes during slow cooling. As a result, if the original pyroxene composition can be reconstructed from the intergrowths found today, it may be used for geothermometry. At very high-grade conditions, orthopyroxene also exhibits incomplete solid solution with garnet and this results in significant Al-contents for some natural orthopyroxenes. This solid solution is favoured by high pressures; the Al-content of orthopyroxene co-existing with garnet is therefore a useful geobarometer for extreme conditions (Müller *et al.* 2013; Wood & Banno 1973).

3.3.8   Mineral Composition Geothermometers

Some geothermometers are based simply on the composition of a specific mineral. However, other minerals must be present in equilibrium with these minerals in order to constrain variation in composition for reasons other than simply temperature.

Trace Element Geothermometers Even where co-existing minerals do not show extensive solid solution, they may influence each other's composition in a way that is dependent on temperature. For example, many rocks contain a Ti-rich mineral, such as rutile, ilmenite or titanite and these have the effect of saturating other minerals with Ti. In thermodynamic terms, the presence of rutile buffers the chemical potential of TiO_2 in the system (Chapter 2, Section 2.8), and this is reflected in the Ti concentrations of all the phases present. For example, biotite can contain significant amounts of Ti substituted into the structure and the amounts increase with temperature, providing a geothermometer (Henry *et al.* 2005).

Less obviously, quartz can also accommodate small amounts of elements such as Ti and Zr, especially at very high temperatures, and attempts have also been made to use their concentrations in quartz as geothermometers (Wark & Watson 2006). This is an attractive approach because it involves very common minerals, but some studies have found significant discrepancies between temperature estimates from trace elements in quartz and other geothermometers (Morgan *et al.* 2014).

Chlorite Geothermometer Chlorite is ubiquitous in low-grade metabasites. The first empirical chlorite thermometer was calibrated on natural chlorites sampled from geothermal fields in basaltic rocks where the temperature of growth (150–300 °C) was measured directly down-hole (Cathelineau 1988). The thermometer is based on the concentration of Al in tetrahedral sites (see Chapter 4, Box 4.2), which increases with temperature. The chlorite geothermometer has since been revised a number of times, primarily with the aim of taking better account of the compositional parameters that can also influence the amount of tetrahedral Al. The most recent revision is based on a range of well-constrained natural chlorites including those from deep diagenetic sedimentary sequences as well as volcanic geothermal fields (Bourdelle & Cathelineau 2015).

3.3.9 Computational Methods: Obtaining Metamorphic Conditions From an Entire Assemblage

Many metamorphic rocks contain a diverse range of minerals (and mineral compositions), and so in principle it should be possible to identify a number of mineral equilibria that are potentially able to record the metamorphic conditions under which that assemblage equilibrated. Each of these equilibria give rise to a curve on a *P–T* diagram and these should intersect at the conditions under which the rock formed. As more experiments were carried out in the latter part of the twentieth century, and computers became more powerful, it became possible to extract thermodynamic data for minerals by combining results from different experiments on different equilibria to produce an **internally consistent dataset.** This is the dataset whose thermodynamic data best reproduce all the experimentally-determined equilibrium curves which went into producing it. For example, the line shown passing between the data points in Figure 3.7 is the statistical best fit, but slightly different lines could also fit these points. If these data points were to be combined with results of other experiments involving garnet and biotite in an internally consistent dataset, it would be possible to calculate a line through the data points on Figure 3.7 that was also consistent with all the other experimental data. This would be preferable to a fit based on one dataset alone as shown in Figure 3.7.

The first comprehensive dataset for mineralogy and petrology (Helgeson 1978) began a revolution in theoretical petrology. More recent datasets (e.g. Holland & Powell 2011) underpin software such as THERMOCALC (Powell *et al.* 1998), which can perform a wide range of phase equilibrium calculations. Other software packages have also been developed, including PerpleX (Connolly 2005) and Theriak-Domino (de Capitani & Petrakakis 2010) and these offer different advantages for different problems. Powerful graphics interfaces are making the software packages increasingly user-friendly.

The initial focus of thermodynamic software packages for metamorphic petrology was to enable calculation of metamorphic P–T conditions for a rock whose mineral assemblage and mineral compositions were specified by the user. The codes use thermodynamic datasets that include solid-solution activity models to calculate conditions of formation for any particular rock. The software generates P–T curves for multiple mineral equilibria, using the mineral compositions actually present (Figure 3.9a). For this approach to work, at least some parts of each mineral must still have the same composition now as when they grew together. For rocks metamorphosed at high temperatures, this may not be the case: diffusive exchange between minerals can modify their compositions during cooling so that the calculated temperatures may be the temperature when diffusive exchange ceased, rather than peak temperatures (Chapter 7). Nevertheless, this approach is successful over a wide range of grades and rock types, and is able to show the uncertainties in the P–T curves to aid in their interpretation.

3.3.10   Computing Mineral Assemblages From Rock Compositions as a Function of P and T

The approaches to estimating P–T conditions outlined thus far use the measured compositions of minerals in a rock to calculate the conditions at which they could have co-existed in equilibrium. An alternative approach is to calculate the mineral assemblage that a specific rock will develop at equilibrium under different P–T conditions in order to compare the results with what is actually found. The analysed composition of the rock as a whole (termed the **bulk composition**) is the input and the software produces a petrogenetic grid for that specific rock composition, including information on mineral compositions and modal abundances, so that the

(a)

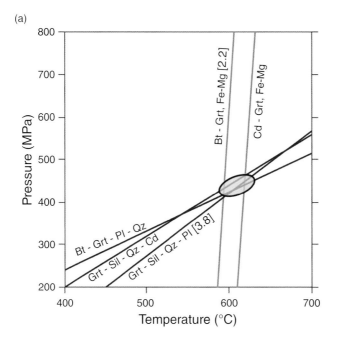

(b)

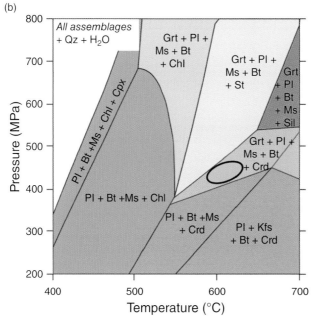

Figure 3.9 Comparison of alternative approaches to determining the *P–T* conditions of formation of a garnet–sillimanite–biotite–cordierite–plagioclase–quartz schist. (a) Intersecting *P–T* curves calculated for cation-exchange geothermometers (orange) and markedly continuous reactions which serve as geobarometers (blue), using the analysed mineral compositions and an internally consistent thermodynamic dataset. The most likely conditions of metamorphism are where the curves intersect (green circle). (b) Isochemical phase diagram constructed for the same rock as part (a), showing mineral assemblages predicted for the rock composition with the green circled area from part (a) superimposed to show the agreement between the two approaches. Simplified from de Capitani & Petrakakis (2010).

petrologist can identify the *P–T* range where the observed mineral assemblage should be stable. This type of diagram is most accurately known as an **isochemical phase diagram**, since it is constructed for a rock of constant composition (de Capitani & Petrakakis 2010). However, these diagrams are also often referred to as 'pseudosections' in the literature (Powell & Holland 2008). Figure 3.9b is an example of an isochemical phase diagram constructed for the same sillimanite-garnet–biotite schist as was used for calculating Figure 3.9a. The results calculated from mineral chemistry are clearly consistent with the isochemical phase diagram.

The isochemical phase diagram is particularly valuable for studying rocks formed at high temperatures where mineral compositions may well have been modified during cooling (as discussed above, this makes peak temperature estimates based on mineral composition invalid). Mineral compositions are normally homogeneous at the highest grades of metamorphism and this means that the composition of the rock mass that was able to interact is simply the same as the bulk rock composition and can readily be determined. It is less easy to use isochemical phase diagrams for rocks metamorphosed at lower grades, because zoned minerals are often present; the diagram is valid only for the composition of those parts of the rock which are able to come to equilibrium, known as the **volume of equilibration** (Powell & Holland 2008). This is because the diagram shows the lowest energy way of combining the constituent atoms in the rock at different *P–T* conditions, *assuming that all the atoms in the rock are able to interact*. Where this was not the case, it is necessary to estimate the composition of those parts of the rock that interacted effectively during each stage of metamorphism from what we see today. In rocks where some grains preserve old, chemically distinct cores, isolated from the rest of the rock (e.g. Figure 2.10a), this will be very difficult. Here, the **effective bulk composition** of the rock changes as minerals react; some chemical constituents become unavailable through being incorporated into grain cores, others are released as grains break down. In a rock whose effective bulk composition changes during metamorphism, isochemical phase diagrams should ideally be calculated for each successive effective rock composition. Most of the codes noted in the previous section are able to calculate a range of types of diagram so that users can investigate these issues.

3.3.11   Non-Equilibrium Approaches to Geothermometry

All the approaches to estimating *P–T* conditions that we have described thus far are based on the requirement that phases have formed at equilibrium, so it seems counter-intuitive that failure to reach equilibrium could be used to constrain temperatures! However, some phases that form at low temperatures grow with structures that do not correspond to their ideal equilibrium structure, because kinetic barriers prevent this happening. For example, the ordering of some atoms in the structure may be somewhat irregular. If these grains are subsequently heated and begin to recrystallise, the atoms reorganise more regularly. There are two thermometers of this type that have been widely used in studies of low-grade metamorphic rock such as slates.

Sheet silicates have been used as indicators of thermal history for many years, in studies of both diagenesis and low-grade metamorphism. Illite is a precursor phase for muscovite during initial metamorphism of muddy sediments. X-Ray Diffraction (XRD) scans of illite–muscovite show a strong reflection from the basal (001) plane of the lattice, but the shape of the peak changes with metamorphic grade. Very low-grade rocks have relatively broad peaks but with increasing grade, the peaks become sharper (Figure 3.10a). The peak shape is quantified by measuring the width of the 001 peak at half its height (the 'Kübler Index') and is commonly referred to as **illite crystallinity**. Illite crystallinity has been widely used to evaluate metamorphic grade below the greenschist facies (Kisch 1991). **Chlorite crystallinity** is defined similarly using the 'Árkai Index' (Árkai 1991), and calibrations exist that tie the two indices together (e.g. Warr & Cox 2016).

Another increasingly-used non-equilibrium technique is based on the recrystallisation of organic matter in sediments to well-crystallised graphite. Optical reflectivity of carbonaceous material has been used for many years as an approximate indicator of metamorphic grade, but their different Raman spectra provide a more quantitative approach (Beyssac *et al.* 2003). Raman spectrometry measures the

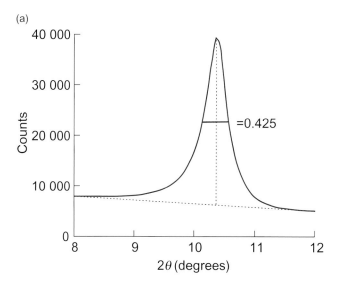

(a)

Figure 3.10 Spectra illustrating crystallinity as a guide to metamorphic grade. (a) Demonstration of the measurement of the 'Kübler Index' of illite crystallinity from an XRD trace of white mica, after Warr (2018). The peak width measured half way between the top of the peak and the sloping baseline is the Kübler Index. (b) Raman spectra for carbonaceous materials from metasedimentary rocks of South Island, New Zealand from Beyssac *et al.* (2016). Spectra from successive samples from low to high grade have been stacked above one another.

(b)

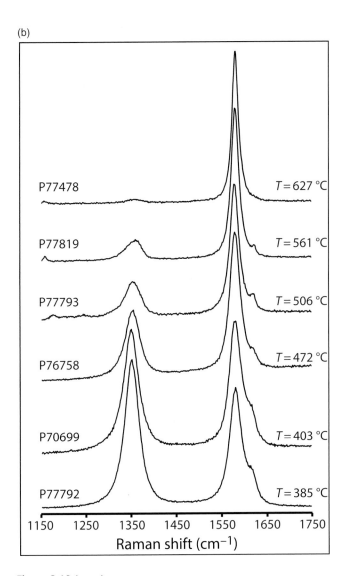

Figure 3.10 (*cont.*)

change in wavelength of laser light that has been scattered by molecules in a sample and records a spectrum of light showing the intensity of scattering as a function of wavelength. Carbonaceous materials give rise to specific peak patterns which become simpler, and sharper, as they recrystallise to graphite. Figure 3.10b provides an example of a series of Raman spectra from carbonaceous metamorphic rocks from New Zealand that range from the zeolite facies to the amphibolite facies. This study showed that temperatures determined by RSCM (Raman Spectroscopy of Carbonaceous Material) were internally consistent and in good agreement with mineral geothermometers.

SUMMARY

Where assemblages of minerals that co-existed in equilibrium during meta-morphism are present in a rock, they can be used to estimate the pressures (and thus burial depths) and temperatures at which metamorphism occurred. Relative pressures and temperatures can usually be readily determined by assigning rocks to metamorphic facies, and a petrogenetic grid can often be used to assign approximate numerical values of pressure and temperature. More-precise estima-tion of equilibrium conditions requires chemical or isotopic analyses of the minerals. These can then be interpreted using geothermometers or geobarometers specific to particular reactions, or input into computer codes that estimate $P–T$ conditions by simultaneously solving for multiple equilibria. Alternatively, if the rock composition is known, it is possible to compute a phase diagram for that specific rock so that its actual assemblage can be compared with those predicted at different conditions.

The most suitable approach depends on how many different minerals are present in the rock. Rocks with few minerals usually have assemblages with a large number of degrees of freedom and so the more sophisticated computational approaches are less effective than using a small number of appropriate individual geothermometers or geobarometers.

It can be difficult to determine the conditions of formation of low-grade rocks because they often contain complex minerals such as clays, for which thermo-dynamic data are very sparse. For these rocks, there are alternative, empirical, approaches based on the evolution of sheet silicates and carbonaceous materials towards equilibrium structures with increased heating.

For most rock types, estimation of metamorphic conditions is now limited only by the degree to which it is possible to interpret textural relationships between minerals and determine which minerals co-existed at different metamorphic stages. The next three chapters deal with the metamorphism of specific rock types, and each includes a discussion of the most useful approaches to the determination of their $P–T$ conditions.

EXERCISES

1. What are the characteristics of a reaction that will make a good geotherm-ometer? Why are cation-exchange reactions particularly well-suited?
2. What are the characteristics of a reaction that will make a good geobarometer?
3. A sample of pelitic schist contains the assemblage biotite, garnet, plagioclase, kyanite, quartz. The garnet and plagioclase are solid solutions containing Ca while garnet and biotite similarly both contain Mg and Fe. How might you set

about determining the P–T conditions at which this assemblage equilibrated? Assume that analyses of both the rock and its constituent minerals are available to you.

4. A sample of schist contains garnet with the composition $Fe_{2.08}Mg_{0.13}Ca_{0.76}$-$Mn_{0.03}Al_2S_3iO_{12}$ co-existing with biotite of composition $K_2Fe_{2.68}Mg_{2.13}$-$Mn_{0.01}Al_{3.34}Ti_{0.24}Si_{5.60}O_{20}(OH)_4$. Calculate the distribution coefficient (K_D) for the partitioning of Fe and Mg between these two phases and hence deduce the temperature of equilibration using the garnet–biotite geothermometer as outlined in the worked example (Box 3.2). Note that you do not need to use all the compositional data provided.

5. The pressure of metamorphism in the Bergell aureole (Figure 2.4) is estimated to be approximately 350 MPa. Use the petrogenetic grid provided in Figure 3.3 to calibrate the isograds on Figure 2.4 where possible, and so construct a profile of peak metamorphic temperature along a traverse approximately at right angles to the granodiorite contact near the middle of the area shown in the map. You will have to estimate the position of the anthophyllite isograd from the assemblage information on the map.

FURTHER READING

Powell, R. & Holland, T. (2010). Using equilibrium thermodynamics to understand metamorphism and metamorphic rocks. *Elements*, **6**(5), 309–14.

Spear, F. S. (1993). Metamorphic phase equilibria and pressure–temperature–time paths. *Mineralogy Society of America Monographs*, **1**, 799.

Spear, F. S., Pattison, D. R. & Cheney, J. T. (2017). The metamorphosis of metamorphic petrology. *The Web of Geological Sciences: Advances, Impacts, and Interactions II*, **523**, 31.

SOFTWARE PACKAGES FOR GEOTHERMOMETRY AND GEOBAROMETRY

A number of software packages are currently in use by metamorphic petrologists wishing to estimate the conditions at which particular rocks have formed. All are based on an assumption of chemical equilibrium, but they may differ in the way in which thermodynamic calculations are performed. There are also differences in the ways in which results can be presented between different packages. All software packages access databases for the thermodynamic properties of minerals and for activity–composition relationships, and these are periodically updated and extended. The most widely-used packages at the time of writing are briefly introduced below. Links to access and download the packages are ephemeral but current links are available at this book's website: https://www.cambridge.org/IMP2e

The interested reader should also search for current information on packages by name.

THERMOCALC is the name used for both the code developed by Roger Powell (University of Melbourne) and Tim Holland (University of Cambridge) for thermo-dynamic calculations, and their extensive thermodynamic dataset (Holland & Powell 2011). The software can be used to calculate isochemical phase diagrams or generate average P–T estimates, but for some applications it compares the stability of assemblages defined by the user, rather than searching for the most stable assemblage possible. The THERMOCALC database was one of the first to be truly internally consistent and has been developed over many years; it is also used in other software packages. If searching for information on THERMOCALC, be aware that there is also commercial software for materials science called Thermo-Calc.

THERIAK-DOMINO is a collection of programs developed by Christian de Capi-tiani (University of Basel) which is particularly effective at plotting isochemical phase diagrams. It predicts the most stable assemblages and mineral compositions for a wide range of P–T conditions and is compatible with datasets developed for THERMOCALC.

PERPLEX is a thermodynamic calculation package developed by Jamie Connolly (ETH Zurich), which is also focussed on generating phase diagrams. It is able to work with a variety of thermodynamic databases and activity models for solid solutions.

4 Metamorphism of Pelitic Rocks

Metapelitic rocks are derived from clay-rich sediments. They are of particular importance for studies of metamorphism because they develop a wide range of distinctive minerals that help constrain peak P-T conditions.

The term 'metapelite' is commonly used loosely as a field term for metamorphic rocks with a high proportion of micas. However, the distinctive minerals that characterise metamorphic zones in metapelites develop only in bulk compositions that are relatively rich in Al and poor in Ca. Originally, these would have been mudrocks rich in clay and with little calcite. This chapter will outline the metamorphic changes that take place in rocks of this 'true' pelite composition under different metamorphic conditions. Mineralogical changes in other compositions, e.g. semi-pelites, will be noted where common or pertinent.

Clay-rich sediments may undergo extensive changes during diagenesis, and, as noted in Chapter 1, there is no sharp distinction between diagenetic and metamorphic processes. During the advanced stages of diagenesis, many sedimentary clays become unstable and pelitic sediments are converted to mixtures of chlorite and secondary clay minerals.

At the lowest metamorphic grades, very fine white mica forms. Because of the difficulty in identifying such small grains, the textural term sericite is used to describe this mica, which is commonly a mixture of muscovite and/or paragonite and/or illite. As we saw in the two field examples described in Chapter 1, metapelitic rocks develop new minerals and become coarser grained as they progressively metamorphose. Except at the lowest grades, they contain a sufficient range of minerals for their assemblages to be used to interrogate their metamorphic history.

In this chapter we will first explore how the assemblages that develop in meta-pelites can be analysed by using phase diagrams to identify the mineralogical differences that reflect changes in metamorphic conditions. This approach will be applied to the metamorphic conditions experienced by the Scottish Highlands example outlined in Chapter 1, Figure 1.2, then extended to look at other areas in which different metapelite assemblages are found.

4.1   The AFM Projection

We saw in Chapter 2 that compositional phase diagrams allow us to understand whether different mineral assemblages in rocks that occur near to each other were formed because rocks of different bulk composition were metamorphosed under the same P-T conditions or because rocks of similar bulk composition experienced different P-T conditions. For metapelites, the most useful type of phase diagram for this purpose is the **AFM projection**.

In chemical terms, reactions in pelitic rocks principally involve the components K_2O, FeO, MgO, Al_2O_3, SiO_2 and H_2O (referred to by the acronym KFMASH from the initial letters). These are the essential components in the metamorphic index minerals introduced in Chapter 1, Section 1.2, and most theoretical and experimental attempts to model metapelitic rocks use this simplified system. Other components, especially Fe_2O_3, TiO_2, MnO, CaO, Na_2O and CO_2, may also be present; however, they do not usually play a major role in the reactions that produce the key metamorphic index minerals.

Considerable simplification is required in order to represent even this ideal-ised six-component system on a compositional phase diagram, and this is done by means of projection, in much the same way as was done to represent phases in the CaO – MgO – SiO_2 – H_2O system on a triangular phase diagram in Chapter 2, Figure 2.6. As in Chapter 2, we make the assumption that an aqueous fluid phase is always present during progressive heating, since most of the reactions release H_2O. Additionally, almost all metapelites contain quartz. If we restrict our study to quartz- and H_2O-bearing rocks, we can then project from these phases to reduce the number of phases that need to be plotted. The remaining components, Al_2O_3, K_2O and FeO and MgO, then plot at the corners of a three-dimensional tetrahedron (known as the **AKFM tetrahedron**) that encompasses the minerals that we want to investigate

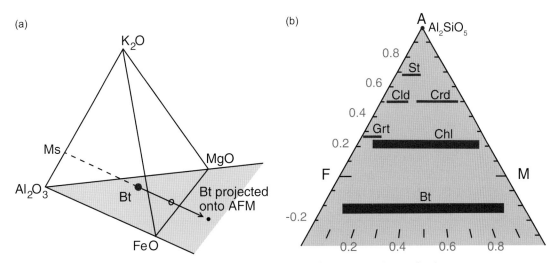

Figure 4.1 The AKFM tetrahedron and projection. (a) AKFM tetrahedron showing the graphical representation of the projection of a biotite composition within the tetrahedron onto the AFM plane from muscovite. (b) The AFM projection showing the usual compositional variation of common metapelite minerals. The numbers along the left hand side are values of the A coordinate; those along the base are the M coordinate (explained in Box 4.1).

(Figure 4.1a). Most minerals plot on the Al_2O_3 – FeO – MgO face of the tetrahedron, but biotite, and natural rock compositions, plot inside it. This tetrahedron is not the easiest of diagrams to work with, since specific compositions cannot be plotted on it uniquely. The difficulty is resolved by a further projection from the common metapelite mineral, muscovite, which is present in most metapelites except at the highest metamorphic grades.

Muscovite plots along the Al_2O_3 – K_2O edge of the AKFM tetrahedron (Figure 4.1a). Projecting from the muscovite point allows most KFMASH compositions to disperse across the base of the diagram, onto a triangle that is delimited by the Al_2O_3 – FeO – MgO (AFM) components.

The plotting procedure is illustrated in Figure 4.1a, which shows a biotite composition projected onto the AFM face by drawing a line from muscovite through biotite (which plots somewhere in the middle of the tetrahedron) until it intersects the AFM plane. Note that biotite, despite being a very common mineral in metapelites, is actually a tricky mineral to plot, as its projected composition on the AFM diagram lies outside the original AKFM tetrahedron. Box 4.1 outlines how the coordinates of different minerals for an AFM plot are calculated.

Remember that AFM diagrams can only be used to understand mineral reactions in rocks that contain both muscovite and quartz, *and* that can be assumed to have had an aqueous fluid phase during metamorphism. All of these phases (muscovite, quartz and fluid) should be taken into account when attempting to balance reactions deduced from AFM projections.

BOX 4.1 Calculating plotting coordinates for an AFM diagram

Despite the greater complexity of the AFM projection, the principle is the same as for projection of the $MgO - SiO_2 - H_2O$ system in Chapter 2, Figure 2.6.

The numerical procedure for calculating where a rock or mineral composition lies has to take into account the fact that muscovite contains Al, and therefore the AFM coordinates cannot be given simply by the relative (molecular) proportions of Al_2O_3, FeO and MgO. Instead, for analyses containing K, the analysis must be recast to give its composition in terms of the components $K_2Al_6Si_6O_{20}(OH)_4$ (muscovite), Al_2O_3, FeO and MgO. The amount of muscovite component is dictated by the amount of K in the analysis, and, because the muscovite formula can be broken down to $K_2O + 3Al_2O_3 + 6SiO_2 + 4OH$, three moles of Al_2O_3 must also be assigned for every mole of K_2O assigned to muscovite. This obviously reduces the remaining Al_2O_3 component to plot on the AFM triangle. The AFM coordinates of any analysis of an AKFM mineral (or rock) on the AFM projection are therefore given by:

$$A = \frac{([Al_2O_3] - 3[K_2O])}{([Al_2O_3] - 3[K_2O] + [MgO] + [FeO])}$$

$$M = \frac{[MgO]}{([MgO] + [FeO])}$$

Square brackets denote the number of moles of the oxide in the analysis. To plot a bulk rock analysis, the A value must be further reduced to allow for Al combined with Na and Ca in other minerals such as plagioclase or epidote.

The obtained A values define the position of a horizontal line in terms of its distance from the F–M join towards A, while the M values define a line radiating from A, lying between the A–F join (M = 0) and the A–M join (M = 1). The analysis plots at their intersection. The compositional ranges of the common metapelite minerals are shown plotted on an AFM projection in Figure 4.1b.

4.2   Metamorphism at Moderate Pressures and Temperatures

The classic zonal sequence of changing index minerals found in metapelites in the Scottish Highlands, outlined in Chapter 1, is also found in metapelitic rocks of various ages in many other parts of the world. It reflects metamorphism under conditions where the crustal heat flow is neither very high nor very low (Figure 1.6)

Table 4.1 Characteristic assemblages of metapelitic rocks from the zones of the south-east Scottish Highlands

Zone	Typical assemblage
Chlorite	Chlorite + phengitic muscovite + quartz + albite ± calcite ± stilpnomelane ± paragonite
Biotite	Biotite + chlorite + phengitic muscovite + quartz + albite ± calcite
Garnet	Garnet + biotite + chlorite + muscovite + quartz + albite ± oligoclase (chloritoid is very rarely present in the Scottish Highlands, but is common at comparable grades elsewhere)
Staurolite	Staurolite + garnet + biotite + muscovite + quartz + plagioclase (rare chlorite at the lowest grades)
Kyanite	Kyanite ± staurolite + garnet + biotite + muscovite + quartz + plagioclase
Sillimanite	Sillimanite ± staurolite + garnet + biotite + muscovite + quartz + plagioclase ± kyanite relics
Common accessories	Ilmenite (lower grades) or rutile (higher grades), magnetite, pyrite or pyrrhotite, tourmaline, zircon, monazite, apatite, graphite

and provides an excellent starting point for showing how different minerals grow as temperature and pressure increase in rocks of pelitic composition. Table 4.1 presents a summary of the characteristic assemblages found across the different metamorphic zones in the Scottish Highlands. These assemblages are developed at temperatures and pressures that, as a whole, are intermediate between more extreme high and low pressures estimated for some other types of regional metamorphism. Rocks of similar bulk composition would be expected to develop different assemblages at higher or lower P–T conditions (examples of such assemblages are described later in this chapter).

The mineralogy of metapelites in each of these metamorphic zones in turn may be usefully plotted onto AFM projections to show the mineralogical changes that took place as metamorphic grade increased. In some cases, subtle differences in bulk composition between samples collected from the same location can have a significant influence on the mineral assemblages that develop as P and T increase. The strength of the compositional phase diagrams is that they show whether these differences in mineralogy are due to differences in the bulk composition or in P–T conditions.

The compositional phase diagrams also show changes in mineral composition as metamorphic grade changes. These changes are particularly great in the sheet silicates. Box 4.2 outlines the common substitutions in the sheet silicates during prograde metamorphism.

BOX 4.2 Common substitutions in sheet silicates during metamorphism

The sheet silicates, a group that includes the micas, chlorite, clay minerals, talc and serpentine, are formed of sheets of six-membered rings of SiO_4^{4-} tetrahedra (Figure 4.2). An OH^- ion commonly lies in the centre of the ring, in what is known as the **hydroxyl** site.

A layer of cations (e.g. Fe^{2+}, Mg^{2+}, Al^{3+}) in **octahedral coordination** lies between these sheets of **tetrahedrally-coordinated** rings, and bonds the sheets together. The bonds between the cations *within* the sheets are strong, but *between* the sheets are weak. This is the reason why micas cleave easily into sheets but do not break across the sheets easily: like pages in a book. Two different layered structures can form from this arrangement. **Trioctahedral** sheet silicates, such as biotite, have a structure in which each O or OH is surrounded by three divalent ($^{2+}$) cations. **Dioctahedral** sheet silicates, such as muscovite, have a structure in which each O or OH ion is surrounded by two trivalent ($^{3+}$) cations.

These two types of sheets have different charge balances, so there is little solid solution between them. However different divalent or trivalent cations can exchange for each other, e.g. Mg^{2+} and Fe^{2+} (as covered in Chapter 3 for other mineral groups), or Al^{3+} and Fe^{3+} (a similar process). In addition, F^- and Cl^- can swap for OH^-. Mineral formulae for all the minerals and end-members discussed below can be found in Appendix 1.

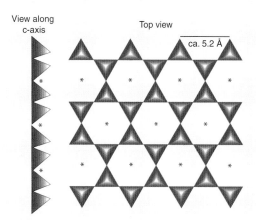

View along c-axis   Top view   ca. 5.2 Å

Figure 4.2 Schematic diagram showing how sheets of six-membered SiO_4^{4-} tetrahedra are structured. In the centre of each ring sits an OH^-, or hydroxyl, ion, marked with a *. An O atom sits at the corner of each tetrahedron and an Si sits in the centre. Not to scale.

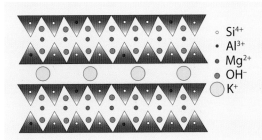

Figure 4.3 Schematic view down the c-axis of the trioctahedral biotite structure. Note that three-dimensional aspects are not shown. Not to scale.

○ Si^{4+}
• Al^{3+}
• Mg^{2+}
● OH^-
◯ K^+

Biotite

Biotite is formed of stacked trioctahedral layers (Figure 4.3) and forms in a wide variety of rock compositions under a wide range of temperatures. A common compositional change in biotite during metamorphism is $Fe^{2+} - Mg^{2+}$ exchange with other Fe – Mg silicates. We saw in Chapter 3, Section 3.3.5, that such exchange between biotite and garnet is the basis for a simple geothermometer for metapelites. Overall, biotite is generally Fe-rich at low grades and becomes more Mg-rich as temperatures and pressures increase.

Another important substitution in biotite involves the exchange of two ions of Al^{3+} for one ion of Mg^{2+} or Fe^{2+} on an octahedral site and one ion of Si^{4+} on a tetrahedral site:

$$^{VI}(Al^{3+}) + {}^{IV}(Al^{3+}) \rightarrow {}^{VI}(Mg^{2+}) + {}^{IV}(Si^{4+})$$ [4.1]

This is known as **Tschermak's substitution** and leads towards biotite end-members rich in Al. These two independent substitutions mean that there are four biotite end-members, but the most important ones for metamorphic studies are phlogopite, the Mg-rich member of the Al-poor biotites, which is found in marbles, and the Al-rich end-members siderophyllite (Fe-rich) and eastonite (Mg-rich) which dominate metapelite biotites. Granites and semi-pelites may have an Fe-rich biotite which is poor in Al, known as annite.

Ti^{4+} may also exchange for Si^{4+} in biotite. This substitution is temperature-dependent (with higher concentrations of Ti found in biotite at higher temperatures) and forms the basis of the Ti-in-biotite thermometer outlined in Chapter 3, Section 3.3.8.

Muscovite

Muscovite is formed of stacked dioctahedral layers (Figure 4.4). The main compositional variations with P and T are exchange of alkali cations leading to solid solution between muscovite and paragonite, its Na-equivalent, and the Tschermak's substitution (reaction [4.1]). Tschermak's substitution allows muscovite to form a solid solution towards an end-member that is poorer in Al, known as aluminoceladonite. This end-member can separately show some substitution of Fe^{3+} for Al, towards the Fe^{3+}-rich mica celadonite (found in nature as the green clay in some hydrothermally altered volcanic rocks).

Solid solutions between muscovite, aluminoceladonite and celadonite are known as 'phengite' (Rieder *et al.* 1998). Although phengite is mainly discussed in the context of high-pressure rocks, phengitic muscovite is ubiquitous in low-temperature rocks such as slates.

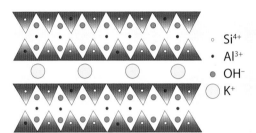

∘ Si^{4+}
• Al^{3+}
• OH^-
○ K^+

Figure 4.4 Schematic view down the c-axis of the dioctahedral muscovite structure. Note that three-dimensional aspects are not shown. Not to scale.

Chlorite

Chlorite is formed of alternating dioctahedral and trioctahedral layers (Figure 4.5). The most common compositional variation under regional metamorphic conditions is due to $Fe^{2+}-Mg^{2+}$ exchange, between the Fe-rich end-member chamosite and the Mg-rich end-member clinochlore. Chlorite in metapelites is generally richer in Fe at low grades and becomes more Mg-rich as temperatures and pressures increase.

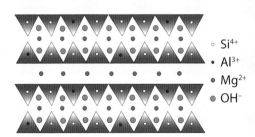

∘ Si^{4+}
• Al^{3+}
• Mg^{2+}
• OH^-

Figure 4.5 Schematic view down the c-axis of the chlorite (clinochlore) structure. Note that three-dimensional aspects are not shown. Not to scale.

Chlorite composition may also vary via the Tschermak's substitution [4.1] and by di/trioctahedral exchange:

$$3(Mg, Fe)^{2+} = \square + 2Al^{VI} \qquad\qquad [4.2]$$

where $\square$ is a vacancy.

As in the other micas, these exchanges are sensitive to $P-T$ conditions and rock composition; as with biotite, some of the highest Al-contents are found in metapelite chlorites.

4.2.1 Chlorite Zone

In the Scottish Highlands, pelitic rocks in the chlorite zone are fine-grained slates that commonly contain graphite. Their fine grain size makes it difficult to study these rocks, even under the microscope, but they typically contain chlorite and muscovite with variable amounts of quartz, albite and opaque accessories such as pyrite or magnetite (Figure 4.6). Some samples may contain K-feldspar, stilpnomelane (an Fe-rich phyllosilicate) or minor calcite. In other parts of the world, chlorite-zone rocks are sometimes coarser-grained than in Scotland (schists rather than slates); in these rocks, the constituent minerals are more readily identifiable. Because chlorite is often the only AFM mineral that these rocks contain, plotting their compositions on an AFM diagram is not very informative.

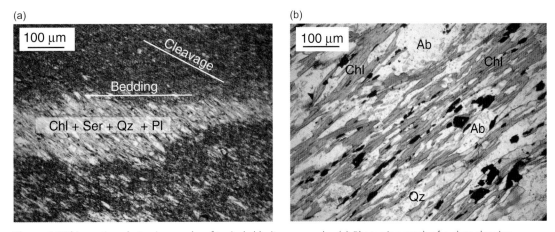

Figure 4.6 Thin-section photomicrographs of typical chlorite-zone rocks. (a) Photomicrograph of a slate showing mineralogical variations associated with original bedding (horizontal) at an angle to the slaty cleavage, now delineated by aligned micas. Ser denotes sericite. (b) Photomicrograph of a chlorite schist. Note the coarser grain size compared to the slate.

4.2.2 Biotite Zone

Biotite flakes that have developed in phyllites and metasandstones in the Scottish Highlands can often be seen with the aid of a good hand lens. A typical biotite-zone metapelite contains chlorite + biotite + phengitic muscovite + quartz + albite (Figure 4.7a). Calcite and graphite may also be present. Biotite appears in metapelites at relatively low temperatures via the reaction:

$$\text{Si-rich muscovite} + \text{chlorite} \rightarrow \text{biotite} + \text{Al-rich muscovite} + \text{quartz} + H_2O \qquad [4.3]$$

Although it appears from reaction [4.3] that two compositions of muscovite are present in the rock, in reality there is only one. As this continuous reaction proceeds, the white mica changes from being relatively richer in Si (more phengitic at low temperatures) to becoming relatively richer in Al (less phengitic at higher temperatures; Box 4.2). The temperature at which reaction [4.3] commences depends on the extent of the phengite substitution in the initial mica, and the X_{Mg} values (Chapter 2, Section 2.4.3) of the chlorite. A higher Si content of muscovite and a low X_{Mg} of chlorite favour biotite growth at lower temperatures. This is because, at equilibrium, $X_{Mg}^{Chl} > X_{Mg}^{Bt}$. As a result, Fe is fractionated into biotite and Fe-rich chlorite breaks down to Fe-rich biotite under conditions where Mg-rich chlorite is still stable. The association of chlorite + muscovite + biotite is stable over a wide temperature interval, emphasising the markedly continuous nature of this reaction (Chapter 2, Section 2.4.2). As an aside, it is worth noting that biotite sometimes appears at even

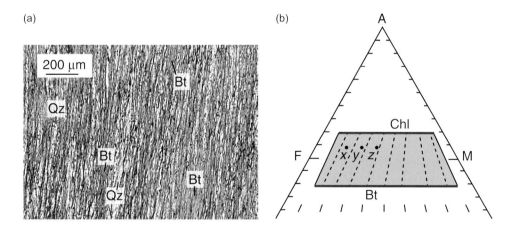

Figure 4.7 Biotite-zone rocks. (a) Photomicrograph of a typical biotite-grade schist from Sikkim, India, containing fine-grained intergrown biotite, chlorite, muscovite, quartz and plagioclase. Sample courtesy of Catherine Mottram. (b) AFM projection for the biotite zone of the Scottish Highlands. The dashed lines are representative tie-lines connecting the compositions of the Fe–Mg minerals that co-exist. The points *x*, *y* and *z* are three possible rock compositions discussed in the main text. All assemblages include muscovite, quartz and H_2O.

lower grades in metasandstones, where it can be formed by the reaction of K-feldspar with chlorite.

An AFM plot of biotite-zone assemblages can shed light on how the equilibrium compositions of chlorite and biotite are linked. The characteristic biotite-zone assemblage muscovite + quartz + biotite + chlorite, with its three degrees of freedom, is represented in Figure 4.7b by a polygonal field bounded by chlorite and biotite. You will notice that a wide range of chlorite and biotite compositions are possible in terms of their relative proportions of Mg and Fe (the F and M corners of the AFM diagram).

Remember from Chapter 2, Section 2.4.3, that Fe and Mg exchange between chlorite and biotite varies in a systematic way (via the distribution coefficient K_D that is dependent on temperature but is independent of mineral composition). This implies that at any fixed P–T value, any specific X_{Mg} composition of biotite can only co-exist with one X_{Mg} composition of chlorite. On an AFM diagram, constructed for a fixed value of P and T, tie-lines can be drawn between chlorite and biotite that express this. Although an infinite number of tie-lines are possible, they must all slope in such a way that:

$$K_{D Fe-Mg}^{Bt-Chl} = \frac{(Mg/Fe)_{Bt}}{(Mg/Fe)_{Chl}} = constant$$

A consequence of this constraint is that no two tie-lines, however closely drawn, may actually cross over one another if they represent equilibrium at the same pressure and temperature. In terms of the phase rule, we have specified two degrees of freedom (P and T) and only one remains. If we specify the X_{Mg} composition of biotite, the X_{Mg} composition of chlorite cannot be an independent variable, and vice versa.

4.2.3 Garnet Zone

In the Scottish Highlands, as in most terranes that have experienced regional metamorphism at moderate pressures, the garnet isograd is easy to trace in the field because garnet appears in a wide range of rock compositions and commonly forms conspicuous porphyroblasts. Metapelites in this zone are typically schists (e.g. Chapter 1, Figure 1.3), whose principal minerals are coarse enough to be readily identified both in the field and in thin section, even though the actual grain size and textures vary between different areas.

A typical garnet zone metapelite contains garnet + biotite + chlorite + muscovite + quartz + albite (Figure 4.8). Garnets are almandine (Fe)-rich; typified by their deep red colour in hand sample (this is important to note because paler pink spessartine (Mn)-rich garnets may develop at lower grades in sediments that are rich in Mn).

Oligoclase, ilmenite or magnetite and minor sulphides may be present also. Epidote may form in rocks that contain higher concentrations of calcium. Common

Figure 4.8 Photomicrographs of garnet-zone metapelites, Scotland. (a) Sample from the lower-grade part of the garnet zone. (b) Sample from a higher-grade part of the garnet zone.

accessory minerals that may be coarse enough to be identified in thin section include apatite, tourmaline, monazite and zircon.

Fe^{2+}-rich (almandine) garnet grows by a continuous reaction such as:

$$\text{chlorite} + \text{muscovite} \rightarrow \text{garnet} + \text{biotite} + \text{quartz} + H_2O \qquad [4.4]$$

Chlorite, biotite and garnet can all vary considerably in their Fe–Mg ratios. The co-existing mineral compositions can be plotted on AFM diagrams to help determine whether changes in mineral composition between different samples imply a change in bulk composition or a change in grade.

We saw in the section above that a typical biotite-zone assemblage has three degrees of freedom. That means that at any fixed P–T value, any specific X_{Mg} composition of biotite can co-exist with only one X_{Mg} composition of chlorite (different examples are shown as tie-lines in Figure 4.7b). A typical garnet-zone assemblage of muscovite + quartz + chlorite + biotite + garnet has one more mineral, and so only *two* degrees of freedom; as a result, co-existing mineral compositions are fixed for any specific P–T conditions (P and T are the independent variables). This assemblage is shown in Figure 4.9a–c as a triangle whose corners represent the unique compositions of the phases that can co-exist at the P–T conditions for which each diagram is drawn. We saw in Chapter 2, Section 2.4.3, that where chlorite, garnet and biotite co-exist, Fe and Mg are distributed between them so that $X_{Mg}^{Chl} > X_{Mg}^{Bt} > X_{Mg}^{Grt}$. Because garnet has a marked preference for Fe compared to biotite and chlorite, it develops first in the most Fe-rich rocks (Figure 4.9a, bulk composition *x*).

How does the mineral assemblage of this Fe-rich bulk composition *x* change as P–T conditions increase? Initially, in the biotite zone, rock *x* contains chlorite and biotite

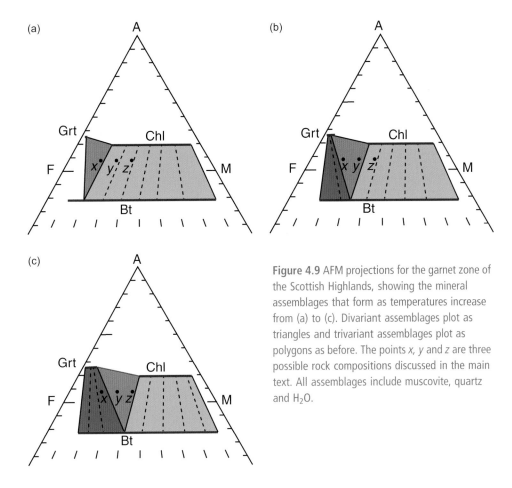

Figure 4.9 AFM projections for the garnet zone of the Scottish Highlands, showing the mineral assemblages that form as temperatures increase from (a) to (c). Divariant assemblages plot as triangles and trivariant assemblages plot as polygons as before. The points x, y and z are three possible rock compositions discussed in the main text. All assemblages include muscovite, quartz and H_2O.

whose compositions are fixed by a tie-line of constant $K_{D\,Fe-Mg}^{Bt-Chl}$ value (Figure 4.7b). On heating, the first garnet to appear has an Fe-rich composition (Figure 4.9a) and, with further heating, garnet becomes more abundant as the biotite–garnet tie-line tracks towards bulk rock composition x. All three Fe–Mg minerals (garnet, bitiote and chlorite) become progressively richer in Mg (Figure 4.9b) despite the rock composition remaining constant, because the Mg-rich chlorite is becoming less abundant while Fe-rich garnet becomes more abundant. Eventually, the chlorite–biotite tie-line sweeps past rock x and the rock loses chlorite, leaving the trivariant assemblage muscovite + quartz + garnet + biotite (Figure 4.9c).

In contrast, a more Mg-rich rock with bulk composition z will not grow garnet until the P-T conditions are such that rock x has lost chlorite completely (Figure 4.9c). The assemblages of rock y are intermediate between those that form in rocks x and z. At the conditions showing in Figure 4.9b, rocks with

compositions *x* and *y* both have the same mineral assemblage, and the compositions of chlorite, biotite and garnet will be the same in both. What varies is the
proportion of the minerals in both rocks: rock *y* will contain more chlorite and less
garnet than rock *x*. Figure 4.9a–c demonstrates two major points: (i) the compositions of garnet, biotite and chlorite in rocks of the same bulk composition all
become more Mg-rich as metamorphic grade increases, and (ii) rocks of different
bulk composition will grow the index minerals at different conditions and so can
have different assemblages at the same *P*–*T* conditions. Plotting the compositions
of mineral phases in two nearby rocks on an AFM diagram helps to determine
whether any differences between them are due to different bulk compositions or
differences in metamorphic grade.

Chloritoid is another mineral that can form in garnet-grade metapelites, although
it is rare in the Scottish Highlands because of the bulk compositions present there. It
only forms in rocks that are relatively rich in Al (Figure 4.10); rocks whose
compositions plot above the garnet–chlorite tie-line on the AFM projection. Rocks
containing less Al will grow biotite instead. Under the garnet-zone conditions that
existed in the Scottish Highlands, chloritoid cannot co-exist with biotite and garnet
and chlorite form instead.

Metapelites commonly show a change in plagioclase compositions within the
garnet zone. This change is also observed in other metasediments and in meta-
igneous rocks. At lower grades, albite (Na-rich plagioclase) is present. Any Ca in
the rock is incorporated in garnet, accessory calcite, epidote or other phases.
Within the garnet zone, oligoclase (10–30% anorthite component) appears.
Because the plagioclase solid-solution series is not continuous at low temperatures, two separate co-existing plagioclase phases, albite *and* oligoclase, may be
present. By the upper part of the garnet zone, the plagioclase gap has closed and
the rock contains only a single plagioclase composition. This mineralogical

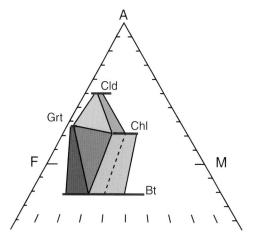

Figure 4.10 AFM diagram showing phase relationships involving chloritoid.

change is not a particularly useful field isograd in metapelite-rich areas, as plagioclase compositions may be difficult to identify in hand specimen and other more identifiable index minerals exist. However, in regions where there are abundant meta-igneous rocks and few metapelites, this plagioclase isograd can prove useful (Chapter 5, Section 5.2.1).

4.2.4   Staurolite Zone

Staurolite is more limited in its occurrence than garnet and mainly grows in Al-rich, Ca-poor metapelite compositions. Staurolite schists are often interbedded with garnet–mica schists that lack staurolite.

Typically, staurolite-zone metapelites contain staurolite + garnet + biotite + muscovite + quartz + plagioclase ± chlorite (Figure 4.11a). Minor opaque and accessory phases are similar to those present in the garnet zone. The appearance of the accessory phosphate mineral monazite has also been linked to the staurolite-in-isograd. This is significant because monazite is an important metamorphic chronometer, as we shall see in Chapter 9.

Depending on the bulk composition, staurolite may co-exist with a number of phases including garnet, biotite, chlorite or a combination (Figure 4.11b). Note that the tie-line between garnet and chlorite, which existed at lower grades (Figures 4.9 and 4.10) has gone. This 'tie-line flip' suggests that staurolite may be produced by the reaction:

$$\text{garnet} + \text{muscovite} + \text{chlorite} \rightarrow \text{staurolite} + \text{biotite} + \text{quartz} + H_2O \qquad [4.5]$$

Since reaction [4.5] is a discontinuous reaction, it takes place at a fixed temperature for any given pressure and proceeds until one of the three reactants has been

(a)   (b)

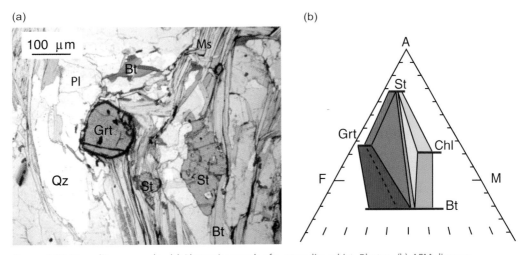

Figure 4.11 Staurolite zone rocks. (a) Photomicrograph of a staurolite schist, Bhutan. (b) AFM diagram showing a variety of staurolite-bearing assemblages that may form depending on bulk composition.

consumed. When this reaction has ceased, further staurolite may be produced by continuous reaction involving the remaining phases, for example in relatively Mg-rich rocks:

$$\text{chlorite} + \text{muscovite} \rightarrow \text{staurolite} + \text{biotite} + \text{quartz} + H_2O \qquad [4.6]$$

This reaction drives chlorite towards more Mg-rich compositions, in a similar way to the continuous reactions that were discussed earlier. It has the effect of gradually eliminating the chlorite–biotite field in Figure 4.11.

Some staurolite schists contain inclusions of chloritoid in minerals such as garnet or plagioclase. These inclusions provide evidence that the bulk composition of these rocks allowed chloritoid to form during metamorphism at lower grades but that chloritoid was no longer stable at the highest temperatures reached. A comparison between Figure 4.9 and Figure 4.11 shows that chloritoid plots within a triangle between staurolite, garnet and biotite. This suggests that staurolite may form in rocks of suitable composition via a reaction that consumes chloritoid (and involves the disappearance of chloritoid from the AFM projection), such as:

$$\text{chloritoid} + \text{muscovite} \rightarrow \text{staurolite} + \text{garnet} + \text{biotite} + \text{quartz} + H_2O \qquad [4.7]$$

4.2.5   Kyanite Zone

The kyanite zone is typified by assemblages such as kyanite + staurolite + biotite, or kyanite + biotite, both of which co-exist with muscovite + quartz (Figure 4.12a). Staurolite zone assemblages such as staurolite + garnet + biotite may still persist in rocks of suitable composition but, at this grade, chlorite is no longer present. The minor and accessory phases are similar to those found in the garnet and staurolite zones.

The AFM projection suggests that the disappearance of chlorite at the same time as the appearance of kyanite is due to a tie-line flip reaction involving the replacement of the staurolite–chlorite tie-line with a kyanite–biotite tie-line (Figures 4.11b and 4.12b). This corresponds to the reaction:

$$\text{muscovite} + \text{staurolite} + \text{chlorite} \rightarrow \text{biotite} + \text{kyanite} + \text{quartz} + H_2O \qquad [4.8]$$

Staurolite can also start to react out at this grade in rocks without chlorite, through the continuous reaction:

$$\text{staurolite} + \text{muscovite} + \text{quartz} \rightarrow Al_2SiO_5 + \text{biotite} + H_2O \qquad [4.9]$$

While both these reactions couple growth of kyanite to breakdown of staurolite, it is not uncommon to find composite grains of intergrown kyanite and staurolite, suggesting that more-complex natural rocks are not exactly modelled by the idealised KFMASH system.

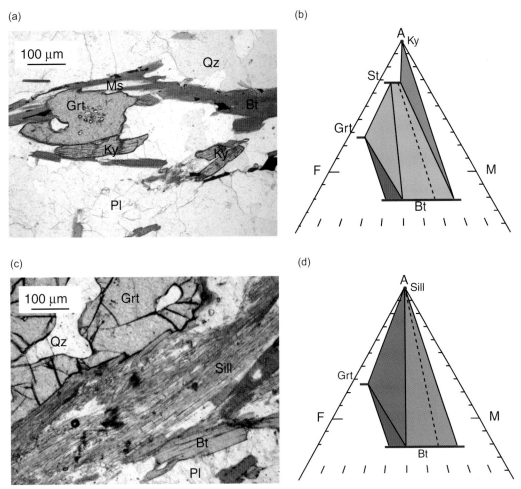

Figure 4.12 Kyanite and sillimanite zone rocks. (a) Photomicrograph of a garnet–kyanite schist, Bhutan. (b) AFM projection showing kyanite zone assemblages co-existing with muscovite, quartz and H_2O. (c) Photomicrograph of a sillimanite gneiss, Bhutan. Very little muscovite is left in this rock, although remnants do still exist. (d) AFM projection showing sillimanite zone assemblages co-existing with muscovite, quartz and H_2O.

4.2.6   Sillimanite Zone

This zone differs from the kyanite zone only by the presence of sillimanite as the Al-silicate phase. Kyanite may also still be present. The sillimanite normally occurs in the form of very fine needles, which may be matted together or penetrate grains of biotite or quartz, and is known as **fibrolite** (Figure 4.12c). The transition from the kyanite zone can nominally be represented by the polymorphic transition:

$$\text{kyanite} \rightarrow \text{sillimanite} \qquad\qquad [4.10]$$

The fact that some kyanite commonly remains in sillimanite-bearing rocks suggests that this reaction is very sluggish. The two minerals have different crystallographic coordination of Si and Al, which makes it difficult for sillimanite to replace kyanite directly (covered further in Chapter 7, Section 7.5.2). Rather than forming directly from the breakdown of kyanite, sillimanite is instead largely produced from the breakdown of other minerals. For example, reaction [4.9] takes place over a temperature range that spans the boundary between the stability fields of kyanite and sillimanite; at higher temperatures, sillimanite is generated instead of kyanite.

Within the sillimanite zone, staurolite commonly disappears from muscovite and quartz-bearing metapelite assemblages. The AFM projection suggests that this may take place as a terminal reaction, producing garnet, for example:

$$\text{staurolite} + \text{muscovite} + \text{quartz} \rightarrow \text{garnet} + \text{biotite} + \text{sillimanite} + H_2O \qquad [4.11]$$

This terminal reaction is shown on the AFM diagram in Figure 4.12d. In metapelites that initially contained less muscovite, and which have lost it all by this grade, staurolite still reacts out by a discontinuous reaction but does not produce biotite:

$$\text{staurolite} + \text{quartz} \rightarrow \text{garnet} + \text{sillimanite} + H_2O \qquad [4.12]$$

Evidence for this reaction has been documented in Connemara, Ireland (Yardley *et al.* 1980). It does not show up on an AFM diagram because it takes place only in rocks from which muscovite is absent.

4.2.7 Limitations on the Applicability of the AFM Diagram

It ought to be a straightforward practical exercise to identify the assemblages of any set of metapelites and to deduce their relative grades by plotting different assemblages on AFM projections. However, many natural assemblages appear to have too many phases, i.e. four rather than three of the phases plotted on the projection. This common circumstance cannot result just from the fortuitous sampling of rocks that were metamorphosed at the precise P–T conditions of a discontinuous reaction such that both reactants and products are preserved. Instead, the phenomenon must therefore result either from persistence of earlier-formed minerals beyond the conditions under which they were stable, or from the presence of additional components (e.g. Mn) in the natural system serving to stabilise a larger number of phases than the KFMASH system suggests would co-exist.

The occurrence of kyanite and sillimanite in the same sample is most likely to result from persistence of the first-formed polymorph outside its own stability field. Both minerals occur as nearly pure Al_2SiO_5 and so their stability is most unlikely to be affected by additional chemical components in the rocks. Instead, the persistence of kyanite outside its stability field is likely due to kinetic factors (covered in more detail in Chapter 7).

In contrast, one of the common assemblages of the kyanite and sillimanite zones is garnet + staurolite + Al-silicate + biotite + muscovite + quartz, while garnet + staurolite + chlorite + biotite + muscovite + quartz may occur in the staurolite zone. In both these cases it is notable that the garnet contains Mn and Ca, often in significant amounts (>15% grossular component and exceptionally >40% spessartine), but these elements do not readily substitute into the other KFMASH system phases. Here, garnet is likely stabilised as an 'extra' phase by the presence of these additional elements: the rock cannot be adequately represented by six components, and since it has additional components it can also contain additional phases. This stabilisation of garnet also accounts for the common co-occurrence of garnet and kyanite (Figure 4.12a) which is not predicted by Figure 4.12b.

4.3   Metamorphism at High Temperatures

Further metamorphic changes take place at temperatures higher than those of the sillimanite zone described in Section 1.2.1. In part, these involve stabilising new metamorphic mineral assemblages, but melting reactions play an increasingly important role as temperature rises. The composition of this melt can vary considerably, but is invariably different from the original metapelite since it is dominated by quartz and feldspar. If melts move away from their source then the composition of the rock that is left behind will inevitably be different from the original material.

How much melting takes place during high-grade metamorphism is largely dictated by the availability of heat and water. Melting reactions are strongly endothermic (they require lots of heat). The presence of water lowers melting temperatures since it dissolves readily in silicate melts. During melting, the rock may therefore cease to have excess water present; this 'drying out' of the rock has significant implications for the reactions that can take place in the remaining rock.

4.3.1   Upper Sillimanite Zone

The most widespread isograd above the first appearance of sillimanite in metapelitic rocks is a second reaction that creates additional sillimanite from the breakdown of muscovite:

$$\text{muscovite} + \text{quartz} \rightarrow Al_2SiO_5 + \text{K-feldspar} + H_2O \qquad [4.13]$$

The upper sillimanite zone is therefore characterised by the co-existence of sillimanite and K-feldspar (rather than just by the presence of sillimanite) and by the absence of muscovite, except as a retrograde phase. This 'second sillimanite isograd' is a particularly useful indicator of metamorphic grade because it develops in a very wide range of rock types, not just 'true' meta-pelites; it may even be traced in quartzite in some instances. In rocks where muscovite has already reacted out but where staurolite remains, the staurolite-out reaction [4.12], which also produces sillimanite, takes place at very similar temperatures.

4.3.2   Partial Melting

At higher metamorphic grades, pelitic rocks begin to melt, and form migmatites (Chapter 1, Figure 1.12d and e, Box 4.3). The earliest signs of melting occur in rocks which contain Na–K (alkali) feldspars as well as quartz and muscovite (see Section 4.3.4). To a good approximation however, we can describe the onset of melting purely in terms of the KASH system (K_2O, Al_2O_3, SiO_2, H_2O; i.e. with no Na involvement). In both cases, the melts that form are granitic in composition.

Rocks which contain water as a free fluid phase melt at the lowest temperature. At relatively low pressures, in rocks where muscovite has already broken down by reaction [4.13], the first melting reaction will be:

$$\text{K-feldspar} + \text{quartz} + \text{Al-silicate} + H_2O \rightarrow \text{melt} \qquad [4.14]$$

At moderate pressures, however, reaction [4.14] is replaced by:

$$\text{muscovite} + \text{quartz} + H_2O \rightarrow \text{sillimanite} + \text{melt} \qquad [4.15]$$

These reactions are shown by the blue line on Figure 4.13a. Both these reactions require water to be present and therefore they are known as **wet melting** reactions. As the reaction proceeds, water dissolves into the melt, and water becomes more soluble with increasing pressure. As a result, the temperature of wet melting is lower at high pressures than at low ones and the curves for reactions [4.14] and [4.15] have negative slopes on Figure 4.13a. They cross the muscovite dehydration reaction [4.13] at a high angle. These reactions correspond closely to the combined 'wet granite solidus' discussed in igneous petrology (see also Section 4.3.4).

Wet melting can only take place to a significant degree in places where a sufficient external supply of water is available, e.g. Weinberg & Hasalová (2015). Such fluids may be sourced from nearby crystallising magmas, or from nearby dehydrating metapelites in regions with strong temperature and permeability gradients. An example of the first scenario is provided by migmatites in Connemara, Ireland, that formed when fluids released from a crystallising hornblende-rich magma induced melting in nearby metapelites (Yardley & Barber 1991). An example of the second scenario is provided by the Central Steep Belt in the Alps where the fluids released by the dehydration of an underthrusted metapelite-rich unit are

BOX 4.3 Migmatites

Melting reactions give rise to the development of migmatites (Chapter 1, Figure 1.12d and e). Migmatites vary greatly in their nature and mode of occurrence (see Further Reading). The bulk of the Earth's migmatites occur in extensive and fairly uniform (often Precambrian) high-grade shield terranes. These terranes do not generally exhibit a gradational transition from non-migmatised to migmatised rocks. In other (commonly younger) areas of both regional and contact metamorphism, such transitions do occur. These latter regions have proven to be central to our understanding of how migmatites form in different environments, for example during regional heating in continental collision zones or during local (contact) heating associated with the intrusion of magmas from mantle sources. There is considerable mineralogical variation between different migmatite terranes, related to different starting materials, different pressure and temperature conditions, the availability of water and the presence or absence of deformation.

Migmatites arise because the melt produced during **anatexis** (the process of partial melting) commonly migrates on at least an outcrop scale before it solidifies (Figure 1.12d). Owing to the high viscosity of the melt, migration of melt on more than about a cm scale requires deformation, which is itself significantly aided by the overall weakening that anatexis provides (Rosenberg & Handy 2005). On cooling, the melt will crystallise feldspar and quartz. Segregated melt patches form the granitic leucosome component of the migmatite. Sometimes, especially if very little melt has formed, the melt does not move and instead crystallises in the grain boundaries of the parent schist. In this case, it may be difficult to determine if a rock containing sillimanite + K-feldspar formed by reaction [4.13] or by one of the melting reactions. Reactions [4.14] and [4.15] produce sillimanite directly from the reaction and K-feldspar indirectly during (later) melt crystallisation. Careful observation of thin sections can provide textural clues to constrain which reaction has taken place e.g. (Dyck *et al.* 2020).

considered to have facilitated anatexis of metapelites in the overlying rock unit (Berger *et al.* 2008).

Once any available water has reacted, further heating results in the breakdown of the hydrous micas to produce melt without an aqueous fluid phase. Such **dehydration melting** takes place over a temperature interval that depends on both the

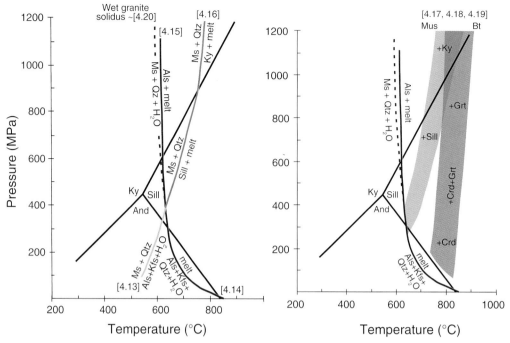

Figure 4.13 Petrogenetic grids to illustrate the conditions under which melting can take place in pelitic rocks. Reaction numbers correspond to those in the text. (a) Reactions involving the breakdown of muscovite and quartz. Data from Huang & Wyllie (1973); Johannes (1984); Johannes & Holtz (1992); Thompson (1982); Weinberg & Hasalová (2015). (b) Fields for the dehydration melting of muscovite (orange) and biotite (red), with the wet melting curves from (a), indicating the solids produced by the incongruent dehydration melting reactions in a typical metapelite composition. Data from Weinberg & Hasalová (2015).

mineral that is breaking down and the pressure. Muscovite dehydrates at the lowest temperatures:

$$\text{muscovite} + \text{quartz} \rightarrow \text{K-feldspar} + \text{sillimanite} + \text{melt} \qquad [4.16]$$

In natural rocks, dehydration melting of muscovite takes place over a range of temperatures, indicated by the orange band in Figure 4.13b.

Both reactions [4.13] and [4.16] produce sillimanite + K-feldspar directly. On Figure 4.13a both reactions have similar slopes, meeting where they both intersect the wet melting reactions [4.14] and [4.15]. The difference between them is that *water* is produced in reaction [4.13] whereas *melt* is produced in reaction [4.16]. Reactions [4.16] and [4.15] are both examples of **incongruent melting**, where melting produces a new solid phase as well as melt. All dehydration melting reactions are incongruent. Such reactions are also known as **peritectic reactions** in igneous petrology, and the resulting solid phases are called **peritectic phases**. In reaction [4.15] the peritectic phase is sillimanite (or, at high pressures, kyanite), in reaction [4.16] sillimanite and K-feldspar are both peritectic phases.

4.3.3   Cordierite–Garnet–K-Feldspar Zone

After the onset of anatexis, and at higher temperatures, further incongruent melting of pelitic rocks at moderate pressures is marked by the production of cordierite and/or garnet. Al-silicates and K-feldspar are also often present. These minerals appear as a result of dehydration melting of biotite and are typically very different in size, morphology and chemistry from occurrences of the same minerals at lower grades where they were produced by solid-state reactions.

There are a range of possible melting reactions involving biotite, some including sillimanite as a reactant and some involving sillimanite as a product. Important examples include the continuous reactions:

$$\text{biotite} + \text{sillimanite} + \text{quartz} \rightarrow \text{K-feldspar} + \text{cordierite} + \text{melt} \qquad [4.17]$$

$$\text{biotite} + \text{sillimanite} + \text{quartz} \rightarrow \text{K-feldspar} + \text{garnet} + \text{melt} \qquad [4.18]$$

Whether cordierite or garnet develops depends partly on pressure (cordierite is produced at lower pressures, garnet at higher pressures) and partly on the Mg/Fe ratio of the rock (garnet will form more readily in rocks with a more Fe-rich bulk composition, cordierite in ones richer in Mg, since $X_{\text{Mg}}^{\text{Grt}} < X_{\text{Mg}}^{\text{Crd}}$, Chapter 2, Section 2.4.3).

Reaction relationships in this zone are not as straightforward as they are at temperatures below the **solidus** (the line in *P–T* space below which a rock is completely solid) because the composition of the melt phase is very variable. For this reason, the approximate range of conditions over which biotite dehydration melting takes place has been indicated by a red field on Figure 4.13b.

The final biotite dehydration melting reaction in sillimanite-bearing metapelites is discontinuous in the KFMASH system:

$$\text{biotite} + \text{sillimanite} + \text{quartz} \rightarrow \text{cordierite} + \text{garnet} + \text{K-feldspar} + \text{melt} \qquad [4.19]$$

The garnet–cordierite–K-feldspar assemblage is typical of high-grade pelitic migmatites in the granulite facies. Reaction [4.19] is often taken to mark the boundary between the amphibolite and granulite facies in pelitic rocks.

4.3.4   Partial Melting Involving Plagioclase

All of the reactions that we have focussed on up to this point involve only the KFMASH components. Plagioclase, which contains additional Ca and Na, has been ignored, because Ca and Na are not important components of the index minerals we have been concentrating on. However, plagioclase does enter into the melt phase and so also participates in melting reactions. In practice, plagioclase is often a more important component of the melt phase than K-feldspar. For example, the first wet melting reaction encountered by rocks of suitable mineralogy is:

$$\text{muscovite} + \text{Na-plagioclase} + \text{K-feldspar} + \text{quartz} + H_2O \rightarrow \text{melt} \qquad [4.20]$$

This reaction is known as the wet granite solidus in igneous petrology and is shown in Figure 4.13a as a dashed blue line. Note that it occurs at very similar conditions to reaction [4.16].

4.3.5   Beyond Melting: Ultra-High-Grade Zones

Until relatively recently, there was a long-held understanding that, away from mantle-derived intrusions, the continental crust would never reach temperatures greater than 800–850 °C. However, more than 40 crustal localities have now been described where continental crust appears to have reached 'ultra-high' temperature (*UHT*) conditions of >900 °C for geologically-significant periods of time (see Further Reading). In practice, the field defined as *UHT* metamorphism is limited at its upper boundary only by the crustal **liquidus** (i.e. the temperature at which the crust melts completely). Continental crustal rocks that have reached *UHT* conditions are granulites rich in **refractory** minerals (resistant to melting), and these have been the focus of many experiments to constrain their conditions of formation.

The mineral assemblages that are diagnostic of *UHT* metamorphism in 'true' metapelites include orthopyroxene + sillimanite ± quartz, and sapphirine + quartz. The former assemblage is the most common – described from around two thirds of known *UHT* localities – and is stable at 900 °C at $P > 0.8$–0.9 GPa. The formation of orthopyroxene with sillimanite can be related to the more common cordierite–garnet assemblages through the reaction:

$$\text{cordierite} + \text{garnet} \rightarrow \text{sillimanite} + \text{orthopyroxene} \qquad [4.21]$$

The association of sapphirine + quartz, documented from a third of all known *UHT* localities, is the most extreme diagnostic assemblage for *UHT* metamorphism, requiring temperatures of c. 1000 °C. It forms through the reaction:

$$\text{orthopyroxene} + \text{sillimanite} \rightarrow \text{sapphirine} + \text{quartz} \qquad [4.22]$$

Other mineral indicators of *UHT* metamorphism in metasediments include combinations of cordierite, sapphirine and spinel as well as more exotic minerals such as osumilite, kornerupine and corundum (Kelsey 2008). Many of these minerals are only found in complex reaction textures, covered in more detail in Chapter 7.

4.4   Metamorphism at Low Pressures

As we saw in Chapter 1, different zonal sequences of metapelites form during metamorphism at different pressures. Lower pressures promote the formation of andalusite rather than kyanite, and of cordierite rather than garnet. The metamorphic sequence that has developed in the aureole around the Bugaboo granite-granodiorite intrusion of British Columbia, Canada (Figure 1.5) provides a good example for demonstrating the sequence of new mineral appearances and the development of distinctive rock textures that form during metamorphism at low pressures.

4.4.1   Metamorphic Zones Around the Bugaboo Batholith

The country rocks that host the Bugaboo batholith and its metamorphic aureole are regionally-metamorphosed greenschist-facies slates and phyllites. Like similar rocks in the Scottish Highlands, their fine grain size makes it difficult to identify constituent minerals without a high-powered microscope. They typically contain muscovite, chlorite, quartz and ilmenite with some paragonite (sodic white mica) and detrital plagioclase. The micas provide a greenish-silvery sheen to the cleavage planes of the rock (see Chapter 1, Figure 1.11b). Their regional metamorphism took place prior to the emplacement of the intrusion that led to the contact metamorphism.

Andalusite–Cordierite–Chlorite Zone About 1 km from the contact with the batholith, mineralogical changes are apparent, modifying the country rocks. Porphyroblasts of new metamorphic cordierite, andalusite and plagioclase appear, and paragonite disappears. Note that this latter change can only be detected in thin section with advanced petrographic techniques due to the optical similarity of paragonite and muscovite. The cordierite and plagioclase (albite) appear on cleavage planes in the rock as pale, millimetre-sized, rounded spots, whereas the andalusite appears as grey-coloured millimetre-sized elongate grains. The chemical balance of these mineral changes suggests the following reactions:

$$\text{paragonite} + \text{quartz} \rightarrow \text{andalusite} + \text{albite} + H_2O \qquad [4.23]$$

$$\text{paragonite} + \text{chlorite} + \text{quartz} \rightarrow \text{cordierite} + \text{albite} + H_2O \qquad [4.24]$$

As in the Scottish Highlands example described earlier, local bulk composition plays an important role in determining which metamorphic minerals formed during heating. In the Bugaboo aureole, subtle variations in the country rock composition mean that not all rocks contain all of these new minerals in this zone: some contain only andalusite, some only cordierite, some both and some neither. The AFM

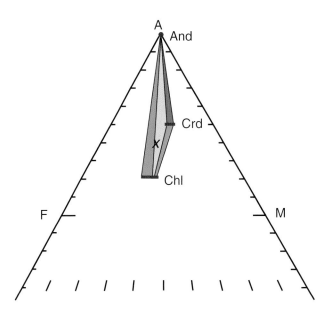

Figure 4.14 AFM diagram showing the lowest-grade assemblages formed in the metamorphic aureole around the Bugaboo batholith. The point marked *x* is the bulk composition of a typical rock in the area. Modified from Pattison & DeBuhr (2015).

diagram in Figure 4.14 shows how narrow the mineral stability triangles are, such that subtle variations in Fe/Mg of the bulk composition *x* will determine whether cordierite is stable.

Andalusite–Cordierite–Biotite Zone Closer to the intrusion, the rocks change from 'spotted' slates and phyllites to silvery, fine-grained schists. At this point tiny crystals of biotite and plagioclase start to appear, and within 50 m of biotite appearing, chlorite disappears. The rocks now look less shiny, with more densely-packed porphyroblasts.

Still closer to the intrusion, the grain size of the porphyroblasts decreases and matrix grains coarsen and become randomly oriented. These rocks are now hornfelses rather than schists. An example of a similar rock from another metamorphic aureole is shown in Figure 4.15a. The amount of biotite relative to muscovite increases, as does the amount of andalusite relative to cordierite.

The mineral assemblage of andalusite, cordierite and biotite is plotted on an AFM digram in Figure 4.15b. From inspection of Figures 4.14 and 4.15b, the disappearance of chlorite and the increase in amount of andalusite, cordierite and biotite suggest a reaction such as:

$$\text{muscovite} + \text{chlorite} + \text{quartz} \rightarrow \text{andalusite} + \text{cordierite} + \text{biotite} + H_2O \qquad [4.25]$$

The gradual decrease in the amount of muscovite and cordierite and increase in the amount of andalusite and biotite after chlorite has reacted out suggest a further continuous reaction:

(a)

(b)

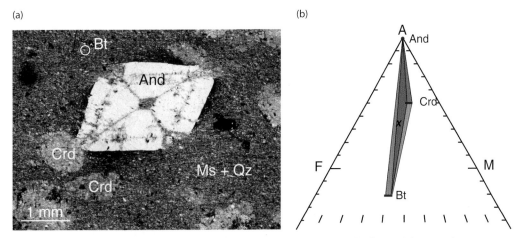

Figure 4.15 Andalusite–cordierite-zone rocks. (a) Thin section photomicrograph of an andalusite cordierite hornfels, Halifax, Canada, cross polarised light (XPL). Photo: Becky Jamieson. (b) AFM diagram showing Bt–Crd–And assemblages formed in the metamorphic aureole around the Bugaboo batholith. The point marked x is the same as in Figure 4.14. Modified from Pattison & DeBuhr (2015).

$$\text{muscovite} + \text{cordierite} \rightarrow \text{andalusite} + \text{biotite} + \text{quartz} + \text{H}_2\text{O} \qquad [4.26]$$

Muscovite–K-Feldspar–Sillimanite Zone Sparse fibrolitic sillimanite starts to appear about 300 m from the igneous contact. Where it first appears, it is weakly associated with biotite or quartz. Within the next 40 m or so, metamorphic K-feldspar also starts to appear, and primary muscovite eventually disappears. The final mineral assemblage in many samples close to the contact (except for those which are migmatitic, see below), is cordierite + biotite + andalusite + sillimanite $\pm$ K-feldspar.

As in the Scottish Highlands example described earlier, there is little textural evidence that the sillimanite formed from a direct polymorphic transition from andalusite via the reaction:

$$\text{andalusite} \rightarrow \text{sillimanite} \qquad [4.27]$$

which is chemically equivalent to reaction [4.10]. Instead the textural evidence suggests that sillimanite first appeared as the andalusite-producing reaction [4.25] continued into the sillimanite stability field.

The development of K-feldspar porphyroblasts in conjunction with the disappearance of muscovite but in the absence of any obvious melt, suggests muscovite breakdown by reaction [4.13]. This zone is therefore equivalent to the second sillimanite zone described in Section 4.3.

Migmatite Zone Close to the contact with the igneous intrusion, the country rocks contain millimetre to centimetre-scale segregations and veins of quartzofeldspathic material, interpreted as crystallised melt. Evidence for a local origin of this melt includes mantles of biotite-rich rock around coarse segregations of quartz, plagioclase and/or K-feldspar (the leucosome). These mantles formed during the back-reaction between the melt and the minerals that did not take part in the melting reaction. The K-feldspar in these segregations is anhedral and inclusion-poor, in contrast to the discrete, equant and inclusion-rich K-feldspar porphyroblasts that occur in the adjacent, lower-grade zones.

The close association of migmatitic and non-migmatitic rocks, the loss of muscovite before the melt-present zone, the absence of K-feldspar in many lithologies and the similarities in overall mineral assemblage suggest that the most likely melting reaction was a wet melting reaction involving feldspar and quartz with a fluid:

$$\text{feldspar (plagioclase or K-feldspar)} + \text{quartz} + H_2O \rightarrow \text{melt} \qquad [4.28]$$

The fluids required for this melt reaction may have been sourced from the nearby crystallising batholith or from metamorphic dehydration reactions taking place deeper underground in the local area.

4.5   Metamorphism at High Pressures

Some metapelites contain assemblages which differ from any of those present in the Scottish Highlands and Bugaboo aureole examples. Instead, they contain a range of distinctive, dense, mineral assemblages which are indicative of metamorphism at relatively high pressures. In a similar way to metamorphic changes at moderate and low pressures, a wide variety of mineral assemblages and a range of distinctive mineral compositions are produced during metamorphism under high pressures.

4.5.1   High-Pressure Metapelites: an Alpine Example

The Western European Alps expose one of the world's best-characterised regions of high-pressure (*HP*) metamorphism, where metasediments preserve characteristic and diagnostic mineral assemblages. A number of major metapelite-bearing terranes preserve evidence for different grades of *HP* metamorphism: in the west, the Briançonnais terrane has been metamorphosed to the lowest grades; in the middle, the Schistes Lustrés complex contains intermediate-grade rocks; and towards the east, the Piemonte Zone, Internal Crystalline Massifs (Dora Maira, Gran Paradiso

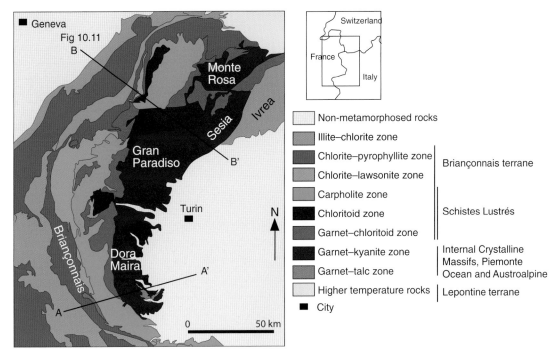

Figure 4.16 Metamorphic map of the Western Alps showing the high-pressure metamorphic zones recorded in metapelitic rocks during the Alpine continental collision. Simplified from Bousquet *et al.* (2012), with additional information from Beltrando *et al.* (2010) and Goffe & Chopin (1986).

and Monte Rosa) and Austroalpine units have been metamorphosed at the highest grades (Figure 4.16). It is important to note that the protolith history, metamorphic structure and tectonic evolution of the Alps are complex, and rocks of 'true' pelite composition are rare – we only provide a very simplistic overview here. Parts of the Alps expose rocks that were also first metamorphosed during the Permian Variscan orogeny and were only lightly overprinted during the Tertiary Alpine orogeny (e.g. the Ivrea Zone reached granulite facies metamorphism in the Permian but only illite–chlorite grade in the Tertiary). The following zones are presented from west to east, along line A–A' on Figure 4.16.

Illite–Chlorite Zone At the western extremity of the western Alps, the lowest-grade metamorphic rocks (above the baseline diagenesis) consist of fine-grained slates that contain quartz, chlorite, and the clay minerals illite and kaolinite. As in the Scottish Highlands, their fine grain size makes it difficult to study the mineralogy of these rocks under an optical microscope.

Chlorite–Pyrophyllite Zone Moving east, across a major tectonic break, the metasediments in the Briançonnais terrane are phyllites and schists, with a larger

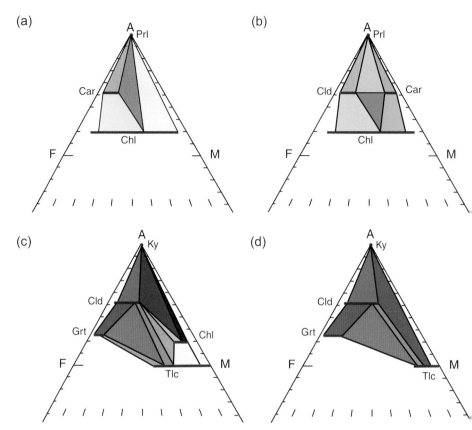

Figure 4.17 AFM projections for high-pressure metamorphism of metapelites. (a) Assemblages involving chlorite and pyrophyllite. (b) Assemblages involving chloritoid and carpholite. (c) Assemblages involving garnet, kyanite and talc with chlorite. (d) Assemblages involving garnet, kyanite and talc without chlorite. All assemblages contain phengitic muscovite and quartz. Modified from Goffe & Chopin (1986).

grain size than in the slates. Careful analysis of the phyllosilicates has indicated that they contain a mixture of chlorite and an aluminosilicate called pyrophyllite. Whilst pyrophyllite is not a high-pressure mineral as such (for example, it is also associated with ore mineral formation in other contexts), its presence in metasediments in the Alps helps to confirm the metamorphic progression along a low-temperature, high-pressure metamorphic trajectory; see Figure 4.17a.

Carpholite Zone The metapelites of the westernmost exposures of the Schistes Lustrés contain the iron-rich end-member of a rare mineral called carpholite, commonly associated with pyrophyllite. These minerals form in part from chlorite but also co-exist with it; see Figure 4.17b.

Chloritoid Zone At somewhat higher pressures and temperatures, grain size in the Schistes Lustrés increases, carpholite becomes more Mg-rich and Fe-rich chloritoid appears (Figure 4.17b). This change in mineralogy suggests a continuous reaction such as:

$$\text{carpholite1} \rightarrow \text{carpholite2} + \text{Fe-chloritoid} + \text{quartz} + H_2O \qquad [4.29]$$

where $X_{\mathrm{Mg}}^{\mathrm{Car2}} > X_{\mathrm{Mg}}^{\mathrm{Car1}} > X_{\mathrm{Mg}}^{\mathrm{Cld}}$.

This dehydration reaction is strongly dependent on temperature, and therefore the compositions of the carpholite and chloritoid in this area are very useful for deciphering thermal evolution.

At higher temperatures within this zone, pyrophyllite disappears and is replaced by kyanite (both plot at the A corner on the AFM projection, Figure 4.17b and c). Note that throughout this sequence, biotite, a ubiquitous low-grade mineral at similar temperatures in moderate- and low-pressure metapelites, is uncommon or absent.

Garnet–Kyanite Zone At the eastern side of the Schistes Lustrés terrane, and into the western side of the Dora Maira Massif (across a further tectonic boundary), the metasedimentary rocks become increasingly coarse-grained. Garnet appears in association with increasing Mg-rich chloritoid (Figure 4.17c and d). A key difference between the garnet which forms in the Alps and the garnet which forms in the Scottish Highlands is that the former is enriched in the Mg-end-member (pyrope) relative to almandine. In the field, these pyrope-rich garnets tends to be more of a purple colour compared to the brighter red of almandine.

Further east in the Dora Maira Massif, kyanite replaces pyrophyllite and chlorite starts to disappear, to be replaced by talc. Interrogation of the AFM diagram (Figure 4.17c and d) suggests the following divariant reactions depending on the rock bulk composition:

$$\text{chlorite} + \text{quartz} \rightarrow \text{garnet} + \text{talc} + H_2O \qquad [4.30]$$

$$\text{chlorite} + \text{quartz} \rightarrow \text{chloritoid} + \text{talc} + H_2O \qquad [4.31]$$

$$\text{chlorite} + \text{quartz} \rightarrow \text{kyanite} + \text{talc} + H_2O \qquad [4.32]$$

Garnet–Talc Zone At nearly the highest pressure conditions recorded in the Alpine metasediments in the Dora Maira Massif, chloritoid disappears from the schists, to leave a garnet + talc + kyanite assemblage. This suggests a reaction such as:

$$\text{chloritoid} + \text{muscovite} \rightarrow \text{garnet} + \text{talc} + \text{kyanite} + H_2O \qquad [4.33]$$

The association of talc and kyanite forms rocks known as 'whiteschists', due to the talc being colourless while the kyanite is very pale blue. In the classic locality in the Alps, the pyrope-rich garnets are also extremely pale in colour, but so large that originally they were misidentified as clasts and the rock was thought to be some kind of strange conglomerate. Unfortunately, there is very little of this iconic rock left now, despite the outcrop being protected in law, as a result of indiscriminate sampling by too many 'geo-tourists' over the years.

4.5.2   High-Pressure Greywackes

As we have already seen, rock bulk composition plays an important role in the mineral assemblages that form during metamorphism. High-pressure terranes are no exception to this. 'True' metapelites that have been metamorphosed to *HP* conditions are quite rare; many *HP* metamorphic belts are instead characterised by immature greywacke sediments, which are richer in Ca. The higher Ca content of the greywacke compositions means that Ca–Al-rich minerals such as plagioclase, epidote, lawsonite and Ca-rich garnet are present, and preclude the formation of many of the Fe–Mg–Al silicates typical of 'true' metapelites. In the Alps, for example, lawsonite forms towards the eastern side of the Briançonnais terrane near the tectonic contact with the Schistes Lustrés units (marked on Figure 4.16).

Metasediments described from some of the high-pressure circum-Pacific regions such as New Caledonia or Japan contain Ca-rich amphibole or epidote in significant quantities. In both regions, phengite, paragonite, chlorite, albite and quartz occur throughout the terrane, with pyrope-rich garnet appearing at higher grades. Some of these assemblages, especially where they contain minerals such as lawsonite or glaucophane, are similar to those that form in metabasites (see Chapter 5).

A common reaction in Ca-rich metasediments at high pressures, roughly equivalent to the pressures at which talc forms in 'true' metapelites, is one in which anorthite (Ca-rich) plagioclase reacts to form grossular (Ca-rich) garnet plus kyanite:

$$\text{anorthite} \rightarrow \text{grossular garnet} + \text{kyanite} + \text{quartz} \qquad [4.34]$$

This reaction is equivalent in metasediments to the reaction that forms eclogites in basic rocks (Chapter 5, Section 5.5.2).

4.5.3   High-Pressure–High-Temperature Rocks

High-pressure rocks do not always form at relatively low temperatures: high pressures may also be associated with high temperatures. One example is exposed near the Manicouagan Impact Crater (the effects of the impact are distinct from regional metamorphism) in eastern Canada. The Tschenukutish Terrane exposes kyanite-bearing migmatitic metapelites containing quartz, plagioclase, K-feldspar, garnet, kyanite rutile and some biotite (Indares & Dunning 2001). About 20% of the rocks comprise granitic leucosomes containing quartz, K-feldspar, garnet and kyanite. The close association of the leucosome and the metapelite, and the absence of sillimanite, suggest that they formed via the dehydration of micas in the kyanite field.

4.6 Summary: the Effect of Pressure on Metamorphic Zoning Patterns

The differences in zoning patterns between the Scottish Highlands, the Bugaboo aureole and the metapelites in the Western Alps are quite distinctive. The first two contain chlorite, muscovite and biotite at low grades, whereas biotite is conspicuously absent in the Western Alps. Similarly, sillimanite is present in both the Scottish Highlands and the Bugaboo aureole at the highest grades, but only kyanite occurs in the third example. Garnet is prevalent in the Scottish and Alpine examples, but cordierite is present in the Bugaboo example instead. Andalusite is also only present in Bugaboo, whereas kyanite forms in both the other two examples.

There are also subtler differences between the moderate-pressure regional and lower-pressure contact types of metamorphism. Staurolite overlaps with kyanite in the Scottish Highlands but in some areas reacts directly to sillimanite. At Bugaboo, muscovite breaks down before there is any evidence of melting, whereas in the Scottish Highlands, melting takes place in rocks with stable muscovite. Figure 4.18 illustrates the relationships between the principal dehydration and melting reactions of metapelites and the Al-silicate phase diagram at moderate to high temperatures (the reactions that take place in the Western Alps are off the scale of this figure). The intersections between them define the distinctive sequences of reactions that characterise the different zonal schemes.

In the Scottish Highlands, kyanite is the first Al_2SiO_5 phase to appear and is replaced by sillimanite at higher temperatures. Although partial melting is not recorded in the area of the Scottish Highlands that we have been looking at, staurolite appears in the rocks well before sillimanite appears. These observations suggest that the Scottish Highlands followed a heating path illustrated by B–B' on Figure 4.18. This is a typical mineral reaction sequence for regional metamorphic terranes.

Along the section through the Bugaboo aureole that was looked at in detail earlier in this chapter, the Al-silicate polymorph that forms furthest from the pluton is andalusite, succeeded by sillimanite closer to the contact. Furthermore, the upper sillimanite zone assemblage K-feldspar + sillimanite develops before the rocks start to melt. Both of these observations fit with a heating path at lower pressures: best represented by D–D' on Figure 4.18.

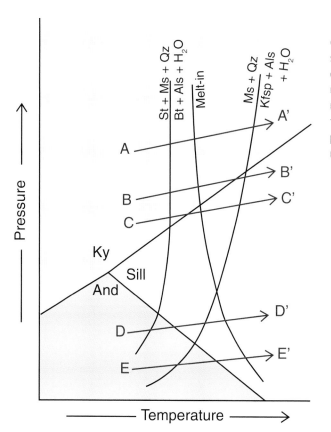

Figure 4.18 Schematic *P–T* diagram to illustrate the effect of stylised heating paths at different pressures on the mineral assemblages and mineral assemblage sequences that form in metapelites. Higher-pressure reactions are not shown.

Along path E–E', andalusite is predicted to remain stable up to the temperatures of muscovite breakdown, which will result in the growth of andalusite + K-feldspar through reactions [4.13] and [4.17]. This distinctive low-pressure–high-temperature assemblage is present in some sections through the Bugaboo aureole (Pattison & DeBuhr 2015). This variation in zoning patterns around the same aureole suggests that exhumation of the aureole has not been uniform, with some exposed sections of the aureole around the batholith now recording different depths to others.

Metamorphic heating along the C–C' path also predicts sillimanite to be the first Al_2SiO_5 phase produced. However, along this path, the rocks are predicted to melt before the second sillimanite isograd is reached. This type of heating path has been suggested for metamorphism in Western Maine, USA, where primary muscovite only disappears a few hundred metres past the first appearance of migmatites (Johnson *et al.* 2003). Along path A–A', however, the rocks are predicted to melt in the kyanite field, without ever forming sillimanite. Kyanite-bearing migmatites have been reported from a number of localities, including the Himalaya (Iaccarino *et al.* 2015).

Although the examples of low-pressure metamorphism that we have described in detail are from thermal aureoles, the combination of andalusite followed by sillimanite can also form in regionally-metamorphosed areas. This is the case in Connemara, Ireland, for example, where regional metamorphism and deformation is temporally and spatially associated with the emplacement of syn-tectonic intrusions. Such 'regional contact metamorphism' is discussed further in Chapter 10, Section 10.2.2.

4.7 Determining *P–T* Conditions for Metapelite Metamorphism

In the preceding sections we have seen that metapelites form a number of distinctive index minerals during metamorphism that allow us to constrain *relative* metamorphic grade. However, as we saw in Chapter 3, we rely on experimental data to constrain the absolute conditions at which different assemblages are stable. With the aid of thermodynamic datasets, a range of options are available for estimating pressures and temperatures of formation of metapelite assemblages more precisely.

4.7.1 Petrogenetic Grid

Figure 4.19 is a petrogenetic grid that illustrates the approximate stability limits of some of the key assemblages discussed in this chapter. The phase diagram for the three Al_2SiO_5 polymorphs, discussed in Chapter 2 (Section 2.3.1, Figure 2.1), provides a key focus point for being able to make broad subdivisions on the basis of both pressure and temperature. Many metapelite equilibria are represented by steep curves on the petrogenetic grid, indicating that they are predominantly sensitive to changes in temperature and make potentially useful geothermometers. For example, the presence of staurolite provides a good indicator of metamorphic temperature since it is stable only over a narrow temperature range which does not change much with pressure. Another reaction that provides relatively good constraints on temperature is the muscovite breakdown reaction (reaction [4.13]) that gives rise to the second sillimanite isograd. Conversely, the flatter the equilibrium curve on the petrogenetic grid, the better it is for geobarometry. For example, the ilmenite to rutile, sillimanite to kyanite and cordierite to garnet transitions are all important for estimating pressure.

4.7.2 Geothermometers and Geobarometers

There are a wide range of geothermometers and geobarometers that are usefully applied to metasedimentary rocks, including most of the types outlined in Chapter 3. At low (e.g. chlorite zone) grades, non-equilibrium approaches such as chlorite or illite crystallinity, spectroscopy of carbonaceous materials (RSCM; Chapter 3, Section 3.3.11), as well as oxygen isotope thermometry (Chapter 3, Section 3.3.6), provide useful constraints on temperature (Árkai 1991; Beyssac *et al.* 2003; Bourdelle *et al.* 2013; Kisch 1991; Lacroix & Vennemann 2015). These approaches

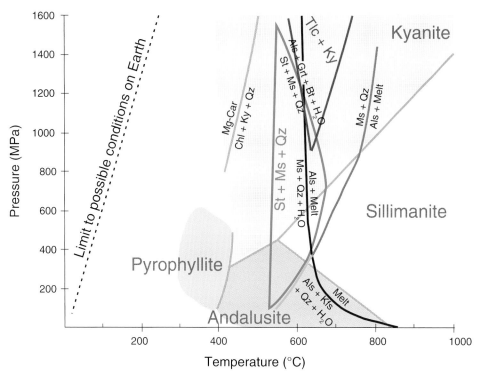

Figure 4.19 Petrogenetic grid for metapelites, including the wet solidus (blue) and muscovite breakdown curves (yellow, orange) from Figure 4.13. Shading denotes the stable Al-silicate mineral under water-saturated conditions. To the left of the wet solidus, conditions are all water-saturated. Green lines define the stability of staurolite with muscovite + quartz (*MT–MP* metamorphism), purple lines outline the talc + kyanite field (*HP–LT* metamorphism) in wet, Mg-end-member systems. Data from Kerrick (1968); Spear & Cheney (1989); Vidal *et al.* (2001); Wei & Powell (2006), plus references in the caption for Figure 4.13.

generally require access to specialist equipment but calibrations and the understanding of how temperature affects these minerals is improving all the time.

At higher grades, and especially where garnet is present, cation-exchange reactions have proven useful but are now mainly being replaced by computational multi-equilibria approaches (see below). The most commonly applied reactions in metapelites include Fe–Mg exchange between garnet and biotite (Chapter 3, Section 3.3.5), garnet–muscovite and garnet–cordierite. Calibrations of these thermometers are constantly being revised but at the time of publishing the most recent include Kaneko & Miyano (2004); Wu & Cheng (2006); Wu & Zhao (2006). Exchange of Fe and Mg between the reacting minerals may also occur as rocks are cooling, so temperatures yielded by these thermometers may commonly only be minima.

Trace element thermometers such as Ti-in-biotite (Chapter 3, Section 3.3.8) are also increasingly useful in rocks where Ti-rich minerals such as rutile, ilmenite or titanite are present to buffer the concentrations of Ti in co-existing phases (Wu &

Chen 2015). As the Ti content of biotite is less easily perturbed than the Fe/Mg ratio, this thermometer provides useful complementary information about the thermal evolution of the rock when used alongside the garnet–biotite thermometer.

The continuous reactions that lead to exchange of calcium between garnet and plagioclase provide the basis for some of the most useful barometers for metapelites, as calcium-rich garnet is denser than calcium-rich plagioclase and thus is preferentially formed at higher pressures. A commonly applied garnet–plagioclase barometer is the Garnet–Aluminosilicate–Plagioclase (GASP) reaction (Chapter 3, Section 3.3.4, reaction [3.8]) (Newton & Haselton 1981; Wu & Cheng 2006). Another useful geobarometer is the GRAIL reaction involving garnet, rutile, aluminosilicate, ilmenite and quartz (Koziol & Bohlen 1992). Further reactions involving grossular, anorthite, muscovite, biotite and/or cordierite have also been suggested as useful for metapelite barometry (Wu 2015).

Computational approaches such as isochemical phase diagrams (Chapter 3, Section 3.3.10) are increasingly being used to constrain the detailed *P-T* history recorded in metasedimentary rocks, e.g. (McDade & Harley 2001; Wei *et al.* 2004; Wei & Powell 2003). As described in Chapter 3, these approaches combine empirical and experimental data from a variety of systems, thus allowing more accurate constraints on conditions. Remember from Chapter 3 that these computational phase-diagram approaches are particularly useful for studying mineral assemblages at higher temperatures, where individual mineral compositions may have been modified during cooling. The zoned minerals that are present at lower grades require significant assessment to determine the volume of equilibration before interpretations of the phase diagrams are valid.

SUMMARY

Metasedimentary rocks that are rich in Al are common in the sedimentary record in the form of mudstones. Rocks of this bulk composition form a number of readily identified index minerals in close succession during metamorphism. Such metapelite mineral assemblages can therefore be used to provide a rough estimate of metamorphic conditions in the field.

In this chapter we have explored how the equilibrium assemblages that develop in metapelites can be analysed using AFM diagrams to help to determine whether different assemblages arose under the same peak metamorphic conditions because of variations in bulk composition or whether they reflect different metamorphic conditions affecting rocks of similar composition. AFM diagrams can be plotted for any metapelite assemblage so long as it contains muscovite and quartz.

Distinct sequences of index minerals develop in rocks of similar bulk composition at different metamorphic grades. A typical 'moderate-pressure–moderate-temperature' sequence develops mineral assemblages that contain chlorite, biotite, garnet, staurolite, kyanite and sillimanite up-grade, eventually leading to melting and the formation of migmatites. At higher temperatures, continued dehydration promotes the formation of anhydrous minerals such as orthopyroxene and sapphirine. At lower pressures, andalusite and cordierite form instead of kyanite and garnet, and the sequence of growth of different index minerals can help provide clues about subtle variations in pressure (and thus depth) of formation. At higher pressures, rocks of metapelitic composition grow minerals such as carpholite and talc, the latter in combination with garnet and kyanite.

Whilst the index mineral sequence is a useful rough guide to the pressure and temperature conditions of metapelite metamorphism, there are also a number of experimentally-calibrated reactions that are useful for determining absolute $P–T$ conditions in metapelites. These include cation-exchange reactions such as the one which underpins the garnet–biotite thermometer, and the continuous reaction between plagioclase and garnet that produces aluminosilicate and quartz, which underpins the GASP barometer. The concentrations of titanium (a trace element) in biotite is temperature-dependent, as is the crystallinity of chlorite at very low temperatures.

EXERCISES

1. A colleague brings a pile of rocks (labelled *a* to *e*) back from the field. The rocks contain the following mineral assemblages:
   (a) quartz, garnet, plagioclase, biotite, staurolite, muscovite
   (b) kyanite, garnet, quartz, muscovite, biotite
   (c) plagioclase, biotite, muscovite, quartz
   (d) sillimanite, K-feldspar, garnet, biotite, quartz
   (e) garnet, quartz, muscovite.
   On the basis of their mineral assemblages, place rocks *a* to *e* in order of increasing metamorphic grade as far as is possible.

2. Predict the mineral assemblages that grow in rocks of bulk composition *y* and *z* (Figure 4.9) through the staurolite, kyanite and sillimanite zones (Figures 4.11 and 4.12).

3. Why do some metapelites grow chloritoid during prograde metamorphism and others not?

4. Sketch reaction [4.7] on an AFM diagram. Describe how this reaction is similar or different to the anorthite-out reaction investigated in Chapter 2, Exercise 6 (Figure 2.2).

5. Sketch reaction [4.5] on an AFM diagram. Describe how this reaction is similar or different to reaction [4.7].

6. Suggest two distinct mineralogical features of metapelites metamorphosed under relatively low pressures, moderate pressures and high pressures.
7. Using the petrogenetic grid in Figure 4.19, what are:
   (a) the maximum P–T conditions for a rock containing sillimanite, biotite and garnet;
   (b) the minimum P–T conditions for a rock containing kyanite and talc?

FURTHER READING

Baxter, E. F., Caddick, M. J. & Ague, J. J. (2013). Garnet: Common mineral, uncommonly useful. *Elements*, **9**(6), 415–19.

Clark, C., Fitzsimons, I. C., Healy, D. & Harley, S. L. (2011). How does the continental crust get really hot? *Elements*, **7**(4), 235–40.

Sawyer, E. W. (2008). *Atlas of Migmatites* (Vol. 9). NRC Research Press.

Sawyer, E. W., Cesare, B. & Brown, M. (2011). When the continental crust melts. *Elements*, **7**(4), 229–34.

Wei, C. & Powell, R. (2003). Phase relations in high-pressure metapelites in the system KFMASH (K_2O–FeO–MgO–Al_2O_3–SiO_2–H_2O) with application to natural rocks. *Contributions to Mineralogy and Petrology*, **145**(3), 301–15.

Wei, C. J., Powell, R. and Clarke, G. L. (2004). Calculated phase equilibria for low- and medium-pressure metapelites in the KFMASH and KMnFMASH systems. *Journal of Metamorphic Geology*, **22**(5), 495–508.

5 Metamorphism of Basic Igneous Rocks

Metamorphosed lava flows, and their related minor intrusions and volcanogenic sediments, are frequently found interspersed with metamorphosed sedimentary rocks. Larger intrusive bodies are sometimes also metamorphosed, especially where old crystalline crust is caught up in a later orogeny and **remobilised** (subjected to later metamorphism and/or deformation). This chapter will outline the metamorphic changes that take place in igneous rocks of basic composition, such as basalts and gabbros, to produce **metabasites**. Metabasites differ from the metapelites described in Chapter 4 in that their mineral assemblages tend to vary progressively, often without distinct zones marked by index minerals. Despite this we shall see that their assemblages continue to vary with metamorphic conditions over a very wide range of pressures and temperatures. In particular, amphiboles are stable in metabasites over a very wide range of conditions, and we shall see how the composition of the amphibole varies considerably with pressure and temperature.

Igneous rocks often experience regional and contact metamorphism alongside interbedded sediments, but they differ fundamentally from sediments in their metamorphism, because of the contrasting role of water. The initial igneous assemblage is made up of mostly anhydrous minerals stable at high temperatures, and may include glass. The first changes that take place during their metamorphism normally involve the formation of hydrous minerals stable at low temperatures. This requires the *addition* of water, not its loss. The extent of such retrograde reaction depends on the amount of water that is able to penetrate the rock, and the stage at which metamorphism begins depends in part on the physical properties of the rock unit. Permeable tuffs often react extensively at very low temperatures, whereas massive lava flows and intrusions may retain relics of igneous minerals and textures to moderate grades of metamorphism.

Hydrothermal metamorphism is important in shallow volcanic systems, especially below the sea bed, and is a process we can study in progress today in geothermal fields. We will use examples of hydrothermal metamorphism to introduce the study of metasomatism and the characteristic features of metasomatic rocks.

5.1   The Breakdown of Primary Igneous Mineral Assemblages

The transition from an igneous rock to a metamorphic rock is one that can take place in very different ways according to the nature of the original igneous rock and the depth at which it was emplaced. While volcaniclastic sediments rich in glass can be completely hydrated to low-temperature assemblages under near-surface conditions, coarse-grained igneous intrusions can survive to lower crustal depths and temperatures before they begin to recrystallise (e.g. Figures 1.8 and 5.1c). The primary reason for these different patterns of behaviour is the availability of water. Water must be added to the rock for metamorphism to take place and, once metamorphism begins, different primary igneous minerals react with water at different rates so that some are more likely to be preserved than others. For example, olivine is usually much more readily hydrated than clinopyroxene, altering to serpentine in all but the freshest igneous rocks. One of the most widespread minerals to form from hydration of basic igneous rocks is chlorite. This is significant because chlorite has a lower silica content than most of the igneous minerals that it replaces. As a result, the growth of chlorite in metabasites is normally accompanied by the appearance of quartz.

Water can most readily access an igneous rock if it is porous, with interconnected pores throughout, and volcaniclastic sediments are therefore particularly susceptible to metamorphism. These rocks, together with highly-vesicular flow tops, often act as aquifers in geothermal fields (Figure 5.1a). More-massive lava flows and intrusions depend on fractures for their permeability, and are much less readily accessed by fluids. Metamorphism can only begin when deformation provides flow paths, but both brittle fractures and shear zones provide access for fluids. In outcrop, hydrous minerals are commonly found along sharp planar fractures or around zones of ductile deformation (e.g Figure 5.1b). Basalts on the ocean floor are altered by sea water at low temperatures soon after eruption, but even here, beneath the ocean, alteration is patchy. Long drillcore sections

(a)

(b)

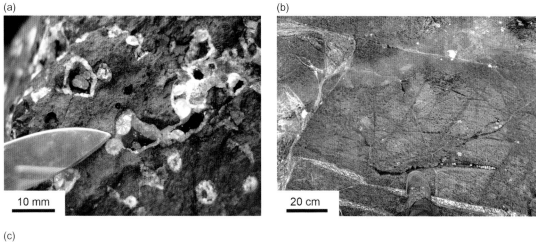

10 mm

20 cm

(c)

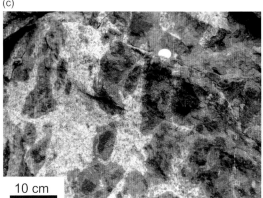

10 cm

Figure 5.1 Examples of the metamorphism of basic igneous rocks. (a) Amygdaloidal basalt with extensive development of hydrothermal minerals in vesicles, Iceland. (b) Remnant gabbro (brown) with extensive development of hornblende (greenish black) along fractures, Connemara, Ireland. (c) Coarse pyroxene–plagioclase pegmatite metamorphosed to produce rims of red garnet around dark pyroxene under granulite facies conditions, Gaupas, west Norway.

extending up to 500 m through the western Atlantic oceanic crust showed extensive development of calcite, clays and other hydrous minerals at flow tops and in volcaniclastic sediments, but relatively little alteration in massive flows and at greater depths (Staudigel *et al.* 1996). Thus, while parts of the upper oceanic crust are partially hydrated during mid-ocean ridge processes, further hydration of more-massive basalt takes place in other settings.

At medium metamorphic grades, water driven off from surrounding sediments can cause hydration of the margins of igneous bodies and hydration becomes extensive. In deformed, regionally-metamorphosed terranes such as the south-east Highlands of Scotland (Chapters 1 and 4), original sills and lava flows tens of metres thick typically develop hydrated metamorphic mineral assemblages with only rare igneous relics by the time they reach greenschist-facies conditions. Similarly, in high-pressure terranes, sills and lavas are almost entirely composed of metamorphic minerals by the time they reach blueschist facies. With further heating, such assemblages begin to dehydrate, undergoing prograde metamorphism and releasing fluid in a similar way to the pelitic metasediments described in

Chapter 4. In some circumstances, massive igneous intrusions can retain their primary features to very high metamorphic grades. This is particularly the case where intrusions were emplaced into the deep continental crust and encountered very little water (Figure 5.1c).

5.2 The Facies Classification

The mineral assemblages of metabasites are less sensitive to changes in temperature than those of metapelites, except at very low metamorphic grades. The zones that they define therefore represent a broader range of possible conditions of formation. In this sense metabasites are less useful than metapelites as metamorphic indicators. However, metabasites are found in most metamorphic belts irrespective of the setting in which the accompanying sediments were deposited, and the zones identified in most other rock types can be correlated with metabasite assemblages. Additionally, metabasites provide good indicators of high-pressure metamorphism.

For these reasons Eskola based his scheme of metamorphic facies, introduced in Chapter 3, Section 3.2, on the mineral assemblages of metabasites. The facies classification is a useful qualitative approach to evaluating the conditions of metamorphism, especially in the field, but is superseded by quantitative approaches once rocks and minerals can be analysed. Table 5.1 summarises the distinctive mineral associations of metabasites in each of the facies.

The facies shown in Figure 3.2 can sometimes be subdivided into zones on the basis of additional phases such as garnet or epidote but the details of whether such zones develop are dependent on rock composition. Hacker *et al.* (2003) have calculated the equilibrium assemblages for a rock with MORB (mid-ocean ridge basalt) composition over the full range of metamorphic conditions, and this provides a valuable reference point.

5.2.1 Mineralogical Changes Defining the Facies

As in metapelites, a number of mineralogical features of metabasites are useful for constraining changes in the intensity of metamorphism as *P–T* conditions change. The difference in metabasites, however, is that in general it is the *composition* of the phase that is important, rather than its presence or absence. The most useful mineralogical changes that are used to distinguish between the different metamorphic facies in metabasites include:

- changes in the composition of amphibole;
- changes in the composition of plagioclase feldspar;

Table 5.1 Characteristic metabasite mineral associations of the metamorphic facies

Zeolite	laumontite (most typical), analcite, heulandite, wairakite. Relic igneous grains widespread
Prehnite – Pumpellyite	prehnite + pumpellyite (lower grades), pumpellyite + actinolite (higher grades), chlorite, albite common, ± epidote. Relic igneous grains common
Greenschist	actinolite + albite + chlorite ± epidote (lower grades), at higher grades hornblende + actinolite; oligoclase + albite is possible. Relic igneous grains usually rare
Amphibolite	hornblende + plagioclase ± epidote ± garnet
Granulite	orthopyroxene + clinopyroxene + plagioclase ± hornblende ± garnet ± olivine
Blueschist	glaucophane (or glaucophane–riebeckite) + albite ± lawsonite. Relic igneous grains possible
Eclogite	omphacite + garnet without plagioclase

- the formation of hydrous Ca–Al silicates, including epidotes, lawsonite, Ca-zeolites, prehnite or pumpellyite;
- formation of pyroxenes, including diopside–augite, omphacite or jadeite;
- the formation of garnet, ranging in composition from almandine to pyrope, often with a significant grossular content.

Of these, the most important are the composition of amphibole and the type of plagioclase feldspar, if present.

Changes in Amphibole Composition Amphiboles are formally classified into a large number of varieties, which are periodically reviewed by the International Mineralogical Association (Hawthorne *et al.* 2012). Amphiboles can form both monoclinic and orthorhombic structures, and it is the compositions of monoclinic amphiboles which are the main focus in this book. The main varieties of amphibole we will cover are actinolite, hornblende and glaucophane (Box 5.1, compositions in Appendix 1). A detailed review of the amphibole compositions present in metamorphic rocks can be found in Schumacher (2007).

Broadly speaking, the lower-temperature facies are characterised by actinolite. At higher temperatures hornblende appears, and at higher pressures, glaucophane is stable. Although intermediate amphibole compositions between these three main groups are found in nature, breaks in composition are commonly quite abrupt. There has been considerable investigation into the extent to which this may result from miscibility gaps in the amphibole solid-solution series and the existence of such miscibility gaps is now generally accepted (Schumacher 2007). Despite such

BOX 5.1 Amphiboles – a brief introduction

The amphiboles are a group of double-chain silicates, in which Si–O tetrahedra are joined at their corners to create a potentially infinite double chain or ribbon, four tetrahedra in width and with all the tetrahedra facing the same way (Figure 5.2). These ribbons are stacked together so that they are all parallel but with the tetrahedra pointing in opposite directions in adjacent ribbons. The ribbons are bonded together by strips of octahedrally-coordinated cations between inward-pointing ribbons and by distinct octahedral sites between the bases of ribbons pointing in opposite directions. Alkali metal cations are also found here, if needed to maintain charge balance (Figure 5.2).

The general formula of an amphibole can be written as: $A_{0-1}B_2C_5T_8H_4$. Here, 'A' represents the site where optional alkali cations can occur, to compensate for charge imbalance due to substitutions on other sites, 'B' represents the octahedral sites between the bases of the ribbons of tetrahedra (M4 in Figure 5.2), 'C' represents the octahedral sites between inward-facing ribbons of tetrahedra (M1, M2, M3 in Figure 5.2), 'T' represents the tetrahedrally-coordinated sites making up the ribbons themselves, and 'H' represents the hydroxyl (OH) site, which lies within the rings of tetrahedra in the ribbons. The most common ions present in the sites of the general formula are:

A – Na^+, K^+ (but is empty in actinolite and glaucophane);
B – Ca^{2+} or Na^+;
C – Mg^{2+}, Fe^{2+}, Fe^{3+} Al^{3+};
T – Si^{4+} or Al^{3+};
H – OH^-, Cl^-, F^-.

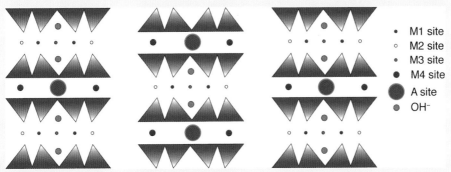

- • M1 site
- ◦ M2 site
- • M3 site
- ● M4 site
- ⬤ A site
- • OH⁻

Figure 5.2 The essential elements of amphibole structure. Si–O tetrahedra link together to form a continuous double chain or ribbon, here effectively perpendicular to the page. These chains are parallel to the c-axis of amphibole crystals, and where these are elongate, this is also the long dimension. The ions that most commonly occur in the sites illustrated here are described in the text.

In detail, the octahedral sites designated here by 'C' are not all identical, and some are more suitable for substitution by Al. The large number of sites, and the potential for different positive charges at each according to the atoms present, helps to explain the enormous diversity of the amphibole group. The major amphibole types discussed in this chapter are:

actinolite   – $Ca_2(Fe, Mg)_5Si_8O_{22}(OH)_4$
hornblende   – $(Na, K)_{0-1}(Ca, Na)_2(Fe, Mg, Al)_5(Si, Al)_8O_{22}(OH)_4$
glaucophane – $Na_2(Fe, Mg)_3Al_2Si_8O_{22}(OH)_4$.

Of these, hornblende is the most variable, with multiple end-members. Some Al is always present in hornblende, and the extent to which it is present in tetrahedral or octahedral sites is a reflection in part of the conditions under which it grew. The Tschermak's substitution, described for sheet silicates in Box 4.2, is also important in hornblendes. Glaucophane commonly contains some Fe^{3+} substituted for Al; this increases the intensity of the distinctive blue and lilac colours seen in thin section, and also extends the range of conditions over which it is stable to lower pressures (see Figure 5.3). Riebeckite, $Na_2Fe^{2+}_3Fe^{3+}_2Si_8O_{22}(OH)_2$, the Na–$Fe^{3+}$ amphibole end-member, is stable to low temperatures and pressures; it occurs as veins in banded iron formations, for example.

25 μm

Figure 5.3 Actinolite grains in a greenschist from the chlorite zone of the greenschist facies, New Zealand, with relic cores of blue amphiboles (glaucophane–riebeckite series). The growth is evidently sequential and so this texture cannot be used to argue for the stable co-existence of actinolite and glaucophane.

complications, activity–composition models (Chapter 3, Box 3.1) have been developed for some amphiboles, to enable thermodynamic modelling of metamorphic conditions (Diener & Powell 2012) (Chapter 3, Section 3.3.9).

In addition to the more-common Ca- or Na-bearing monoclinic amphiboles, Ca-poor amphiboles (cummingtonite and the orthorhombic amphiboles, anthophyllite and gedrite) are also found in some metabasites, especially at low pressures.

Changes in Feldspar Composition   Albite is the only plagioclase feldspar stable at low temperatures, and is replaced by an intermediate plagioclase at higher grades in much the same way as was described for metapelitic schists in Chapter 4, Section 4.2. Albite can co-exist with oligoclase over a narrow temperature range around 450 °C, but at slightly higher temperatures only oligoclase is present. In this book, we have taken this transition to mark the transition from the greenschist facies to the amphibolite facies because it is seen in a range of rock types (Section 5.3.2). As temperatures increase further, plagioclase tends to have higher anorthite contents. At very high pressures, plagioclase reacts out completely, with Na incorporated into clinopyroxene as the jadeite component in omphacite and Ca into garnet (grossular component) or, at lower temperatures, lawsonite.

5.3   Metamorphism at Low Pressures and Temperatures

Igneous rocks metamorphosed at shallow depths and modest temperatures often retain many physical characteristics of their precursors, and are not greatly deformed. Often, they are studied for their igneous features rather than any metamorphic ones, but some volcanic rocks are very susceptible to alteration at low grades, owing to their interconnected porosity (Section 5.1).

Low-grade metamorphism of volcanic rocks takes place in two distinct settings: hydrothermal metamorphism associated with the development of geothermal fields at active volcanic centres (both on land and under water) and burial metamorphism associated with the accumulation of thick sequences of volcaniclastic sediments in subsiding basins. Figure 5.4 shows a field photograph of a typical zeolite facies

Figure 5.4 Steeply-dipping bed of coarse volcanogenic greywacke with a finer tuff bed to the right. The greywacke contains rip-up clasts of similar ash layers together with coarse fragments of volcanic rocks and minerals. Only in thin section is it apparent that this rock has undergone zeolite facies burial metamorphism. Kaka Point, New Zealand.

metagreywacke; depositional features are preserved in fine detail but metamorphic minerals now dominate. Lack of pervasive deformation is characteristic of burial metamorphism, although these rocks are folded. Both types of low-grade metamorphism may involve a degree of chemical change (metasomatism) as circulating warm waters introduce or remove components in solution. Metasomatic change is so integral to metamorphism in geothermal systems that it is described separately in Section 5.7.

5.3.1   Zeolite Facies

In Southland, New Zealand (Figure 5.5), zeolites occur in an extensive belt of Triassic-aged metavolcanogenic greywackes (Figure 5.4) and tuffs (Coombs 1954; Coombs *et al.* 1959).These rocks were particularly susceptible to metamorphism because of their high porosity and presences of unstable glass shards as well as high-temperature igneous minerals.

Mineralogical variation in the zeolites in these rocks is closely related to stratigraphic depth across most of this region (although this is not always the case elsewhere). At the shallowest levels (probably equivalent to around 5 km depth of burial), detrital intermediate plagioclase survives, but glass shards have been

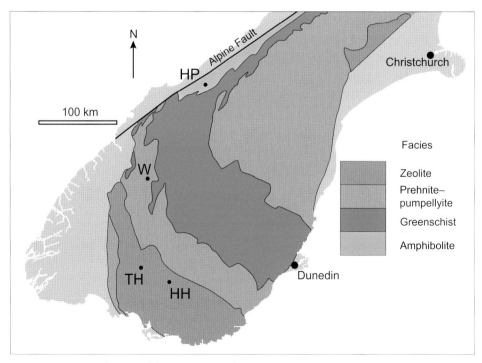

Figure 5.5 Metamorphic map of the South Island of New Zealand, illustrating the distribution of the major facies in rocks metamorphosed during the Cretaceous Rangitata Orogeny. Simplified from Menzies *et al.* (2018). Note the steep increase in metamorphic grade close to the Alpine Fault. Letters denote localities discussed in the text: HH – Hokonui Hills, HP – Haast Pass, TH – Taringatura Hills, W – Wakatipu district.

replaced by the zeolites heulandite and (less commonly) analcite (Figure 5.7a). Secondary quartz and fine-grained phyllosilicates (essentially montmorillonite or celadonite) have also formed by this depth, which is characterised by the hydration of igneous material. Deeper in the sequence, plagioclase is pseudomorphed by albite, and at slightly greater depth the original zeolites have been replaced by laumontite. Chlorite also appears in this zone, and adularia (K-feldspar) forms pseudomorphs after analcite. By this depth, some of the first-formed hydrous minerals are therefore beginning to undergo dehydration metamorphism themselves, even though fresh igneous material is often still present. In the deepest part of the sequence, zeolites become scarce and laumontite is replaced by the hydrous Ca–Al silicates prehnite and pumpellyite; minor epidote also makes its appearance.

The metamorphism across Southland is of regional extent but largely unaccompanied by pervasive deformation. For this reason, and because the zoning patterns in zeolite reflect depth of burial, the term burial metamorphism was coined to describe it. Appreciable metasomatic changes to the chemical composition of the original tuffs have been reported from Southland (Boles & Coombs 1975), and this is probably a common feature of zeolite facies metamorphism. It is likely that as well as the original rock composition, the chemistry of the circulating fluids is a critical factor in determining which zeolites develop.

5.3.2   Prehnite–Pumpellyite Facies

The appearance of pumpellyite near the base of the Taringatura Hills sequence in Southland, New Zealand (Figure 5.6) marks the onset of the **prehnite–pumpellyite facies**. Both pumpellyite and prehnite are further examples of hydrous Ca-aluminosilicates, but pumpellyite contains significant Fe and Mg. Prehnite is chemically simpler but of more limited stability. In the Taringatura Hills, burial, rather than regional, metamorphism produced both these minerals, but prehnite-pumpellyite facies assemblages are also present in low-grade regional metabasites, for example in slate belts, as well as in some geothermal fields. In southern New Zealand, burial metamorphic rocks grade into deformed, regional metamorphic rocks and in the Wakatipu district (Figure 5.5) a transition is seen from regional prehnite–pumpellyite-bearing rocks to pumpellyite–actinolite assemblages (Kawachi 1975), and then to greenschists.

Typical minerals of prehnite–pumpellyite facies metabasites include prehnite, pumpellyite, actinolite, chlorite, epidote, albite, quartz, calcite, sericite, titanite and stilpnomelane. The rocks often have a distinctive blue-green colour imparted by Fe-pumpellyite, but relic igneous phases may also be present (Figure 5.7b). In many areas, distinct prehnite–pumpellyite and pumpellyite–actinolite zones can be recognised. In some regions, including New Zealand, lawsonite also appears in the prehnite–pumpellyite facies where it co-exists with albite and chlorite. This assemblage is mainly found in metamorphosed greywackes with volcanogenic detritus,

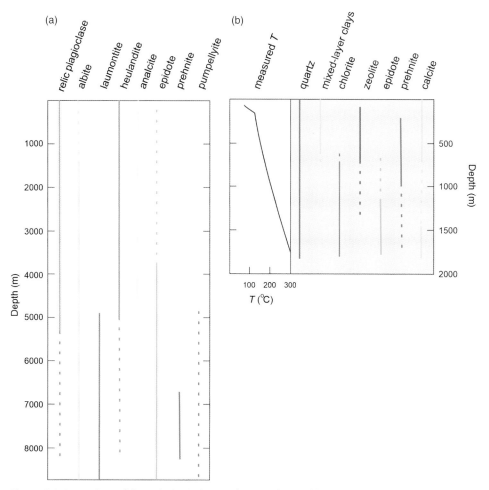

Figure 5.6 Comparison of the variation in low-grade mineral assemblages with depth in burial metamorphism and in geothermal field metamorphism. (a) Burial metamorphic zones of the Taringatura Hills (Coombs 1954) with a wide range of pressure at low temperature. (b) Variations in mineralogy with depth from a representative geothermal well on the Reykjanes peninsula, Iceland (Tómasson & Kristmannsdóttir 1972) with a steep increase in temperature at shallow depths. Note the different depth scales in the two parts. Broken lines denote sparse occurrences.

rather than massive metabasites, and is significant because lawsonite only forms at relatively high pressures (see below, Figure 5.11) and so lawsonite–albite assemblages are transitional to the blueschist facies.

5.3.3   Why are Distinctive Low-Grade Minerals Sometimes Absent?

In many low-grade metamorphic belts, the distinctive phases such as zeolites, prehnite or pumpellyite are rare. Instead, less diagnostic assemblages with chlorite and calcite are present. Minerals such as laumontite, prehnite and pumpellyite are

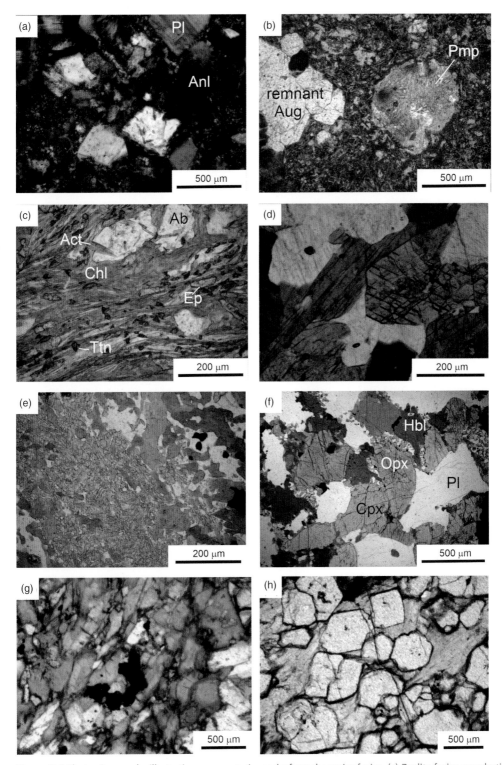

Figure 5.7 Photomicrographs illustrating representative rocks from the major facies. (a) Zeolite-facies pyroclastic rock with igneous fragments, including plagioclase, in a matrix with clays and zeolites,

hydrous Ca–Al silicates that are not stable in the presence of even low levels of CO_2. For example, in the Salton Sea geothermal field (California) zeolites are absent but calcite, epidote and chlorite are common and fluid contains CO_2 (Muffler & White 1969). Where CO_2 is present, calcite grows with minerals such as chlorite, epidote and quartz; these minerals are also widespread in the greenschist facies. Only small amounts of dissolved CO_2 are needed to stabilise calcite and, with sufficient CO_2, the minerals distinctive of the lower-grade facies may never form. The zeolite facies metagreywackes in New Zealand do contain calcite as well as zeolites, but it appears that the CO_2 introduced in pore waters was exhausted while abundant reactive volcanic glass still remained to be hydrated.

5.4   Metamorphism at Low to Moderate Pressures

Many metamorphic rocks have developed along low- or moderate-pressure facies series, and if metapelites are present they can be divided into discrete zones as discussed in Chapter 4. However, metabasites are almost invariably present, even if Al-rich metapelites are not and, while narrow zones are not developed, they do show broad changes between facies.

5.4.1   Greenschist Facies

Greenschist-facies metabasites are among the most common of metamorphic rocks, being widespread in both oceanic and continental sequences. They can be the direct products of deep hydrothermal metamorphism on the sea floor (Section 5.6.2) or the result of prograde regional metamorphism. In addition, retrograde greenschist-facies assemblages are often developed in higher-grade rocks that have been hydrated by water moving along faults or shear zones as they returned to the surface or simply as they resided in the mid-crust.

It is not always possible to distinguish between prehnite–pumpellyite and greenschist-facies rocks, especially in the field. The absence of pumpellyite and

Figure 5.7 (*cont.*) including analcite, (XPL). (b) Pumpellyite metabasalt with radial pumpellyite infilling a vesicle in a matrix that includes a large, unaltered crystal of augite. (c) Greenschist-facies metabasalt with prominent epidote (yellow-green), actinolite (blue-green) and chlorite (green) in a matrix of albite and quartz. (d) Amphibolite dominantly composed of stubby pleochroic hornblende and plagioclase with accessory titanite. (e) Retrograde amphibolite after eclogite. Green hornblende and plagioclase mantle spongy clinopyroxene. (f) Hornblende granulite with orthopyroxene, clinopyroxene, hornblende and plagioclase. Note the small orthopyroxene grains growing at some hornblende grain boundaries. (g) Blueschist with strongly coloured pleochroic blue glaucophane and garnet. (h) Eclogite. Abundant garnet is set in a matrix of omphacite pyroxene.

co-existence of epidote and actinolite are the best indicators of greenschist-facies conditions, but epidote, chlorite, albite, actinolite and calcite can be found in both facies. The transition can be ascribed to the continuous reaction:

$$\text{pumpellyite} + \text{chlorite} + \text{quartz} \rightarrow \text{actinolite} + \text{epidote} + \text{fluid} \qquad [5.1]$$

Low-grade greenschist-facies metabasites, such as those exposed around Lake Wakatipu (Landis & Coombs 1967; Figure 5.5) or in the southern Highlands of Scotland, are associated with chlorite- or biotite-zone metapelites. Typically, some original igneous features remain, such as ophitic textures in the feldspars, and recognisable pillow lavas in outcrop, but the mineral assemblage is entirely metamorphic. Albite commonly pseudomorphs original igneous plagioclase. The dominant minerals are chlorite, epidote, albite, pale- or bluish-green actinolite and quartz, commonly with calcite and biotite (Figure 5.7c). Stilpnomelane is a common accessory mineral and is easily confused with biotite. Typical greenschist-facies metabasites show distinctive dark green chlorite and yellow-green epidote in hand specimen, commonly with pink albite and small amounts of quartz (Figure 1.15d).

Within the greenschist facies, there are some common trends with increasing grade. Both actinolite and blue-green hornblende can be present throughout the greenschist facies, but hornblende is very rare in the chlorite zone, and at higher grades actinolite often occurs only as relics mantled by hornblende. This texture suggests replacement rather than co-existence. Chlorite and epidote become scarcer with increasing grade. In particular, the association of calcite with chlorite becomes rarer as a result of the reaction:

$$\text{chlorite} + \text{calcite} \rightarrow \text{epidote} + \text{actinolite} + CO_2 - H_2O \text{ fluid} \qquad [5.2]$$

Garnet growth in the greenschist facies is favoured by Fe-rich rock compositions and higher pressures. It generally appears in metabasites at somewhat lower grades than in metapelites but is very rare at low pressures, where associated metapelites contain andalusite. As in metapelites (Chapter 4, Section 4.2), garnet first appears where albite is the only plagioclase present but at about this grade oligoclase also appears, occurring as a distinct phase co-existing with albite. With increasing grade, the compositions of co-existing albite and oligoclase converge, until albite disappears and there is only one plagioclase present, which is oligoclase. This change defines the beginning of the amphibolite facies.

5.4.2 Amphibolite Facies

Metabasites that are composed mainly of hornblende, plagioclase and some quartz (amphibolites) are very common. These form in the amphibolite facies, mostly by prograde metamorphism of greenschist-facies metabasites; however, they can also form directly by the hydration of basic igneous rocks (Figure 5.1b), or from retrograde alteration of higher-grade granulite or eclogite-facies metabasites. Figure 5.7e illustrates an amphibolite with remnants of spongy corroded pyroxene, which is typical of retrogressed eclogite.

Metapelites associated with amphibolites commonly contain staurolite, kyanite or sillimanite zone mineral assemblages (Chapter 4, Section 4.2). Epidote and/or garnet are often present in amphibolites, as are trace amounts of biotite. Titanite, ilmenite and apatite are common accessories, and chlorite is sparse or absent. Hornblendes in amphibolites are generally darker in colour than many greenschist-facies amphiboles, and are likely to form stubby prisms rather than the very elongate grains characteristic of actinolite in the greenschist facies (Figure 5.7d).

It is difficult to write specific reactions to account for the changes marking the transition from greenschist to amphibolite facies because the rocks contain relatively few phases (often only four to five) and these are made up of a large number of components, e.g. Na_2O, K_2O, CaO, MgO, FeO, Al_2O_3, SiO_2 and H_2O. Additionally, CO_2 and Fe_2O_3 may also be important. Hence reactions are likely to be continuous, and the reacting assemblage may have several degrees of freedom.

A number of possible reactions between end-members of the naturally occurring solid solutions can summarise the changes that take place. Hornblende growth can be represented by reactions that generate end-members such as tschermakite hornblende $Ca_2(Mg, Fe)_3Al_4Si_6O_{22}(OH)_2$, or edenite hornblende $NaCa_2(Mg, Fe)_5AlSi_7O_{22}(OH)_2$. For example:

$$\text{chlorite} + \text{clinozoisite} + \text{quartz} \rightarrow \text{tschermakite hornblende} + \text{anorthite} + \text{fluid} \quad [5.3]$$

This also leads to the production of the anorthite component which drives the composition of existing plagioclase to more calcic compositions. Other reactions that contribute to the overall changes in mineral abundance and composition include:

$$\text{albite} + \text{actinolite} \rightarrow \text{edenite hornblende} + \text{quartz} \quad [5.4]$$

$$\text{actinolite} + \text{chlorite} + \text{quartz} + \text{clinozoisite} \rightarrow \text{tschermakite hornblende} + \text{fluid} \quad [5.5]$$

The transition from greenschist to amphibolite has recently been investigated via isochemical phase diagrams, and more details of this transition in a specific bulk composition are shown in Figure 5.8 (Elmer et al. 2006). These broadly confirm the relationships derived from field studies, but provide some additional insights, particularly on the effects of pressure. Figure 5.8 shows that chlorite may persist to higher temperatures at low pressures, while epidote (clinozoisite) may be favoured at higher pressures. Although garnet is not predicted for the composition plotted in Figure 5.8, it is a common mineral in moderate- to high-pressure amphibolites. Note that the predicted fields of co-existing actinolite and hornblende in Figure 5.8 conflict with the field evidence that these are rare assemblages

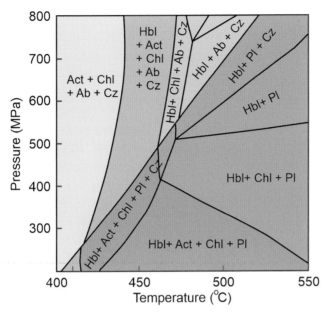

Figure 5.8 Isochemical phase diagram to illustrate the predicted assemblages for a simplified metabasite composition around the conditions of the transition from the greenschist facies to the amphibolite facies, simplified from Elmer *et al.* (2006). Quartz and an aqueous fluid are assumed to be present in every assemblage. The colour shading distinguishes between assemblages with albite versus plagioclase and actinolite versus hornblende. The relationships between these predicted assemblages and natural occurrences are discussed in the text.

in nature. In contrast, the figure predicts a sharp transition from albite to oligoclase-bearing assemblages, whereas these commonly co-exist over a distinct temperature interval in nature. The discrepancies may be a reflection of the incomplete equilibration of most natural metabasites at these grades. However, even at higher temperatures where equilibration might be expected to be reasonably complete, anomalies remain in current thermodynamic models for metabasites (Forshaw *et al.* 2019).

Field-based studies and computer-based modelling both point to some trends between low- and moderate-pressure metamorphism of amphibolite-facies metabasites, including the following.

(a) The presence of garnet at relatively high pressures, with chlorite persisting to higher temperatures at low pressure.

(b) The appearance of plagioclase in place of albite at lower grades in low-pressure metabasites where actinolite is still the dominant amphibole.

(c) The more-common development of Ca-poor amphiboles, notably cummingtonite, in low-pressure rocks.

5.4.3   Granulite Facies

At the highest grades of metamorphism, equivalent to the second sillimanite zone and higher in metapelites, hornblende begins to break down in metabasites and pyroxenes appear. At these grades, metabasite mineralogy becomes closer to that of primary igneous precursors.

A distinctive association of plagioclase, clinopyroxene and orthopyroxene characterises the granulite facies in metabasites (Figure 5.7f); green-brown hornblende and garnet are normally also present, there may be traces of biotite, and quartz is often absent. At the highest grades, hornblende becomes rare and there is often extensive evidence for melting. The transition from amphibolite to granulite facies is marked by dehydration as amphibole is progressively replaced by pyroxene over a temperature interval up to several hundred degrees. This temperature interval is also dependent on rock composition. At low pressures (200 MPa), hornblende breaks down above temperatures of around 750 °C (Spear 1981), with somewhat higher temperatures (810–990 °C depending on rock composition) required for hornblende to break down at mid to lower crustal pressures (Palin *et al.* 2016). The facies boundary between amphibolite and granulite facies is therefore quite indistinct.

At the high temperatures of the granulite facies, amphibole-bearing rocks start to melt via a series of dehydration reactions that are similar to the (amphibolite facies) mica dehydration reactions discussed in Chapter 4, Section 4.3. The role of melting in the transition from amphibolite facies to granulite facies has been the subject of extensive debate. Some granulite-facies metabasites are migmatites, and it is likely that melt is an important product of reactions at the amphibolite–granulite-facies boundary. However, since metabasites melt at higher temperatures than metapelites, it is possible that in some circumstances dehydration of amphibole might release water, which fluxes melting in nearby metapelites.

Many granulites occur in distinct Precambrian Shield terranes, but much younger granulites have also been described, including Miocene examples from Indonesia (Pownall *et al.* 2014). The boundaries of these terranes are often unconformities or faults, and so it is generally difficult to study the progressive metamorphism of amphibolites to granulites. In most cases where amphibolites and granulites occur together, the amphibolites are of retrograde origin due to later reworking and infiltration of fluids.

Some granulites may be the recrystallised products of deeply-emplaced and slow-cooled gabbro bodies with little history of earlier hydration, or even the products of direct crystallisation of basic magmas under high-pressure conditions. In some deep crustal settings, such as now exposed in the Fiordland region of New Zealand, batholithic bodies of intermediate to basic composition exhibit a range from relic igneous textures to foliated metamorphic textures, although all have similar assemblages yielding similar, granulite or amphibolite facies, metamorphic conditions (Bradshaw 1990). They can be referred to as orthogneiss for simplicity. The plagioclase–pyroxene pegmatite from western Norway with metamorphic garnet mantling original pyroxene, illustrated in Figure 5.1c, is an example of the direct transformation from an igneous to a high-grade metamorphic assemblage. In such massive igneous bodies, the absence of water is probably the chief reason why the original assemblages survived unchanged for extended periods. Even once reaction does occur, it is often incomplete and the **reaction rim**, or **corona** texture illustrated in Figure 5.1c is an example of this. These textures are described in more detail in Chapter 7, Section 7.5.3.

The granulite facies embraces a wide pressure range at high metamorphic temperatures, but is distinguished from the higher-pressure eclogite facies by the presence of plagioclase. Additionally, there are variations between granulites formed at different pressures within the granulite facies (Pattison 2003), illustrated in Figure 5.9. These variations include the following.

(a) *Low-pressure granulites* contain a characteristic association of olivine + plagioclase. Under most high-*T* crustal conditions this pair is unstable relative to garnet and/or aluminous pyroxenes and it is normally only found in aureoles around gabbro bodies.

(b) *Moderate-pressure granulites* are characterised by the classic association of clinopyroxene + orthopyroxene + plagioclase, usually with hornblende and often with garnet. Quartz is a possible accessory but does not normally co-exist with both garnet and clinopyroxene.

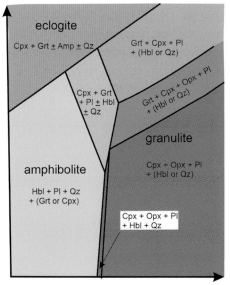

Figure 5.9 Schematic *P–T* diagram to show the relationships between the characteristic metabasite assemblages of the amphibolite and granulite facies, and the differences between granulites formed at high, moderate and low pressures. From Pattison (2003).

(c) *High-pressure granulites* usually contain abundant garnet and lack orthopyroxene. The assemblage garnet + clinopyroxene + plagioclase + hornblende is characteristic, but while quartz is possible in high-pressure granulites it is not normally present together with hornblende. High-pressure granulites may form from eclogites during decompression (O'Brien & Rötzler 2003), rather than from heating of amphibolite-facies rocks.

5.5   Metamorphism at High Pressures

Metabasites are especially important for investigating metamorphism at relatively high pressures and low temperatures because unlike metapelites they undergo conspicuous and diagnostic mineralogical changes under such conditions.

5.5.1   Blueschist Facies

At elevated pressures, lilac-to-blue sodic amphiboles (glaucophane, containing Na–Mg–Fe^{2+}–Al) replace green calcic amphiboles (hornblende and actinolite, containing Ca–Mg–Fe^{2+}; Box 5.1). The sodic amphibole assemblages are often denser, and this points to formation at higher pressures than metabasites of the prehnite–pumpellyite and greenschist facies.

The Ile de Groix, an island off the northwest coast of France, provides a typical example of blueschists in the field. In part of the island, isolated metre-scale boudins, lenses and discontinuous layers of metabasite occur within a matrix of mica schist, albite gneiss and quartzite (e.g. Schulz *et al.* 2001). Elsewhere, more extensive metabasites occur, but they are nevertheless in fault-bounded blocks. These metabasites appear blue or blue-grey in colour (Figures 1.1d, 1.15f and 5.7g) due to the presence of glaucophane rather than actinolite or hornblende. Epidote, albite, chlorite, quartz and white micas also occur here, garnet is common and lawsonite is sometimes present. Note that, in high-pressure rocks, the epidote mineral may be zoisite or clinozoisite; i.e. richer in Al and poorer in Fe^{3+} than the common epidote of the greenschist facies.

Lawsonite, essentially equivalent in composition to hydrated anorthite, is found in many of the Ile de Groix blueschists, but is not present in all blueschist occurrences. Sometimes, including in the Ile de Groix, rocks in which lawsonite is no longer present contain distinctive lozenge-shaped pseudomorphs consisting of epidote, white micas and albite which are diagnostic of its former presence (Figure 5.10). The reason why lawsonite is so often replaced by such pseudomorphs is that its breakdown involves dehydration (Figure 5.11). It can therefore react spontaneously without the addition of water.

The blueschists of the Ile de Groix do not directly preserve evidence of the precursor assemblages from which they developed, but some very-low-grade

Fig 5.10 Blueschist from the Ile de Groix, France, containing what appear to be white rhombohedral porphyroblasts. In fact, these rhomboids are pseudomorphs after lawsonite, now consisting of albite, epidote and white mica. (Photo: Barbara Kunz.)

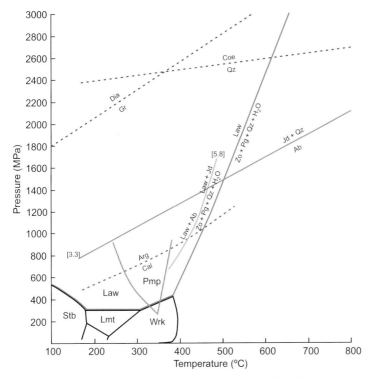

Figure 5.11 Petrogenetic grid applicable to metabasites. This grid shows selected univariant reactions which are close to those that occur in nature, and also indicates the stability limits of selected low-grade minerals. Markedly continuous reactions are not shown. Broken lines are the polymorphic transitions for calcite–aragonite, diamond–graphite and coesite–quartz. Blue lines are the stability limits of some important zeolites: stilbite, laumontite and wairakite. Green lines define limits to lawsonite, brown lines are for pumpellyite. Sources: Day (2012); Frey *et al.* (1991); Johannes & Puhan (1971); Nitsch (1971); Schmidt & Poli (1994). Numbers in square brackets refer to reactions detailed in the text.

blueschists do. Some of the blueschists in the Franciscan Complex of California contain typical high-pressure minerals, including glaucophane, lawsonite, aragonite and jadeitic pyroxene, that directly replace original igneous minerals, as well as replacing prehnite–pumpellyite-facies assemblages. Jadeite is more frequently discussed for its role as a component of omphacite, the characteristic clinopyroxene of eclogites (see Section 5.5.2), but jadeite itself may be present in low-grade blueschist-facies metagreywackes, formed from original albite (see Chapter 3, reaction [3.7] and Figure 5.11). Aragonite is the stable form of calcium carbonate under most blueschist-facies conditions (Figure 5.11), but is only normally found in low-grade blueschists because it rapidly reverts to calcite on exhumation, unless the temperature is below about 250 °C (Carlson & Rosenfeld 1981).

Common blueschist-facies assemblages are chemically equivalent to those found in greenschists and can be related by equilibria such as:

$$\text{actinolite} + \text{chlorite} + \text{albite} + H_2O = \text{glaucophane} + \text{zoisite} + \text{quartz} \qquad [5.6]$$

$$\text{albite} + \text{chlorite} + \text{actinolite} = \text{glaucophane} + \text{lawsonite} \qquad [5.7]$$

$$\text{zoisite} + \text{paragonite} + \text{quartz} + H_2O = \text{lawsonite} + \text{albite} \qquad [5.8]$$

Note that the details of these equilibria, and even the role of water within them, is dependent on the precise composition of the solid-solution minerals involved. These equilibria have been written assuming that the minerals do not exhibit any substitution of Fe^{3+} for Al^{3+}. In practice, many blueschists contain a sodic amphibole which is intermediate between glaucophane, $Na_2(Mg, Fe)_3Al_2Si_8O_{22}(OH)_2$, and riebeckite (or magnesioriebeckite), $Na_2(Mg, Fe)_3Fe_2Si_8O_{22}(OH)_2$, while the epidote group mineral also often contains appreciable Fe^{3+}, which similarly substitutes for Al. As a result, for metabasites with Fe^{3+} present, for example in an Fe-oxide, it is possible to write an equilibrium similar to [5.6] using the Fe-bearing minerals:

$$Fe^{3+}\text{-oxide} + \text{actinolite} + \text{chlorite} + \text{albite} + H_2O = Fe^{3+}\text{-glaucophane} \qquad [5.9]$$
$$+ \text{epidote} + \text{quartz}$$

We saw in Chapter 3 (Section 3.3.4) that where the phases on one side of an equilibrium contain more impurities than those on the other side, the stability field of the less pure phases will be enlarged. It follows that blueschist assemblages (the right-hand side of reactions [5.6] and [5.9]) can form at somewhat lower pressures in Fe^{3+}-bearing rocks than in metabasites which lack an Fe^{3+}-oxide.

In summary, the blueschist facies is defined in rocks of basaltic composition by the presence of a glaucophane-rich sodic amphibole with epidote and/or lawsonite and commonly associated with varying amounts of albite, chlorite, aragonite, pumpellyite and phengitic muscovite. Garnet forms at higher temperatures. The boundaries between the prehnite–pumpellyite, greenschist and blueschist facies are not sharp but are influenced by rock composition.

5.5.2   Eclogite Facies

At very high pressures, metabasic rocks become denser, and turn from an overall blue-purple colour (Figures 1.15f and 5.10) to a green-red colour (Figures 1.15g and 5.7h). Plagioclase has broken down completely, with the albite component becoming incorporated into omphacite and the anorthite component entering into epidote or garnet (Chapter 3, reactions [3.7] and [3.8]). Sodic pyroxenes (omphacite) occur instead of amphibole or igneous augite, and the garnet is relatively rich in Mg (i.e. in the pyrope component) compared to garnets found in other facies. These are all distinctive features of eclogite (Figure 5.7h). Although eclogite is defined by the presence of garnet and omphacite and the absence of plagioclase, other common minerals include epidote, quartz, kyanite, paragonite and phengitic muscovite, and minor lawsonite, rutile and pyrite may be present.

On the Ile de Groix, other metabasic blocks intercalated with the blueschists are of eclogite. Their relationship to the blueschists can be represented by equilibria such as:

$$\text{zoisite} + \text{glaucophane} = \text{garnet} + \text{omphacite} + \text{paragonite} + \text{quartz} + H_2O \qquad [5.10]$$

$$\text{albite} + \text{epidote} + \text{glaucophane} = \text{omphacite} + \text{paragonite} + \text{hornblende} + H_2O \quad [5.11]$$

The Ile de Groix eclogites are typical of many occurrences in a number of mountain belts, and fairly subtle changes in the chemical composition of the protolith, for example in total Ca or Al concentration, may determine the accessory minerals present. For example kyanite develops in more aluminous metabasites (reaction [3.8]).

Lawsonite is rare in eclogites, but lawsonite-bearing eclogites have been described from a number of localities (Tsujimori & Ernst 2014). They require an exceptional combination of low temperature and high pressure, and the conditions in which they form were once considered outside of the realm of what was possible on Earth (Liou *et al.* 2000). Significant cooling during decompression is required to preserve lawsonite during exhumation (Figure 5.11) (Wei & Clarke 2011) and deformation can erase any evidence of their pseudomorphs. Its presence in these rare rocks therefore tells us that very particular *P–T* conditions existed when these eclogites formed, and special conditions were also required for their preservation.

Eclogites are most commonly associated with blueschists and develop at higher grades. However, other eclogites are found closely associated with high-pressure granulites. They may have formed sequentially, but it is also possible that eclogites may develop in some rock compositions at the same time as high-pressure granulites are developing in others, as the reactions that define the granulite–eclogite-facies boundary are highly dependent on bulk composition (De Paoli *et al.* 2012).

5.5.3   Ultra-High Pressure Eclogites

It has long been understood that eclogites form at higher pressures than other metabasites and, while it may be possible for eclogites to form at pressures of around 1.4–1.6 GPa at the lower end of their temperature range, the application of a range of geothermometers and geobarometers has pointed to rather higher pressures of around 2.0–2.3 GPa for a number of localities. Detailed petrographic studies in the late twentieth century turned up a number of mineral occurrences in eclogites which suggested that some of them formed at even higher pressures. The most common of these 'exotic' minerals are coesite and diamond.

The mineral coesite is a dense polymorph of quartz, formed at pressures above about 2.6 GPa (Figure 5.11), and was first found in nature at meteorite impact sites. Coesite inclusions in garnet were reported from eclogites found in the Dora Maira terrane in the western Alps, and in the Western Gneiss Region of Norway in the mid 1980s (Chopin 1984; Smith 1984). A few years later, an occurrence of microdiamonds (average size of around 12 μm, requiring pressures of >3.5 GPa) was reported in eclogites from Kazakhstan (Sobolev & Shatsky 1990), again implying significantly higher pressures than are normally required to form eclogites (Figures 5.4 and 5.11).

In recognition of the exceptionally high pressures that these occurrences imply, coesite- or diamond-bearing eclogites are said to have experienced **ultra-high-pressure (*UHP*) metamorphism** to distinguish them from 'normal' eclogites.

The presence of coesite also documents the conditions required for the preservation of *UHP* minerals as well as their formation. The transformation from quartz to coesite and back again is effectively instantaneous in geological terms at temperatures >400 °C (Perrillat *et al.* 2003). Coesite must therefore remain at high pressures during cooling in order to prevent it from transforming back to quartz. This can only happen where coesite is included in a strong mineral, most commonly garnet or zircon, which acts as a pressure vessel; the inclusion is unable to expand and so pressure does not drop. Typically, coesite inclusions are partially replaced by quartz, and the host mineral has fractures radiating out from the inclusion where it finally burst as a result of the greater volume of the quartz (Figure 5.12). Sometimes, this

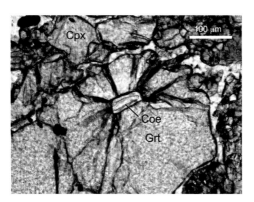

Figure 5.12 Photomicrograph of an inclusion of coesite in garnet from a *UHP* eclogite, Nordøyane, Norway. The radial cracking of garnet around the inclusion is characteristic of the presence of coesite. From Butler *et al.* (2013).

distinctive pattern of radial fractures occurs around inclusions that are of quartz only, suggesting that they may have originally been coesite (other mineralogical clues may also exist to suggest such high pressures). Of course, at the time when coesite present in inclusions was first formed, all the quartz in the rock will have been coesite, but because the retrograde reaction can occur spontaneously without the addition of water, it normally goes to completion. This is similar to the replacement of aragonite by calcite in blueschists (Section 5.3.3).

5.5.4   Field Relationships of High-Pressure Rocks

In many field locations, metabasites of different facies occur in close proximity or are even interlayered. In some, it is evident that greenschist-facies assemblages have replaced blueschist-facies assemblages by retrograde reactions during exhumation, and this is documented for the Ile de Groix (El Korh *et al.* 2013). The retrogression is localised, with large areas of blueschist remaining, and can be linked to fluid infiltration controlled either by lithology or by zones of deformation. This late infiltration of fluid may be demonstrated by using oxygen isotope measurements (e.g. Schliestedt & Matthews 1987). These show that the greenschist-facies assemblages typically have a distinct oxygen isotopic composition arising from interaction with a fluid; the remaining blueschist-facies minerals retain their original oxygen composition.

The common close association of blueschist and eclogite such as is seen in the Ile de Groix can also reflect post-metamorphic processes. For example in parts of the Franciscan Terrane in California, blocks of greenschist, blueschist and eclogite are jumbled up within a scaly mudstone matrix. This chaotic mixture is commonly termed a **mélange** (Silver & Beutner 1980). Metabasite blocks range in size from centimetres to tens of metres and appear to have experienced different peak metamorphic conditions and followed a different exhumation path from their neighbours, but many retain assemblages formed at high pressures. The tectonic processes that can result in the formation of such a mélange are discussed in Chapter 10.

Elsewhere, closely associated blueschists and greenschists may have formed under the same conditions, transitional between the facies, in rocks of different composition. In the Shuksan suite of the North Cascades, Washington State, USA, actinolite-bearing greenschist is interlayered with blueschist on a decimetre scale, although the sodic amphibole of the blueschist contains significant Fe^{3+} which extends its stability field as discussed earlier. The major differences in the bulk compositions of the different protoliths include differences in total iron, Fe^{3+} and the Na/Ca ratio (Dungan *et al.* 1983).

More generally, it is commonly the case that high-P rocks occur in fault-bounded blocks and their assemblages seldom grade progressively into different zones. This reflects the processes that are required to return these rocks to the surface from great depths with their assemblages intact, a topic that we will return to in Chapter 10.

5.6   Fluid Flow and Chemical Change During Metamorphism of Basic Rocks

Many metabasites have compositions that match closely those of unaltered basalts or gabbros, but this is not always the case. Where the initial metamorphism of the igneous rock is by circulating hydrothermal fluids driven by the magmatism itself, the rock composition may be substantially altered, and while subsequent metamorphism may not involve further fluid circulation and metasomatism, the distinctive chemical features created during hydrothermal activity are likely to remain throughout the subsequent metamorphic history. Metabasic rocks can also be changed by circulating fluids within the crust at a later stage, although the effects are generally less dramatic. The retrograde alteration of blueschist to greenschist described in Section 5.5.4 is an example; some chemical change does take place, but the final greenschists are not greatly different in composition from their precursors.

5.6.1   Metamorphism in Sub-Aerial Geothermal Fields

Geothermal fields are of considerable importance for metamorphic petrology, both because some have been extensively drilled and allow us to study active metamorphism and because they provide an analogue for the much more extensive hydrothermal metamorphism that is believed to affect newly-formed oceanic crust on the sea floor. The conditions of formation of the mineral assemblages are known because temperature can be measured down the drill holes where the samples are extracted from.

An important distinction between sub-aerial geothermal systems and those in the deep ocean is that, because of the relatively low pressure, below the critical pressure (Chapter 1, Section 1.4.3, Figure 1.10), boiling is often widespread. One consequence of this extensive boiling is the formation of large volumes of calcite by a process similar to limescale formation in a domestic kettle. Deeper geothermal systems, such as those along mid-ocean ridges, do not experience such extensive boiling because the higher pressures inhibit or prevent it.

One of the best-studied active geothermal fields is the Reykjanes field of southwest Iceland. The rock types affected include hyaloclastic tuffs and breccias as well as lava flows, and alteration principally affects glassy or very fine-grained material. Of the magmatic minerals, olivine is usually completely altered but pyroxene and to some extent plagioclase feldspar are relatively resistant. The temperature profile measured directly down hole is included with the mineral distributions in Figure 5.6b.

The principal silicate alteration products in the upper part of the system are clay minerals, dominated by montmorillonite near the surface (where the temperature is close to 100 °C) but rapidly giving way to mixed-layer clays at depth, while prehnite also appears at quite shallow levels. A number of different zeolite minerals are also found in the upper part of the system, and include mordenite, stilbite, mesolite, analcite and wairakite. In the deeper parts of the system, corresponding to temperatures in excess of 230 °C, epidote and chlorite are found, and plagioclase shows sporadic alteration to albite or, rarely, K-feldspar. Secondary calcite and quartz are found throughout. Elsewhere, drilling in some geothermal fields has penetrated still hotter rocks at temperatures in excess of 300 °C, and depths of around 2 km, and under these conditions actinolite can occur. Note the very small vertical extent of these mineralogical zones in Figure 5.6 when compared with the zones of burial metamorphism from New Zealand, reflecting the steep geothermal gradient. Metamorphism in a geothermal field such as Reykjanes is neither progressive nor isochemical. The metamorphic minerals often develop directly at depth from fresh igneous material once the geothermal system is initiated. Chemical changes can include leaching of the rock by circulating water, cation exchange between minerals and fluid, and precipitation of calcite and other minerals as a direct result of boiling.

5.6.2   Sea-Floor Metamorphism

Convective circulation of sea water through young oceanic crust is widespread along mid-ocean ridges. Cold sea water percolates through relatively large areas of crust and becomes heated, reacting with wall rocks along the way. Pressures are normally too high to permit boiling and so temperatures in excess of 350 °C may be attained (Figure 5.13). As a result, greenschist-facies mineral assemblages are

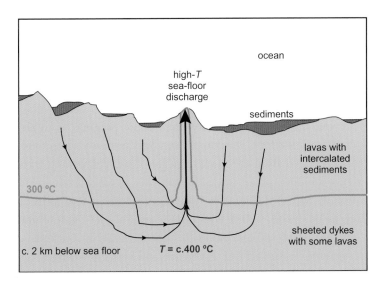

Figure 5.13 Schematic section of hydrothermal circulation beneath a mid-ocean ridge. Water is drawn down through an extensive network of fractures and heated before rising through a more focussed crack network to discharge on the sea bed.

common, often dominated by epidote or chlorite, and zeolites are normally absent. The heated sea water becomes focussed and rises to the surface, giving rise to further reaction with rocks and sometimes to spectacular plumes of mineral precipitates at the sea bed. The focussed flow also results in significant chemical changes to the rocks at depth along the flow path.

Since the first discoveries in the 1970s, sea-floor hydrothermal vents have been found along much of the Earth's mid-ocean ridge network, as well as at other types of volcano built up from the sea bed. Many types of chemical change are possible in the rocks that host the hydrothermal system (Bach & Früh-Green 2010), and these changes make sense of many rocks in the geological record whose origin had previously been controversial.

Sea-floor metamorphism and metasomatism has been coupled to the formation of distinctive precipitates of both sulphides and carbonate minerals at the sea bed (black smokers and white smokers), to the formation of some types of sulphide ore body and more generally to the generation of some distinctive rock compositions. The nature of the metamorphic changes and the accompanying metasomatism depends on temperature and on the extent of fluid circulation in particular, but effects of cation exchange between rocks and sea water predominate while boiling is less important. Important chemical changes include the replacement of plagioclase with albite, the growth of Mg-rich chlorite at the expense of primary Fe–Mg phases, which accounts for the very low Mg content of the oceans at the present time, and the growth of epidote-rich rocks. Beyond the formation of 'smoker' deposits and the metasomatic modification of sea-floor basalts, sea-floor circulation can also cause material leached from them to be precipitated on the sea bed more widely, for example as chert layers, which are commonly associated with hydrothermally altered basalts.

Where the degree of alteration is relatively modest, basalts are altered to rocks rich in albite, chlorite and epidote that retain igneous textures but show extensive alteration of primary minerals. For many years, such rocks, termed spilites, were well known in the geological record and their origin was a source of controversy, with some workers believing they formed directly from sodic magmas with the extensive chlorite a later alteration effect. The controversy was resolved when Cann (1969) found spilitic rocks dredged from the modern ocean floor, confirming their metamorphic origin.

More extreme alteration by focussed fluid flow leads to the development of more extreme rock types such as epidosite, a rock composed essentially of epidote and quartz (Figure 5.14). In contrast to spilites, epidosites are enriched in Ca and Fe (dominantly Fe^{3+}), but depleted in Na and Mg. Epidosites are best known from the sheeted dyke complexes of ophiolites, but similar rocks occur more widely. Epidosites sometimes form bands or 'stripes', typically parallel to dyke walls, and grade into less-altered metabasalt in which other minerals such as chlorite and actinolite are present also (Figure 5.14). Detailed petrographic observations have shown that

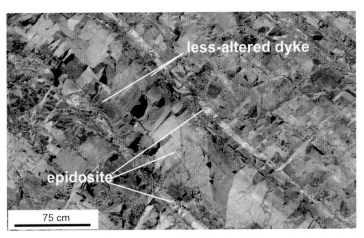

Figure 5.14. Sheeted dyke approximately 0.75 m in width showing bands of alteration to yellow-green epidosite both at the margins and in a central zone of each dyke, Troodos Massif, Cyprus. (Photo: Joe Cann.)

quartz and epidote grains in pure epidosite bands have in part grown into open pore space, implying that the host rock first dissolved and pores were then infilled (Cann *et al.* 2015).

The variation in the number of minerals from just three in epidosites to seven in nearby greenschists is indicative of variation in the amount of reactive fluid that has flowed through rocks with different assemblages: more-extensive fluid flow leads to more alteration but also a decrease in the number of minerals. This rule of thumb is a valuable guide for identifying rocks which are likely to have experienced metasomatism and is encapsulated by the Korzhinskii phase rule, outlined in Appendix 3.

Inevitably, rocks that first experienced metamorphism in a sea-floor hydrothermal system may go on to undergo more conventional regional metamorphism after deep burial. For example, regionally-metamorphosed metabasites sometimes contain large bodies of epidote of possible sea-floor origin. A few rock types have compositions for which there is no known igneous or sedimentary precursor, but which are close to chlorite-rich spilites or other altered sea-floor rocks. Best known of these are rocks rich in cordierite and anthophyllite found in some upper-amphibolite-facies terranes.

5.7   Determining the *P–T* Conditions of Metamorphism from Metabasites

This final section of the chapter is concerned with defining the conditions of pressure and temperature that are required to form the metabasite assemblages characteristic of the different facies. The facies diagram illustrates the general range of conditions of formation of rocks from different facies, but the mineral assemblages of most facies are not readily defined by single reactions which can be determined experimentally. Isochemical phase diagrams are also difficult to apply to metabasites because of the widespread textural evidence for incomplete equilibration. There are, however, a number of the more-specific geothermometers and geobarometers introduced in Chapter 3 that can give useful results for metabasites. Although the emphasis here is on the estimation of temperatures and pressures from measurements of ancient rocks, we have already seen that temperatures of low-grade metamorphism are sometimes measured directly, where deep drilling penetrates geothermal fields in which low-grade minerals are currently forming. These measurements provide important confirmation that our petrological estimates are reasonable.

5.7.1   A Petrogenetic Grid for Metabasites

A few minerals found in metabasite have clearly-defined stability fields, and we have already seen that these provide valuable constraints, especially at high pressures. Examples include aragonite, found in place of calcite at high pressures and low temperatures, and coesite and diamond, found in place of quartz and graphite respectively in *UHP* rocks. Their stability boundaries are shown in Figure 5.11.

Many mineral stability limits are dependent on mineral composition, and while some are defined by continuous reactions which can be quantified, this is not always the case, especially at low grades. Fe-rich pumpellyites, such as occur in some zeolite facies rocks, are stable below 200 °C and break down between 200 °C and 250 °C (Schiffman & Liou 1983), but the more magnesian varieties typical of the prehnite-pumpellyite facies are stable to temperatures between 350 °C and 400 °C, and this provides a temperature for the transition from prehnite–pumpellyite to greenschist facies that is in agreement with the occurrence of sub-greenschist-facies assemblages in deep geothermal wells at temperatures in excess of 300 °C.

5.7.2   Specific Geothermometers and Geobarometers

A wide range of the types of geothermometer and geobarometer outlined in Chapter 3 can be applied to metabasites. These span a wide range of possible metamorphic conditions.

Oxygen isotope geothermometry can be applied to rocks or mineral veins, provided there are mineral pairs that can be shown by their textures to have grown together (Section 3.3.6). Possible co-existing mineral pairs include quartz or albite with calcite, and magnetite with any of the foregoing. Oxygen isotope geothermometers are particularly useful at low grades; for example, this thermometer was used to define the temperatures of formation of blueschist- and greenschist-facies assemblages from the Cyclades, Greece, described in Section 5.5.4 (Schliestedt &

Matthews 1987). This and other studies have indicated that low-grade greenschist-facies rocks often form at around 400 °C.

Another geothermometer for low-grade metabasites is the chlorite geothermometer (Section 3.3.8). This was originally calibrated for metavolcanic rocks and is particularly applicable to low-grade metabasites associated with geo-thermal systems.

Cation-exchange geothermometers are potentially useful for metabasites and the most widely used involves co-existing garnet and hornblende. These minerals co-exist in many medium- to high-grade metabasites, and their Fe–Mg cation-exchange geothermometer has been calibrated using a compilation of experimental data and analyses from natural samples whose conditions of formation had been independently determined (Ravna 2000).

Continuous reactions provide the basis for several important geothermometers (Section 3.3.4) as well as being key to using computational methods to solve for *P–T* conditions from multiple equilibria (Section 3.3.9). Hornblende and plagioclase occur together in metabasites from the greenschist facies to the granulite facies, and are an important target for *P–T* estimates. Two thermometers based on co-existing amphibole and plagioclase are commonly used (Holland & Blundy 1994). A reaction between albite and the simple tremolite end-member of amphibole generates a hornblende end-member, edenite, by partial replacement of Si by Al in tetrahedral sites balanced by the addition of Na in the 'A' site (Box 5.1):

$$\underset{\text{tremolite}}{Ca_2Mg_5Si_8O_{22}(OH)_2} + \underset{\text{albite}}{NaAlSi_3O_8} = \underset{\text{edenite}}{NaCa_2Mg_5Si_4(AlSi_3)O_{22}(OH)_2} + \underset{\text{quartz}}{4SiO_2} \quad [5.12]$$

This geothermometer is only applicable to quartz-bearing rocks, but an alterna-tive equilibrium involving an additional Al-free amphibole end-member, richterite, does not involve free silica and so is not subject to this restriction:

$$\underset{\text{edenite}}{NaCa_2Mg_5Si_4(AlSi_3)O_{22}(OH)_2} + \underset{\text{albite}}{NaAlSi_3O_8} = \underset{\text{richterite}}{Na(CaNa)Mg_5Si_8O_{22}(OH)_2} + \underset{\text{anorthite}}{CaAl_2Si_2O_8} \quad [5.13]$$

This second equilibrium is useful at relatively high grades of metamorphism, where the plagioclase has a significant anorthite content.

The jadeite content of clinopyroxene provides a geobarometer for rocks, such as transitional blueschist–eclogites, where albite persists but co-exists with jadeite or omphacite pyroxene (Section 3.3.4). In the absence of albite, the end-member reaction provides a lower pressure limit for eclogite-facies rocks containing Na-pyroxene without plagioclase. Some kyanite eclogites also contain the Na-white mica, paragonite, and for these the amount of jadeite in the pyroxene is controlled by equilibrium with paragonite and kyanite:

$$paragonite = jadeite + kyanite + fluid \quad [5.14]$$

Reaction [5.14] proceeds to the right with increasing pressure and so the jadeite content of the omphacite can be analysed and used to calculate a pressure of formation for any kyanite–paragonite eclogite.

At high temperatures, there is significant mutual solubility between orthopyroxene and clinopyroxene, and, in granulite-facies rocks, grains of both minerals may show evidence of exsolution of the other. This provides the basis for a solvus geothermometer which, although primarily applicable to igneous rocks, is also appropriate for interpreting granulite-facies temperatures (Wood & Banno 1973). Although the degree of mutual solubility is much more restricted, orthopyroxene co-existing with garnet does contain a small amount of dissolved garnet, is predominantly sensitive to pressure and provides a useful geobarometer (Harley & Green 1982).

SUMMARY

Metamorphosed basic igneous rocks, or metabasites, are found in a very wide range of geological settings and exhibit metamorphic features that may have formed at any time from shortly after the parental magma solidified to many hundreds of millions of years later. This chapter has explained the special significance of water for the metamorphism of igneous rocks: water must be added before igneous minerals can react to form metamorphic ones, and so the onset of metamorphism is strongly dependent on the nature of the original igneous body and the availability of water in the environment into which it was emplaced. Massive lavas and intrusive bodies may retain primary igneous minerals throughout extended histories of burial and heating, whereas pyroclastic deposits often undergo extensive metamorphism at very low temperatures.

The mineral assemblages developed in metabasites vary systematically according to the pressure and temperature of metamorphism. Amphiboles are developed over a wide range of conditions but vary considerably in composition, with glaucophane typical of high pressures and moderate temperatures, actinolite characteristic of relatively low grades and hornblende found at moderate to high temperatures. Systematic changes also take place in plagioclase feldspar and there are a range of hydrous minerals characteristic of low- to moderate-temperature metamorphism. At very high temperatures, assemblages begin to approach those of the primary igneous rock with the growth of pyroxenes in particular. Garnet becomes abundant at high pressures and moderate to high temperatures and may contain distinctive inclusions of coesite or diamond which act as markers of ultra-high-pressure metamorphism.

The mineral assemblages of metabasites do not normally change over such narrow temperature ranges as the assemblages of pelitic metasediments described in Chapter 4 (some very low-grade metabasic rocks provide exceptions to this generalisation, however), and so metabasites are less useful for identifying mappable zones. However, metabasites are present in rock sequences formed in a very wide range of tectonic settings; their assemblages vary with both temperature and pressure, and reflect changing conditions at both low temperatures and high pressures as well as at moderate to high temperatures. As a result, they are very valuable indicators of metamorphic conditions; the recognition of metamorphic facies based on metabasite assemblages is something that can often be done in the field and can locate likely changes in metamorphic conditions. Where laboratory data are available, there is a wide range of approaches to estimating metamorphic conditions of metabasites, but it is important to examine textures carefully because of the common persistence of relic grains formed during original magmatic crystallisation or earlier stages of metamorphism.

EXERCISES

1. List all the metamorphic facies in which each of the following minerals may be present in metabasites: garnet, pumpellyite, albite, plagioclase feldspar, hornblende.
2. Two samples of metabasite have been collected from nearby outcrops. One contains actinolite, albite, chlorite and calcite while the other is composed of plagioclase, orthopyroxene and clinopyroxene. Which facies would you assign them to? What geological inferences could you draw from the fact that they occur close together?
3. Name the distinctive minerals associated exclusively with high-pressure or ultra-high-pressure metamorphism.
4. Why is quartz found in metamorphosed basic igneous rocks but not in their fresh protoliths?
5. A metabasite body contains two distinct amphibole compositions, hornblende and actinolite. Suggest some textural criteria that you might use to determine whether or not these two amphibole types co-existed in equilibrium.

FURTHER READING

Browne, P. R. L. (2003). Hydrothermal alteration in active geothermal fields. *Annual Reviews of Earth and Planetary Sciences*, **6**, 229–48.

Dobrzhinetskaya, L., Faryad, S. W., Wallis, S. & Cuthbert, S. (2011). *Ultrahigh Pressure Metamorphism*. Elsevier, 696 pages.

Ingebritsen, S., Sanford, W. & Neuzil, C. (2006). *Groundwater in Geologic Processes.* Cambridge University Press, 536 pages.

Schiffman, P. & Day, H.W. (1995). Low grade metamorphism of mafic rocks. *Geological Society of America Special Publication,* **296,** 248 pages.

Utada, M. (2001). Zeolites in burial diagenesis and low-grade metamorphic rocks. *Reviews in Mineralogy and Geochemistry,* **45**(1), 277–304.

6 Metamorphism of Limestones – Marbles, Calc-Silicates and Skarns

Metamorphic rocks derived from carbonate-rich sediments, such as limestones and marls, also reflect the temperatures and pressures of metamorphism like metapelites and metabasites, but there is an additional factor that influences their mineralogy. Carbonate minerals release carbon dioxide during metamorphism and the mineral assemblages that form are influenced by the balance between water and carbon dioxide in the metamorphic fluid. In this chapter we will investigate the way in which minerals and fluids interact as limestones undergo metamorphism.

In the sedimentary record, limestones are sometimes very pure, and, because calcite itself is stable under most crustal conditions, these rocks often do not develop new minerals during metamorphism, although they do recrystallise and so become marbles (Figure 6.1a). Many limestone beds, however, contain other constituents, such as detrital grains of quartz or diagenetic dolomite, and these typically react over a range of conditions where pure calcite would remain stable (e.g. Figure 6.1b). Less-pure marbles can develop a range of metamorphic minerals, as we shall see in this chapter.

It is not unusual, especially at medium to high grades, to find rocks rich in Ca-, Mg- or Ca–Mg-silicates (such as zoisite, grossular, amphibole or diopside) but with little or no carbonate. These rocks are known as **calc-silicates**, and are likely to be the products of metamorphism of originally carbonate-bearing sediments, even if no carbonate remains, since calcite and dolomite are the major Ca- and Mg-bearing constituents of sediments.

The metamorphism of carbonate rocks is often accompanied by metasomatism, especially in medium- to high-grade metamorphism. For example, calc-silicates are almost always present at the interface between original limestone and pelite layers

(a)

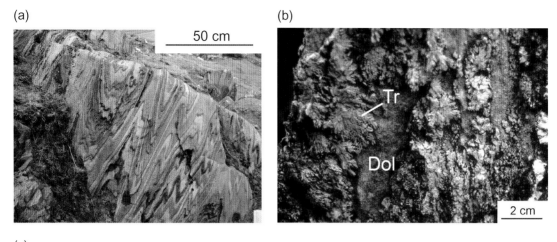

(b)

(c)

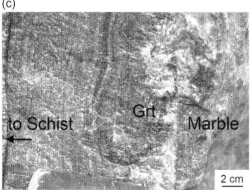

Figure 6.1 Examples of marbles and calc-silicate rocks in the field. (a) Calcite marble with variable amounts of graphite, Connemara. Intense similar-style folding is typical of amphibolite facies marbles, especially when there are no siliceous interbeds (c.f. Figure 1.15b). (b) Tremolite marble, Connemara, Ireland (Yardley & Lloyd 1989). Near-vertical, centimetre-scale layering follows original bedding, with layers of pure dolomite alternating with layers of tremolite rosettes with intergrown calcite which were probably dolomite with quartz sand originally. (c) Calc-silicate layer at the interface between marble (to the right) and schist (metapsammite) beds. Pink garnet dominates close to the marble while dark amphibole is predominant near the schist.

in the upper amphibolite facies (Figure 6.1c), whereas in the sedimentary record such contacts are often very sharp. Sometimes, the metasomatism of carbonate rocks is very extensive and leads to the formation of **skarns,** a variety of calc-silicate rock containing just a small number of minerals; they may be the result of regional metamorphism but are often associated with the release of fluids from cooling plutons.

In this chapter, we will focus first on carbonate-rich marbles and then describe some examples of calc-silicate rocks. By the end of the chapter you will have gained an understanding of mixed-volatile ($H_2O - CO_2$) fluids and how to apply phase diagrams to systems in which they are present. You will also understand how metamorphism can vary according to whether fluid composition was buffered within the rocks by mineral reactions, or controlled externally by influx of fluid from elsewhere.

6.1 Marbles

The term marble is used for metamorphosed calcareous rocks in which carbonate minerals dominate. Marbles may be dominated by calcite or by dolomite (more rarely by other carbonates), and the name may be modified by any other minerals present.

6.1.1 Calcite Marbles

Many marbles are composed predominantly of calcite with minor quartz and phyllosilicates, originally of detrital origin. Graphite derived from organic debris (Figure 6.1a), and/or pyrite are often present. The mineral assemblage in a marble of this type provides few clues as to the conditions of formation, since calcite is stable at all but the highest pressures (Figure 6.2), where it is replaced by aragonite. Note that, even where aragonite does form during metamorphism, it is likely in most cases to change back completely to calcite during exhumation, except at very low temperatures. At very high temperatures and low pressures, calcite reacts with quartz to produce wollastonite (calcium silicate), but otherwise most calcite limestones are remarkably unreactive during metamorphism. Despite this, they are susceptible to extensive textural changes due to recrystallisation of calcite to produce a coarser grain size and often a preferred orientation. As a result, calcite marbles are quite distinct from calcite limestones in appearance, despite having the same mineralogy and similar surface weathering characteristics.

The reaction to form wollastonite provides a simple example of one of the most common types of reaction to occur in carbonate rocks – a decarbonation reaction:

$$\underset{\text{calcite}}{CaCO_3} + \underset{\text{quartz}}{SiO_2} \rightarrow \underset{\text{wollastonite}}{CaSiO_3} + \underset{\text{fluid}}{CO_2} \qquad [6.1]$$

Like H_2O, CO_2 forms a supercritical fluid under metamorphic conditions, usually with a slightly greater density than water.

In the first experimental studies of reaction [6.1] the pressure of the fluid in the experimental capsule (CO_2) was equal to the total applied pressure. This setup is common in high-pressure experiments because the noble metal capsules containing the reacting minerals (the **experimental charge**) are weak and collapse onto the grains in the charge, forcing them together until the pressure on the fluid in the remaining interstices is equal to the applied pressure.

Figure 6.2 demonstrates that, at the pressures encountered at depths of more than a few kilometres below the surface, the temperature required to form wollastonite is beyond the normal range of regional metamorphism. This is consistent with the fact that most wollastonite occurrences are in thermal aureoles formed by contact metamorphism at relatively low pressures.

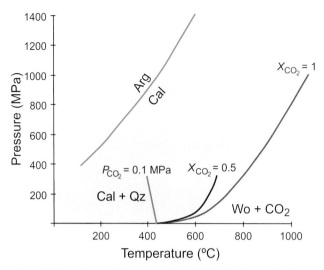

Figure 6.2 *P–T* diagram illustrating the stability of calcite and calcite + quartz under conditions of crustal metamorphism. The upper pressure limit of calcite is in green (Johannes & Puhan 1971) and the upper temperature limit of calcite + quartz (reaction [6.1]) in the presence of a pure CO_2 fluid ($X_{CO_2}=1$) is in purple (Harker & Tuttle 1956). Equilibrium curves for reaction [6.1] at reduced CO_2 pressures are in blues (Greenwood 1962, 1967). Yellow shading denotes the normal stability field of calcite + quartz in marbles, and clearly embraces a wide range of crustal conditions. Terms are defined in Box 6.1.

Nevertheless, wollastonite is occasionally found in regionally-metamorphosed rocks, where it apparently formed without excessive temperatures. To explain these discrepancies between experimental results and field observations we need to consider the composition of the metamorphic fluid phase. The experiments were carried out without water (the fluid phase was pure CO_2), but many metamorphic fluids contain both H_2O and CO_2. Such fluids are known as **mixed-volatile fluids** and are described in more detail in Box 6.1.

If a mixed-volatile metamorphic fluid is a single phase, rich in H_2O with only a small concentration of CO_2, then the partial pressure of CO_2 will be much lower than the total pressure and so decarbonation can occur at relatively low temperatures despite a relatively high total pressure. Experiments to test this idea by Greenwood (1962) were a major breakthrough in experimental petrology and results are included in Figure 6.2. The results confirm that the temperature for the appearance of wollastonite decreases with increasing pressure in the presence of an H_2O-rich fluid and this provides an explanation for the occasional occurrences of wollastonite in regionally metamorphosed marbles. Wollastonite occurs where water produced in adjacent schists was able to infiltrate the marble and give rise to an environment with low P_{CO_2} by flushing away CO_2 as fast as it was produced, allowing continued production of wollastonite.

BOX 6.1 Mixed-volatile fluids

In Chapter 2, Section 2.7, we introduced the principle that higher pressures of H_2O inhibit dehydration reactions, and a similar logic applies to decarbonation reactions as we can see in Figure 6.2. However, if it were possible to apply a high pressure to the solid phases in a quartz-bearing marble by means of a 'leaky piston', which allowed CO_2 to escape at low pressure, then, since the volume of 1 mol of wollastonite is less than the sum of 1 mol quartz + 1 mol calcite, we might expect the decarbonation reaction to take place at *lower* temperatures when the pressure on the solids is raised, because the increased pressure on the solids will facilitate the formation of the assemblage with the smaller solid volume. A leaking piston might seem an improbable device to find in nature, but a similar effect can be produced if the fluid phase in contact with the calcite and quartz is rich in H_2O and poor in CO_2. At greenschist facies and higher temperatures, H_2O and CO_2 are completely miscible supercritical fluids (unless the aqueous fluid contains large amounts of dissolved salts). In a mixed fluid with the total fluid pressure equal to lithostatic pressure, the partial pressures of water and carbon dioxide reflect their concentrations in the fluid. Fluid composition is expressed in terms of the mole fraction of CO_2 or X_{CO_2}:

$$X_{CO_2} = \frac{n_{CO_2}}{(n_{CO_2} + n_{H_2O})}.$$

where n denotes the number of molecules of the subscripted species.

Partial pressures are then given by

$$P_{CO_2} = P_{fluid} \cdot X_{CO_2} \text{ and } P_{H_2O} = P_{fluid} \cdot X_{H_2O}.$$

If the fluid is dominantly H_2O, the partial pressure of CO_2 in a mixed H_2O-CO_2 fluid may be very much less than the total fluid pressure, even if $P_{fluid} = P_{lithostatic}$. As a result, it is possible for a metamorphic rock with a water-rich pore fluid to have a low P_{CO_2} even though both lithostatic pressure and total fluid pressure are high.

The observed effect of adding H_2O to experiments on the equilibrium between calcite, quartz, wollastonite and fluid accords with the phase rule. In the H_2O-absent system there are four phases, three components (CaO, SiO_2, CO_2), and hence one degree of freedom, i.e. the full assemblage can occur stably only along a univariant curve on a P–T diagram. Adding H_2O increases the number of components by one, but does not change the number of phases if it is miscible with CO_2. Hence there are now two degrees of freedom when calcite, quartz, wollastonite and fluid co-exist. Fluid composition is now a variable in addition to T and P, and by specifying one of these three variables, the equilibrium conditions can be represented by a curve on a plot with the other two variables as axes. For example on Figure 6.2, a plot of P versus T, there is a series of univariant curves, each valid for one specific fluid composition.

Another widely used way of representing this sort of equilibrium is on a plot of T versus X_{CO_2}, constructed for some specified constant value of total pressure. An example of such a plot for reaction [6.1], constructed for a total pressure of 200 MPa, is shown in Figure 6.3. Plots of this type are known as **isobaric** $T-X_{CO_2}$ **diagrams** (often shortened to $T-X_{CO_2}$ diagrams). Divariant equilibria, such as reaction [6.1], taking place in the presence of H_2O, plot on such a diagram as a line known as an **isobaric univariant curve**, i.e. when P is fixed, one degree of freedom remains.

The number of phases that can form from limestones composed only of $CaCO_3$ + quartz is clearly limited. Only at exceptionally high temperatures and low pressures do other calcium silicate phases, such as spurrite, scawtite or larnite, appear. These rare calcium-silicate minerals normally result in nature from the heating of limestone by basalt lava flows very close to the surface. They are not of widespread geological importance and so are not considered further in this book.

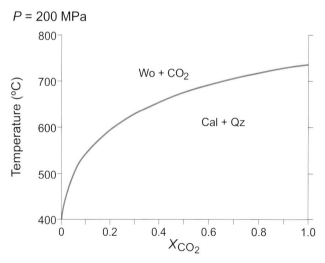

Figure 6.3 $T-X_{CO_2}$ diagram to show the effect of fluid composition on the stability of calcite + quartz. The data are for a total confining pressure of 200 MPa (Greenwood 1962, 1967).

6.1.2   Dolomitic Marbles

Limestones that contain dolomite provide much more useful indicators of metamorphic grade because a range of Ca–Mg-silicates can form in the more usual P–T conditions of metamorphism, notably talc, tremolite, diopside and forsterite. The general sequence of mineral-appearance isograds is:

> talc (not always present)
> tremolite
> diopside *or* forsterite
> diopside + forsterite.

Assemblages with these minerals can be represented on a triangular diagram with CaO, SiO_2 and MgO at the apices (Figure 6.4). CO_2 and H_2O are assumed to be available in excess to produce carbonate or hydrous phases.

Some of the simplest examples of the development of these assemblages are from metamorphism of dolomitic limestones that originally contained chert nodules. One such location is the aureole of the Beinn-an-Dubhaich granite, a Paleogene granite intruded into Cambrian limestones and dolostones on the island of Skye, Scotland. Here, the chert nodules are replaced by Ca–Mg silicates and calcite, with a series of metamorphic zones over a scale of tens to hundreds of metres. Figure 6.5a illustrates how, in the outermost part of the aureole, reaction

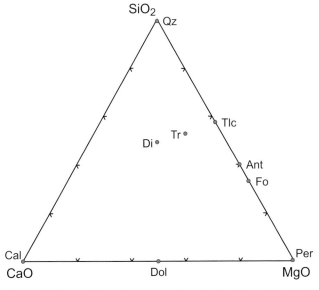

Figure 6.4 Compositions of common minerals of metamorphosed dolomitic marbles in the system $CaO - MgO - SiO_2$, shown projected from H_2O and CO_2. This projection means that carbonates, hydrates and anhydrous minerals are all shown.

(a)   (b)

Figure 6.5 Examples of reaction between chert and dolomitic limestone in the aureole of the Beinn-an-Dubhaich granite, Skye, Scotland. (a) Chert nodule mantled by a thin rind of talc from the outer part of the aureole. (b) A former chert nodule now completely replaced by tremolite with calcite.

has formed a thin shell of talc around the remaining chert. In contrast, closer to the granite, the chert nodules are completely replaced by metamorphic silicates (Figure 6.5b). Tremolite is followed by diopside or forsterite according to the nature of the original limestone (Holness 1992). The presence of readily-weathered calcite intergrown with more resistant silicate reaction products (e.g. reaction [6.2] discussed below), means that the metamorphosed nodules are less prominent than the unaltered chert on weathered surfaces.

Regional Metamorphism of Dolomitic Marbles in the Central European Alps One of the most extensive studies of moderate-pressure regional metamorphism of marbles was undertaken in the Central Alps (Trommsdorff 1966, 1972) and Figure 6.6 is a map showing the metamorphic zoning of dolomitic marbles in the region that he determined. The metamorphic grade increases southwards from low-grade rocks with talc through tremolite marbles to diopside- and forsterite-bearing rocks. The diagnostic assemblages are represented graphically in Figure 6.7, and from this diagram it is possible to suggest reactions to describe the changes between the zones, on the basis of the shift of the pattern of tie-lines.

Figure 6.7a shows the original assemblage before metamorphism and the normal compositional range of siliceous dolomitic limestones. The first metamorphic mineral to appear in calcite-bearing marbles at low grades is talc (Figure 6.7b). The growth of talc results in the replacement of the dolomite–quartz tie-line (Figure 6.7a) by the talc–calcite tie-line (Figure 6.7b), and this change can be represented by the reaction:

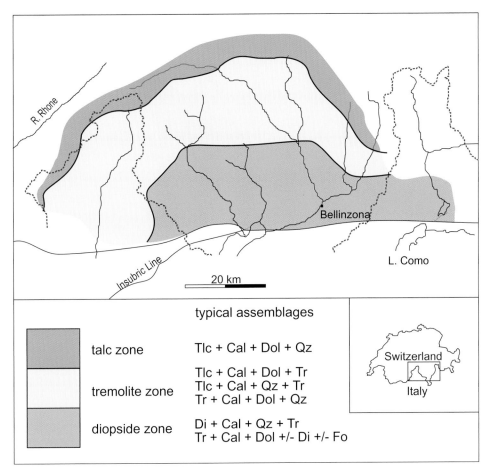

Figure 6.6 Metamorphic zonation of siliceous dolomitic marbles in the Central, Lepontine Alps, based on Trommsdorff (1966, 1972). Broken red line is the International Boundary. Note that dolomite + quartz is a common association in the talc and tremolite zones.

$$3\,\text{dolomite} + 4\,\text{quartz} + 1\,H_2O \rightarrow 1\,\text{talc} + 3\,\text{calcite} + 3\,CO_2 \qquad [6.2]$$

Note that rocks containing all four of these solid phases are relatively common throughout the talc zone. They might be expected where insufficient water was added to the marble to convert all the available reactants to talc. Many other dolomitic marbles show no sign of reaction at this grade, retaining a dolomite + quartz assemblage; presumably this is because they were not infiltrated by water.

The first isograd shown on Figure 6.6, marking the appearance of tremolite, is roughly equivalent to the staurolite isograd in pelitic rocks. To the south of this line, at higher grades, a range of assemblages are possible. Figure 6.7c represents the assemblages found when tremolite first reacts in. The composition of tremolite plots

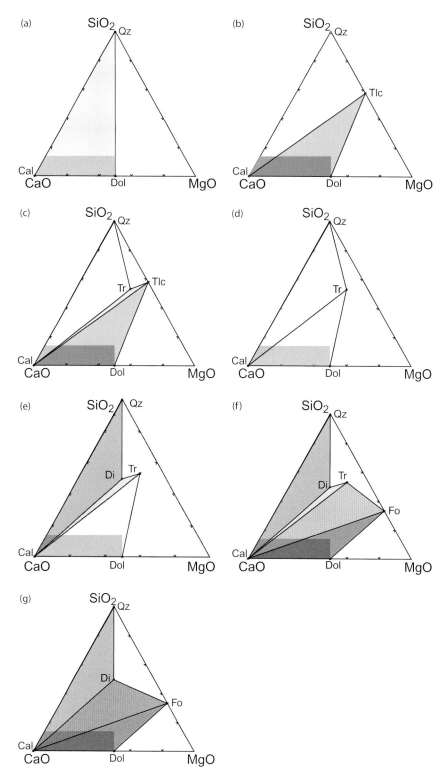

Figure 6.7 Phase compatilibities in the CaO–MgO–SiO$_2$ system projected from H$_2$O and CO$_2$ (see Figure 6.4), illustrating the sequence of phases found in siliceous dolomites of the Lepontine Alps (Figure 6.6). (a) Initial assemblage; the darker blue region is the common range of compositions of dolomitic marbles and recurs through the remaining parts. (b) Talc zone. (c) Upper talc zone. (d) Tremolite zone. (e), (f), (g) Diopside zone. For discussion, see text.

within the talc–calcite–quartz triangle of Figure 6.7b, from which we can deduce that the reaction is:

$$5 \, talc + 6 \, calcite + 4 \, quartz \rightarrow 3 \, tremolite + 2 \, H_2O + 6 \, CO_2 \qquad [6.3]$$

In the case of quartz-poor talc–calcite–quartz rocks, quartz may be completely consumed by this reaction to give the assemblage talc + calcite + tremolite, but in more-siliceous rocks the talc is consumed to give tremolite + calcite + quartz. Dolomite–calcite–talc assemblages are also possible in rocks with a lower silica content, plotting near the base of the triangle. The final disappearance of talc in these rocks, which gives rise to the phase relations illustrated in Figure 6.7d, can be ascribed to the reaction:

$$2 \, talc + 3 \, calcite \rightarrow 1 \, tremolite + 1 \, dolomite + 1 \, CO_2 + 1 \, H_2O \qquad [6.4]$$

At higher grades (approximately equivalent to the pelite sillimanite zone), diopside appears, although tremolite commonly persists. Figure 6.7e shows the appearance of diopside within the tremolite–calcite–quartz field, through the reaction:

$$1 \, tremolite + 3 \, calcite + 2 \, quartz \rightarrow 5 \, diopside + 1 \, H_2O + 3 \, CO_2 \qquad [6.5]$$

In contrast, the association of forsterite + calcite develops in quartz-free dolomitic marbles and implies the replacement of the tremolite–dolomite tie-line (Figure 6.7f) due to the reaction:

$$1 \, tremolite + 11 \, dolomite \rightarrow 8 \, forsterite + 13 \, calcite + 1 \, H_2O + 9 \, CO_2 \qquad [6.6]$$

It is clear from Figure 6.7f that forsterite first appears in rocks whose compositions are silica-poor and lie near the base of the triangle, while diopside appears in relatively silica-rich or dolomite-poor rocks. This observation may account for the discrepancies in their order of appearance in different regions. Diopside and forsterite can co-exist only when the tremolite–calcite tie-line has been removed, due to the reaction:

$$3 \, tremolite + 5 \, calcite \rightarrow 11 \, diopside + 2 \, forsterite + 3 \, H_2O + 5 \, CO_2 \qquad [6.7]$$

This gives rise to the phase relations shown in Figure 6.7g, and results in the final disappearance of tremolite.

The simple reaction sequence outlined here and illustrated in Figure 6.7 is unfortunately more complex in nature. Assemblages involving both reactants and products of these reactions occur over large areas, and apparently low-grade assemblages persist alongside apparently higher-grade assemblages, as listed in Figure 6.6. The reason for this ambiguity is that the triangular diagrams in Figure 6.7 have all been constructed assuming that H_2O and CO_2 are available in excess, but in fact these components occur in a single fluid phase. Applying the phase rule to the marble assemblages we see that the system actually has five components: CaO, MgO, SiO_2, H_2O and CO_2. No assemblages have more than five phases, i.e. four solid phases (Figure 6.6) plus a single mixed H_2O–CO_2 fluid phase. Hence all assemblages have at least two degrees of freedom, even if both the reactants and reaction products co-exist. The possible variables are pressure, temperature and the composition of the fluid phase. So in practical terms, all assemblages can occur over a range of pressures and temperatures according to the composition of the fluid phase. Much of the diversity in the mineral assemblages within the broadly defined zones of Figure 6.6 can be explained if the fluid composition varied between samples. Why this might arise is the subject of the following section.

6.1.3   Controls on the Fluid Composition in Marbles

The *P–T* conditions at which many reactions in carbonate rocks take place are dependent on the composition of the fluid phase. However, the reactions themselves also involve components of the fluid and so as the reaction proceeds the fluid composition changes. If no fluid is introduced from an external source, the fluid composition changes systematically in a way that is determined by rock composition. In other situations, however, as we have seen already in the discussion of regional wollastonite occurrences, water released by reactions in adjacent lithologies, or by the crystallisation of magmas emplaced nearby, may be able to enter marbles and drive reactions. If the fluid composition is controlled by reactions within a rock, without any outside influence, it is said to be **internally buffered**. In contrast, if the fluid composition is being controlled at the source of fluid which is infiltrating the rock as it reacts, it is said to be **externally buffered**.

The Effect of Reaction on Fluid Composition   A marble with calcite + quartz may have a nearly pure H_2O fluid phase present initially, but as it is heated, reaction to produce wollastonite (reaction [6.1]), also releases CO_2 into the fluid. As the fluid becomes enriched in CO_2 the reaction may slow down or stop until the temperature rises further (Figure 6.3).

Reactions that take place in marbles and other carbonate-bearing rocks can be grouped into six types according to how they modify the composition of the metamorphic fluid. Each of these has a distinctively shaped equilibrium curve on an isobaric *T–X* diagram. These are illustrated in Figure 6.8.

Decarbonation Reactions We have already seen that reactions such as [6.1] can be represented by equilibrium curves on the *T–X* diagram which reach a temperature maximum at $X_{CO_2} = 1.0$, and fall to lower temperatures at small values of X_{CO_2}. This is shown schematically by curve *a* on Figure 6.8.

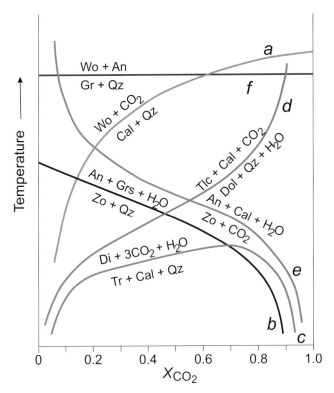

Figure 6.8 Schematic representation of the form of $T–X_{CO_2}$ curves for different types of mixed-volatile reactions, after Kerrick (1974). Note that although specific examples are used, for clarity, they are not shown in their correct relative positions along the temperature axis.

Dehydration Reactions The equilibrium curve for a dehydration reaction is the mirror image of that for a decarbonation reaction (Figure 6.8, curve *b*).

Dehydration–Decarbonation Reactions Many reactions in marbles produce a mixture of CO_2 and H_2O, for example reactions [6.3] to [6.7] above. In the case of dehydration reactions (H_2O release only) or decarbonation reactions (CO_2 release only), the highest temperature on the equilibrium curve on a $T–X$ diagram is attained when the composition of the fluid phase in the system corresponds to that being given off by the reaction, i.e. when it is pure H_2O or pure CO_2 as the case may be. In the same way, an isobaric univariant curve for reactions giving off a mixed-volatile fluid has a temperature maximum on a $T–X$ diagram corresponding to the X_{CO_2} value of the fluid produced. For example, in the case of reaction [6.4] the maximum will be at $X_{CO_2} = 0.5$, for reactions [6.3] and [6.5] it is at $X_{CO_2} = 0.75$ and for reaction [6.6] it is at $X_{CO_2} = 0.9$. The reaction takes place at lower temperatures if the fluid phase present in the rock does not correspond to that given off by the reaction, irrespective of whether it is richer in CO_2 or richer in H_2O (curve *c*, Figure 6.8). This type of reaction can shift the fluid composition towards either CO_2 or H_2O, depending on what it is to start with, but the maximum temperature is reached when the overall pore-fluid composition corresponds to that of the fluid given off by the reaction itself. At this point, CO_2 and H_2O behave as a single component, since their ratio remains constant, and the reaction is therefore univariant.

Hydration–Decarbonation Reactions Other mixed-volatile reactions, for example reaction [6.2], have H_2O and CO_2 on different sides of the reaction. The equilibrium curve for such reactions (e.g. curve *d* on Figure 6.8) displays a particularly large range of equilibrium temperatures according to fluid composition. At low values of X_{CO_2} the equilibrium temperature drops as for other decarbonation reactions, favouring production of CO_2. However, at high X_{CO_2} values it is the behaviour of the aqueous component that dominates the form of the equilibrium curve; as X_{H_2O} diminishes so the hydration aspect of this type of reaction is inhibited by the absence of water, hence the equilibrium temperature approaches infinity as X_{CO_2} approaches 1.

Carbonation–Dehydration Reactions These are similar to hydration–decarbonation reactions except that the effect of increased temperature is to cause CO_2 to be consumed and H_2O to be released. An example is the reaction:

$$\text{zoisite} + CO_2 \rightarrow \text{anorthite} + \text{calcite} + H_2O \qquad [6.8]$$

The form of the equilibrium curve is therefore the mirror image of that for hydration–dehydration reactions (Figure 6.8, curve *e*).

Fluid-Absent Reactions Where no fluid is released or consumed, as in the reaction:

$$\text{grossular} + \text{quartz} \rightarrow 2\,\text{wollastonite} + \text{anorthite} \qquad [6.9]$$

the equilibrium conditions will be unaffected by the composition of any fluid phase that may be present. Such reactions are univariant and are represented by a horizontal line on an isobaric T– X_{CO_2} diagram, indicating a unique equilibrium temperature at each pressure (Figure 6.8, curve *f*).

Buffering We can imagine two end-member scenarios for how fluid composition changes as a marble reacts. In one, no fluid infiltrates from outside and the fluid is internally buffered; in the other, an externally-derived fluid is so effectively introduced that its composition is unchanged by the reactions that it triggers and the fluid is externally buffered. Of course, natural processes may be intermediate between the two. These scenarios are illustrated in Figure 6.9 using the example of the formation of wollastonite according to reaction [6.1].

In the first case, little or no water gains access to the marble after wollastonite starts to form. If the initial fluid composition is represented by X_A on Figure 6.9a,

(a)   (b)

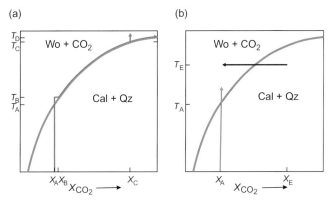

Figure 6.9 Influence of internal versus external buffering of fluid composition on the breakdown of calcite + quartz. (a) Fluid composition is internally buffered; for an initial fluid X_A, reaction takes place over a large temperature interval from T_B to approximately T_D. (b) Externally buffered fluid. If fluid composition remains buffered at X_A during heating, reaction goes to completion at T_A; alternatively if the initial fluid is X_E, an influx of fluid X_A at temperature T_E will drive the reaction to completion without further heating.

wollastonite will first appear at a temperature above T_A, say T_B, and will continue to grow until the fluid composition has shifted to X_B before there is any further rise in temperature. This is because, as we shall see in Chapter 7, reactions cannot start to form new minerals until after the equilibrium conditions have been exceeded.

Further heating is accompanied by further increments of reaction which cause the fluid to become progressively enriched in CO_2 so that the fluid composition and temperature evolve together along the path of the isobaric univariant curve as long as calcite, quartz, wollastonite and fluid are all present. In terms of the phase rule, the assemblage has two degrees of freedom, and so if P and T are independently determined, the fluid composition is fixed, or buffered (see Chapter 2, Section 2.8), by the presence of this particular mineral assemblage, hence the term internally buffered. In this example, suppose reaction proceeds until a temperature T_C is reached, at which point one of the reactants (calcite or quartz) is completely consumed; the fluid composition will remain constant at X_C even if the temperature continues to rise. Suppose instead, however, that the CO_2 released by reaction is able effectively to flush away all the inital H_2O. In this case, the reaction will be inhibited until temperature T_D is reached and the system will then become effectively a three-component system only (no H_2O remains); the reaction is now univariant and will go to completion at T_D.

An important factor in determining whether fluids are internally buffered is the porosity of the marble before and during reaction. It will be easier for a reacting rock to flush away the pre-existing pore fluid, and so internally buffer its pore-fluid composition, if the amount of fluid present initially is very small relative to the amount of new fluid that is released by reaction.

In the alternative process of external buffering, reaction is a response to the infiltration of a fluid derived from elsewhere, and the source region defines the fluid's composition. If a sufficiently large quantity of this fluid passes through the marble, it may be able to flush away the fluid being generated by reactions going on within the marble without having its composition substantially modified. In this case the fluid composition remains more or less constant: it is unchanged by local mineral reactions, and is therefore externally buffered. This type of reaction is represented in two ways on Figure 6.9b: firstly, if influx of fluid of composition X_A continues steadily as the rock is heated, then the heating of the rock is represented by the vertical light blue line. The fluid composition remains constant at X_A and the reaction will go to completion at a temperature close to T_A. Alternatively, fluid infiltration may take place in a single event at a fixed temperature, say T_E. If the initial fluid composition is X_E and the rock is composed of calcite and quartz, then infiltration of fluid X_A will shift the fluid composition at constant temperature, indicated by the horizontal dark blue line at $T = T_E$. This promotes the reaction that produces wollastonite. External buffering is probably responsible for the examples of regional wollastonite, as discussed above.

Internally Buffered Reaction Sequences in Marbles   We have established above that simple mixed-volatile reactions, including reactions [6.1] to [6.8], are divariant. In the absence of fluid being introduced from outside, the fluid composition changes systematically as reaction proceeds and, for rocks with only a small pore volume, large changes in the pore-fluid composition can result from a rather small amount of reaction producing only traces of the solid products.

In most metamorphic rocks, porosity and permeability are very low. This means that most metamorphic fluid compositions are likely to be internally rather than externally buffered (Greenwood 1975). Divariant reactions occurring over a temperature interval therefore make a negligible contribution to the development of new metamorphic assemblages. Instead, distinct isograds develop in response to reactions that can occur without changing the fluid composition and are therefore univariant. The pattern of mineral assemblages shown in Figure 6.6 can readily be explained by internal buffering, with different rock layers undergoing different initial reactions. These local reaction processes overall led to the fluid composition being buffered to different values of X_{CO_2} in each layer (Trommsdorff 1972).

As we have already seen, a mixed-volatile reaction will become univariant if the fluid being released has the same composition as that already present in the pores, since this has the effect of combining H_2O and CO_2 as a single component. Reaction then proceeds at approximately constant temperature until one of the reactants is used up. The growth of wollastonite by reaction [6.1] in the presence of a pure CO_2 fluid is a simple example of this. Dehydration–decarbonation reactions also behave in this way, (e.g. curve c, Figure 6.8) becoming univariant at the maximum temperature point on the curve. Reaction [6.4], which gives rise to an isograd marking the disappearance of the assemblage talc + calcite, is this type of reaction.

An isograd can also arise where divariant equilibria intersect, and this is illustrated by an isobaric T–X diagram in Figure 6.10. The equilibrium curve for reaction [6.2] represents the conditions under which talc, calcite, dolomite, quartz and fluid can co-exist. Similarly, the equilibrium curve for reaction [6.3] in the same system gives the conditions for co-existence of talc, calcite, tremolite, quartz and fluid. These two curves intersect at the point I on Figure 6.10, and so at this point all five solid phases involved in the two equilibria must be able to co-exist. As a result, the assemblage at I has one less degree of freedom and is univariant, which is why on the isobaric T–X diagram it is represented by a point, known as an **isobaric invariant point**. Since talc, calcite, tremolite, dolomite and fluid are all stable at point I, the curve for reaction [6.4] must also pass through this point, as must the curve for another reaction involving four of these five solid phases:

$$\text{dolomite} + \text{quartz} + H_2O \rightarrow \text{tremolite} + \text{calcite} + CO_2 \qquad [6.10]$$

A dolomite–quartz–calcite marble with an initial fluid composition X_A on Figure 6.10, will begin to react on heating to produce talc by reaction [6.2] at temperature T_A; how much talc forms depends on how much water enters the rock. If the fluid composition is internally buffered it will evolve along curve [6.2] until

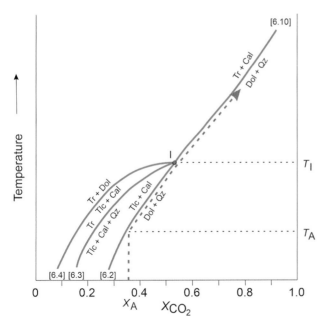

Figure 6.10 Isobaric reactions in siliceous dolomites with the fluid composition internally buffered. A dolomite–quartz(–calcite) marble with initial fluid X_A begins to react at temperature T_A to produce talc. Talc is progressively produced by reaction [6.2] until the isobaric invariant point I is reached at T_I. Here tremolite appears, giving rise to a distinct isograd because reaction continues at this temperature until all talc is consumed.

point I is reached at temperature T_I, and here tremolite will appear. Reaction will proceed at this temperature until one of the minerals talc, quartz, calcite or dolomite is consumed. Note, however, that this is not a simple reaction with a unique stoichiometry, because at different pressures the fluid composition where the curves intersect will be different. In the absence of external buffering, the tremolite isograd in the field corresponds to reaction at point I. To a first approximation, the transition from the talc zone to the tremolite zone in the Lepontine Alps (Figure 6.6) must correspond to conditions close to point I, although the precise form of the phase diagram will depend on pressure.

The temperature and fluid composition of the marble cannot depart from point I as long as all the reactant and product minerals of the intersecting reaction curves are present. However if all the talc is consumed, the remaining assemblage must satisfy the equilibrium conditions for reaction [6.10]. This means that, with further heating, the fluid composition will evolve further along curve [6.10], although again, only a small amount of reaction is likely until the next isobaric invariant point is reached. The tremolite marble illustrated in Figure 6.1b is an example in which the tremolite appears to have grown directly from dolomite + quartz by reaction [6.10], without a talc precursor, because infiltration of water only took place above temperature T_I.

Fluid Infiltration – an Example from Contact Metamorphism Many of the published studies of metamorphism of carbonate rocks are investigations of thermal aureoles around relatively shallow plutons. In some instances these exhibit significant differences from the regional example of the Alps described earlier. In particular, the mineral assemblages and sequence of reactions do not always fit very well to the model of internal buffering. For example around the Beinn-an-Dubhaich aureole (Figure 6.5), a number of features indicative of infiltration of fluid from the adjacent granite have been recognised (Holness 1992).

A well-studied example of infiltration around a pluton is provided by the aureole of the Alta Stock, Utah (Cook & Bowman 2000). The limestone country rocks contain both massive dolomitic limestones with sparse quartz disseminated throughout, and nodular dolomites with chert nodules comparable to those illustrated in Figure 6.5. The distribution of metamorphic zones around a part of the stock is illustrated in Figure 6.11.

In the outermost aureole, the nodular dolomites develop rims of talc and calcite around chert nodules, consistent with reaction [6.2]. The interbedded massive dolomites, in contrast, show little evidence of reaction. Tremolite appears in both types of dolomites at the tremolite isograd, with the most widespread assemblage being tremolite + calcite + dolomite. In the massive dolomites, tremolite was largely produced by reaction [6.10], since generally no talc precursor was present. A key difference from the regional tremolite zone in the Alps is that there is only a very

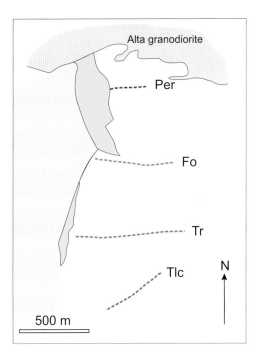

Alta granodiorite

-------- Per

-------- Fo

------- Tr

-- Tlc   N
          ↑

500 m

narrow zone where the four-phase assemblage (tremolite + calcite + dolomite + quartz) is found. In the Alta aureole, the bulk of the tremolite zone comprises rocks with just three out of this set of minerals, whereas in the Alps, assemblages of four minerals are present throughout. Similarly, at the forsterite isograd, tremolite is abruptly replaced by forsterite, through reaction [6.6], giving rise to rocks with forsterite + calcite + dolomite.

The scarcity of rocks in which products and reactants of the simple mixed-volatile reactions co-existed suggests that external buffering of the fluid was important in the tremolite and forsterite zones. It can be seen from Figure 6.9 that, whereas reactants and products co-exist over a significant temperature interval if the fluid is internally buffered, one assemblage replaces the other rather abruptly if the fluid is externally buffered.

Close to the intrusion in some parts of the aureole, the assemblage brucite + calcite is present. Brucite $(Mg(OH)_2)$ is a secondary replacement for original periclase, MgO, which grew with calcite through the reaction:

$$dolomite \rightarrow periclase + calcite + CO_2 \qquad [6.11]$$

Where brucite/periclase + calcite are present, primary dolomite is often absent, although secondary dolomite, formed with the brucite, may occur. The development of periclase is inferred to arise from local infiltration of water, driving the breakdown of dolomite.

Thus far, we have considered the composition of metamorphic fluids only in terms of H_2O and CO_2, but of course other chemical components of rocks are also likely to be present in solution. Where sufficiently large volumes of fluid pass through a rock unit to externally buffer H_2O and CO_2 in the pore fluid, it is reasonable to ask whether other constituents may also be introduced. There is evidence that this happened in the Alta aureole. Limestones in the talc zone and beyond contain very little boron, < 0.5 ppm (parts per million), but samples from the inner parts of the aureole contain up to 40 ppm as well as boron minerals such as tourmaline. Similarly, fluorine is present at higher levels near the intrusion, where the F-bearing mineral clinohumite is sometimes present. The oxygen isotopic composition of marble in the inner zones is also very different from that of the unaltered limestone, demonstrating that it has recrystallised in the presence of an externally-derived fluid (Bowman *et al.* 2009).

In summary, the aureole of the Alta stock provides a clear example of metamorphism being affected by the infiltration of water from an external source, in this case almost certainly the pluton itself. External buffering of the fluid is manifested by both chemical and isotopic changes in the aureole rocks and the mineral assemblages that they develop.

Serpentine in Marbles Serpentine is not normally developed during the prograde evolution of siliceous dolostones, but in many parts of the world marbles contain bright yellow-green coloured serpentine, and are popular decorative stones. These rocks are termed ophicalcite marbles (or more generally ophicarbonate rocks) and their mineralogy is superficially similar to other ophicalcite rocks derived from ultrabasic mantle rocks by hydration and carbonation, although the colour is much brighter. In ophicalcite marbles, the serpentine usually replaces metamorphic forsterite and is developed during retrograde metamorphism.

6.1.4 *P–T* Indicators for Marbles

At first glance the task of relating the assemblages of impure marbles to prevailing temperatures and pressures is a daunting one, because the temperature at which reaction takes place is often strongly dependent on fluid composition and this may be transient. However, if the fluid composition is internally buffered as reaction proceeds, then it becomes possible to assign temperatures to specific isograds, because they correspond to reaction among a univariant assemblage of phases (as we have seen above). It is therefore possible to plot these reactions on a petrogenetic grid, although most of the reactions on it require an assumption that fluid composition is internally buffered.

A Petrogenetic Grid for Marbles Experimental studies in the 1970s provided the data underpinning a series of *T–X* diagrams for different pressures (Skippen 1974; Slaughter *et al.* 1975). Figure 6.12 shows two examples at different pressures that are broadly consistent with the sequence of zones illustrated in Figure 6.6. The talc zone extends up to the temperature of isobaric invariant point A, which gives rise to the tremolite isograd. Diopside may first appear at point B but is likely to become abundant only at the temperature maximum for reaction [6.5]. From the large differences between the diagrams for the two pressures illustrated here, it is clearly difficult to use such diagrams to interpret *P–T* conditions where both pressure and temperature are unknowns.

To arrive at a petrogenetic grid, Connolly & Trommsdorff (1991) used computational methods to calculate the locations of the isobaric invariant points on *T–X* diagrams, such as points A and B in Figure 6.12, across a range of pressures. Figure 6.13, derived from their work, illustrates the fields of three key assemblages which are restricted to specific *P–T* ranges irrespective of fluid composition.

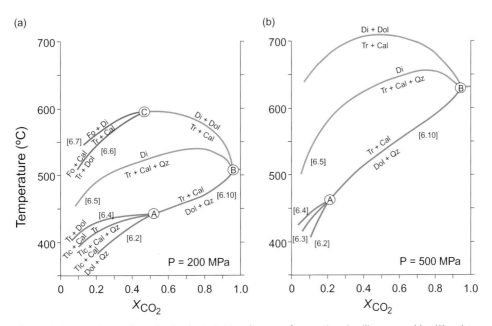

Figure 6.12. Experimentally-derived isobaric T–X_{CO_2} diagrams for reactions in siliceous marbles (Slaughter *et al.* 1975): (a) is for a pressure of 200 MPa, (b) is for 500 MPa. For convenience, the curves are colour coded according to the isobaric invariant points (lettered) that they intersect.

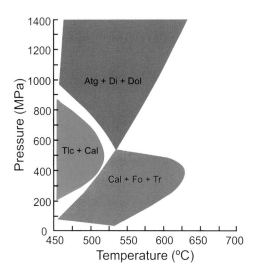

Figure 6.13 Petrogenetic grid for dolomitic marbles (derived from Connolly & Trommsdorff 1991). The coloured regions denote assemblages which are stable only in specific P–T ranges, regardless of fluid composition.

Geothermometers and Geobarometers for Carbonate Rocks There are unfortunately very few geothermometers or geobarometers that are applicable to carbonates. A potentially applicable geobarometer in low-grade rocks is the calcite–aragonite transition. This reaction is one of the only reactions in carbonates that is strongly pressure sensitive and aragonite does occur in prehnite–pumpellyite and low-grade blueschist-facies rocks.

The calcite–dolomite geothermometer is one of the few useful geothermometers. Calcite and dolomite form a restricted solid solution at metamorphic temperatures. The solvus is asymmetrical, and the most reliable way of determining the temperature of formation of co-existing calcite and dolomite is from the Mg-content of the calcite; the dolomite may be almost stoichiometric within the limits of analytical error. A simple formulation for the calculation of the temperature of formation of co-existing calcite and dolomite is:

$$\log_{10} X^{\mathrm{Cal}}_{\mathrm{MgCO_3}} = (-1690/T) + 0.795$$

where $X^{\mathrm{Cal}}_{\mathrm{MgCO_3}}$ is the mole fraction of $MgCO_3$ (*not* dolomite) in calcite co-existing with dolomite, and T is in Kelvin (Rice 1977).

At low temperatures, the limited amount of mutual solid solution between the carbonates makes the calcite–dolomite geothermometer too insensitive to be useful. Oxygen isotope fractionation increases at low temperature, however, and oxygen isotope ratios yield useful temperature estimates for low-grade carbonate rocks provided that there has been no retrograde infiltration of water.

6.2 Calc-Silicates and Skarns

At the beginning of this chapter, calc-silicates were defined as being rocks rich in Ca–Mg-silicate minerals, but with only minor amounts of carbonate. The simplest origin for such rocks is the metamorphism of marly sediments, but as we shall see, many calc-silicates probably arise through metasomatic changes affecting marbles. The example illustrated in Figure 6.1c is of this type.

The first zonal scheme for calc-silicates was developed for the central Highlands of Scotland. Zoisite appears at equivalent conditions to the pelite garnet zone (upper greenschist facies), occurring with a relatively sodic plagioclase (andesine or, at lower grades, albite). Biotite, garnet and hornblende appear in some lithologies. In the zones equivalent to the pelite staurolite and kyanite zones, zoisite is absent and the plagioclase is calcic, typically >70% anorthite component. Hornblende and garnet are also typically present. The breakdown of zoisite is a dehydration reaction (curve *b* in Figure 6.8). In the sillimanite zone, the calcic plagioclase and garnet persist, but hornblende is replaced by clinopyroxene.

6.2.1 Calc-Silicates from the Vassalboro Formation, Maine, USA

One region where changes in mineralogy of calc-silicates following prograde regional metamorphism has been studied in detail is the Silurian Vassalboro Formation of Maine, which was metamorphosed during the Acadian orogeny in the Upper Palaeozoic (Ferry 1983). The formation consists of finely-interbanded semi-pelites, pelites and argillaceous calcareous rocks, typically layered on a scale of only a few centimetres. Metamorphism occurred at low pressures and is spatially related to syn-metamorphic granite stocks (Figure 6.14).

The zonal sequence in Maine is distinctly different from that reported from Scotland, especially in terms of the occurrence of zoisite.

Ankerite Zone   The lowest-grade rocks contain the assemblage ankerite + quartz + albite + muscovite + calcite ± chlorite with accessory pyrite, graphite and ilmenite.

Biotite Zone   This zone is characterised by the co-existence of biotite and chlorite without amphibole. The biotite isograd can be related to the reaction:

$$\text{muscovite} + \text{quartz} + \text{ankerite} + H_2O \rightarrow \text{calcite} + \text{chlorite} + \text{biotite} + CO_2 \quad [6.12]$$

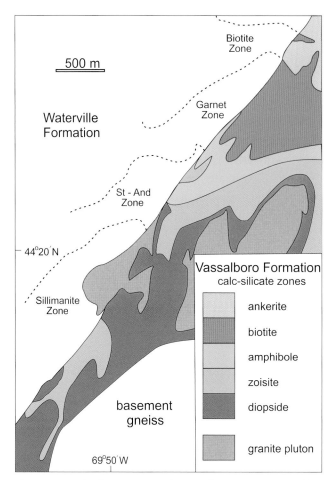

Figure 6.14 Calc-silicate zones in the Vassalboro Formation (north-east USA) showing pelite isograds in the adjacent Waterville Formation. Modified after Ferry (1983).

Within the biotite zone, albite is replaced by an intermediate plagioclase (oligoclase or labradorite) through the reaction:

$$\text{muscovite} + \text{calcite} + \text{chlorite} + \text{quartz} + \text{albite} \rightarrow \text{biotite} + \text{plagioclase} + H_2O + CO_2 \quad [6.13]$$

As a result of this, and similar, reaction(s), muscovite becomes rare and is often absent by the upper biotite zone. Note this stark difference to the occurrence of muscovite in pelitic rocks, where muscovite may persist until the onset of partial melting. The change in plagioclase composition seen in this zone suggests that conditions are comparable to the pelite garnet zone (Chapter 4).

Amphibole Zone   The appearance of calcic amphibole is accompanied by a further shift towards more calcic plagioclase compositions. The characteristic assemblage is Ca-amphibole + quartz + Ca-plagioclase + calcite + biotite $\pm$ chlorite. The isograd reaction is inferred to be:

$$\text{chlorite} + \text{calcite} + \text{quartz} + \text{plagioclase} \rightarrow \text{Ca-amphibole} + \text{Ca-plagioclase} \qquad [6.14]$$
$$+ H_2O + CO_2$$

Zoisite Zone   Zoisite first appears rimming plagioclase at contacts with calcite grains, suggesting growth is due to the reaction:

$$\text{Ca-plagioclase} + \text{calcite} + H_2O \rightarrow \text{zoisite} + CO_2 \qquad [6.15]$$

Within the same zone, the assemblage K-feldspar + Ca-amphibole first appears, due to the reaction:

$$\text{biotite} + \text{calcite} + \text{quartz} \rightarrow \text{Ca-amphibole} + \text{K-feldspar} + CO_2 + H_2O \qquad [6.16]$$

The appearance of K-feldspar and Ca-amphibole is close to the sillimanite isograd in nearby pelites.

Diopside Zone   At the highest grades, the mineral assemblage is diopside + zoisite + Ca-amphibole + calcite + quartz + plagioclase $\pm$ biotite $\pm$ microcline. The growth of diopside results from breakdown of amphibole:

$$\text{Ca-amphibole} + \text{calcite} + \text{quartz} \rightarrow \text{diopside} + H_2O + CO_2 \qquad [6.17]$$

As at Alta (Section 6.1.3), this sequence of zones implies that the fluid composition was externally buffered so that it remained water-rich, even though many of the reactions consumed water and/or released CO_2-rich fluid. For example, zoisite + quartz is not stable in the presence of a CO_2-rich fluid, and at high grades even small amounts of CO_2 will cause reaction [6.15] to proceed to the left (curve *b*, Figure 6.8). The release of CO_2 that accompanies the prograde metamorphism of these rocks should, if the fluid composition were internally buffered, lead to a relatively CO_2-rich fluid phase and preclude the possibility of zoisite being present at the higher-grade zones. The growth of zoisite with increasing temperature can therefore only be accounted for by infiltration of water.

Chemical Changes During Metamorphism of the Vassalboro Formation   The cal-careous layers of the Vassalboro Formation not only change mineralogically with increasing metamorphic grade; their chemical composition also changes. These chemical changes are more marked than was the case for most of the marbles at Alta. Higher-grade rocks are depleted in K and Na in particular (compare the assemblages of the ankerite zone, with abundant albite and muscovite, with those of the higher-grade zones which lack sodic and potassic phases). This depletion may be due to the effects of infiltrating water and is a phenomenon that appears to be quite widespread in calc-silicates.

How much water is required to produce the observed metamorphic zones in the Vassalboro Formation has been the subject of considerable discussion. Large water fluxes have been estimated based on a number of assumptions, including that all chemical change was a result of introduction or removal of material by migrating fluids (Ferry 1994). Later investigations suggested that diffusion between layers could also account for mineralogical differences and so not all of the chemical changes required fluid movement (Penniston-Dorland & Ferry 2006). So far, geochemical estimates of fluid flux have generally not been reconciled with physical models for the sources of fluid and driving mechanisms for flow (Yardley 2009).

6.2.2   Why Do Carbonate Rocks Document So Much Fluid Flow?

The rocks described in this chapter are very different from pelitic and metabasic rocks in that they have commonly changed composition as a result of fluid flow in the course of deep metamorphism, not just during shallow diagenesis or geothermal circulation. What is special about many of the reactions described in this chapter is that the volume of the solid products is much less than the volume of the reactants.

For example, the molar volume of calcite is 36.93 cm^3, that of quartz is 22.69 cm^3 and that of wollastonite is 39.83 cm^3. Reaction [6.1] therefore produces just 39.83 cm^3 of solid products from the breakdown of 59.62 cm^3 of reactants. If the reaction is slow, the reacting rock will collapse as the volume of solids becomes smaller. At faster reaction rates, however, the rock may develop porosity (Figure 6.15). In this case, if there was no collapse at all, the porosity would make

Figure 6.15
Cathodoluminescence photomicrograph of a marble with tremolite (black) and calcite (orange/brown). The marble outcrop is shown in Figure 6.1b. The outer, brighter calcite zones pick out the euhedral form of calcite growing into open space. This transient metamorphic porosity was created by decarbonation through reaction [6.10] (Yardley & Lloyd 1989).

up 45% of the volume of the quartz and calcite that reacted. Calculations by Balashov & Yardley (1998) showed that, where infiltration of H_2O leads to an increase in porosity and permeability, this enhances the infiltration and so speeds up the process still further.

Most reactions that release CO_2 can produce secondary metamorphic porosity if they proceed sufficiently rapidly, such as when there is a fluid of different composition available nearby. Marbles at high temperature readily deform and so secondary porosity is only transient and seldom preserved, but Figure 6.15 illustrates an example. It is often the case, as we shall see in the following section, that the outer beds in limestone units are metasomatised while the interior of the unit still comprises carbonate rock.

6.2.3 Skarns

Metamorphosed limestones are often associated with coarse-grained bodies of silicate and/or oxide minerals which are usually mineralogically rather simple and have bulk compositions unlike any realistic sedimentary precursor. These bodies, which often replace carbonate rocks, are termed skarns, and the reasons for the simple mineralogy of rocks affected by extensive metasomatism are outlined in Appendix 3. A wide variety of skarn types exists, but individual skarns may usually be classified as one of magnesium-, calc- or magnetite skarns (Fettes & Desmons 2007). Many skarns preserve textural evidence, such as the development of minerals with well-developed crystal faces, suggesting that they have grown into open secondary pore space, as in Figure 6.16a. Skarns are particularly well-known from contact metamorphic aureoles, where they may form ore deposits, but can also develop during regional metamorphism. Most skarns develop as replacements of carbonate-rich beds, but skarns can also replace silicate rocks.

Figure 6.16a shows a skarn formed during contact metamorphism by a granite intrusion. The rocks here had undergone earlier regional metamorphism and the folded marble illustrated in Figure 6.1a, and which is from the same region, is probably representative of what this unit looked like prior to the intrusion. Marble beds range up to a few metres in thickness and, in the contact aureole, calcite only remains in the centre of the original beds. The outer parts are replaced by calc-silicate minerals with layers that are made up of c. 90% pale pink, grossular-rich garnet and which probably track individual beds in the original marble. The composition of these layers indicates that Si, Al and some Fe have been added, while some Ca has been removed, and it is reasonable to suppose that the mineral changes are the result of the passage of hot, aqueous fluid derived from the granite. Replacement and chemical change has been most extensive towards the edges of the original marble beds and the adjacent schists are rich in plagioclase, hornblende and clinopyroxene so that overall the skarns are zoned in a way that is reminiscent of the zoned calc-silicate at a marble-schist contact illustrated in Figure 6.1c.

(a)

2 cm

(b)

10 cm

Figure 6.16 Examples of metasomatic calc-silicate rocks or skarns. (a) Calc-skarn developed during contact metamorphism of regionally metamorphosed calcite marble (grey). Layers of pink grossular–andradite garnet parallel bedding; chalky white material is wollastonite. Connemara, Ireland. (b) Coarse diopside skarn (greenish-grey) interlayered with thin quartzite beds. The diopside rock is cut by white K-feldspar veins approximately perpendicular to bedding. Connemara, Ireland.

Figure 6.16b illustrates a rather different skarn where original dolomitic marble with thin quartzite interbeds has been transformed to an almost monomineralic diopside rock (Yardley 2009). The coarse diopside crystals are inferred to have grown initially as a result of the reaction of quartz with dolomite in the absence of water:

$$\text{dolomite} + 2\,\text{quartz} \rightarrow \text{diopside} + 2\,CO_2 \qquad [6.18]$$

In the absence of H_2O, quartz + dolomite remains stable to amphibolite facies conditions, and like reaction [6.1], reaction [6.18] results in a reduction in the solid volume, and so potentially creates porosity. In this bed, near the edge of the marble sequence, it appears that the influx of quartz-saturated water from surrounding schists added enough silica to allow all the remaining dolomite to react to diopside.

6.2.4 Geothermometers and Geobarometers for Skarns and Calc-Silicate Rocks

The widespread evidence for metasomatism accompanying metamorphism of these lithologies makes it very difficult to estimate metamorphic conditions from phase relationships, and in general metamorphic conditions are more readily estimated from interbedded lithologies that have not undergone metasomatism. There are, however, examples of minerals which have grown together from the same fluid and can be used for oxygen isotope thermometry, and in some situations cation-exchange or solvus geothermometers (Chapter 3) may be applicable.

6.3   Final Considerations

A final factor, which has not been considered in most studies of metamorphic fluids in calcareous rocks, is the extent to which CO_2 and H_2O are actually miscible. In the foregoing treatment of calcareous rocks it has been assumed that H_2O and CO_2 can mix to form a single fluid phase. In fact, although pure CO_2 and H_2O are miscible at temperatures above about 275 °C, the miscibility gap between them is greatly extended if dissolved salts are present in the fluid (as is almost invariably the case). Clearly, if two immiscible fluid phases are present, then from the phase rule the number of degrees of freedom will be reduced by one, and, for example, divariant equilibria will become univariant. To date, there is little evidence for the widespread occurrence of immiscible fluids in greenschist-facies or higher-grade marbles, but some examples have been reported. For example, immiscible $H_2O-NaCl$ and CO_2 fluids were found as inclusions in amphibolite-facies marbles in Switzerland and Italy (Trommsdorff *et al.* 1985). Furthermore, co-existing brine and CO_2-rich inclusions in coarse quartz and wollastonite formed during contact metamorphism of chert nodules have also been described (Heinrich & Gottschalk 1995).

Widespread examples of immiscibility in metamorphic fluids will have important implications for our understanding of physical fluid processes as well as for estimating conditions of metamorphism. Not only are the phase diagrams presented in this chapter all based on a single fluid phase, but the way in which fluids move through rocks is also very different if there are immiscible fluids rather than just a single phase.

SUMMARY

The study of calcareous metasediments emphasises the importance of yet another possible variable in metamorphism: the composition of the fluid phase. Despite this additional variable, marbles in regional metamorphic terranes often have mineral assemblages that vary regularly with temperature (and to a lesser extent with pressure) in much the same way as other metamorphic lithologies. Siliceous dolomitic limestones, including nodular limestones, generally develop new minerals in the same sequence, but there is often considerable variation in the full mineral assemblages present. Talc is not always developed, and dolomite + quartz can persist to high grades. This is because the conditions at which the reactions take place are dependent on the composition of the fluid as well as on P and T, and indeed hydrous phases such as talc and tremolite can only develop where water is able to infiltrate the marble.

In some metamorphosed sequences, the distribution of mineral assemblages and isograds can be accounted for if the fluid composition was controlled by the mineralogy of the rock or internally buffered, and did not therefore vary independently of pressure and temperature. This type of behaviour is to be expected if the porosity of the rock is small compared to the volume of fluid being released by reactions.

In contrast, some marbles and many calc-silicate rocks show evidence of introduction of fluid from an external igneous or metamorphic source, or external buffering. In some contact aureoles the mineral assemblages are similar to those that result from internal buffering but are relatively simple and the reactants and products of mixed-volatile reactions only co-exist in narrow zones. Where there has been more extensive infiltration of an externally-derived fluid, the rocks show evidence of metasomatism. Sometimes, calc-silicate rocks have compositions which vary with metamorphic grade and do not correspond to likely sedimentary precursors. In extreme cases of metasomatism associated with original carbonate rocks, the original rock composition has been transformed resulting in skarns composed predominantly of a very small number of minerals.

EXERCISES

1. Sketch an isobaric $T-X_{CO_2}$ curve for equilibrium between calcite + quartz + wollastonite + CO_2-H_2O fluid at a pressure of 100 MPa from the curves in Figure 6.2.
2. Annotate a copy of Figure 6.12a to show the likely range of temperatures of the talc, tremolite, diopside *or* forsterite and diopside *and* forsterite zones at a pressure of 200 MPa.

3. Why might the talc zone be missing in some metamorphosed dolomitic marble sequences?
4. The ferromagnesian minerals in metamorphosed dolomitic marbles are almost always close to the Mg end-member composition. If the original dolomitic limestone contained small amounts of illite clay, what sort of mica might you be likely to find in the equivalent marble after metamorphism?

FURTHER READING

Putnis, A. & Austrheim, H. (2010). Fluid-induced processes: metasomatism and metamorphism. *Geofluids*, **10**(1-2), 254–69.
Yardley, B. W. & Bodnar, R. J. (2014). Fluids in the continental crust. *Geochemical Perspectives*, **3**(1), 1–2.

7 Mineral Growth and Textures in Metamorphic Rocks

In the previous chapters, the emphasis has been on the attainment of chemical equilibrium in metamorphism, because it is only by identifying assemblages of minerals that have co-existed together in equilibrium that the pressures and temperatures of their formation can be determined. Equilibrium studies alone tell us only about the P–T conditions prevailing when a particular assemblage formed, they cannot tell us anything about the rock's history before or after the assemblage grew.

The study of the textures of metamorphic rocks provides a complementary line of evidence about how they developed. Textures are very important to the study of metamorphism because they often indicate deviations from equilibrium that allow us to see the way in which a rock was evolving during metamorphism. For minerals that are stable over a wide range of conditions, such as garnet, detailed investigation of their textures and chemical zonation can provide a treasure trove of information about metamorphic processes (Ague & Carlson 2013).

Metamorphic textures can be broadly divided into two types: those that preserve information about the metamorphic reactions that have taken place, and those that are related to deformation that occurred during metamorphism. The study of the first type is the main subject of this chapter, and it can reveal information about how mineral grains have formed and the factors that controlled their growth, all part of understanding the history of metamorphic conditions that a rock has experienced. Metamorphic textures reflect both the crystallisation of new minerals and the recrystallisation of existing ones. Recrystallisation can proceed independently of metamorphic reactions but can be triggered by them, so that for minerals with a wide stability field, recrystallised grains may combine new mineral matter with that which was present before.

In this chapter, we will treat metamorphic rocks as solid materials using approaches developed as much in materials science as in Earth science. This helps us to understand how the textures of metamorphic rocks reflect processes that took place during metamorphism. You will learn about the factors that control the growth of mineral grains and use this understanding to learn how to interpret metamorphic history, and the processes by which a metamorphic rock has developed, from its textures. Finally, we will consider what we can learn from textures about the factors which control the rates of metamorphic reactions.

7.1 Crystal Shapes and Alignment

The term 'metamorphic textures' covers all aspects of the shape and alignment of mineral grains in metamorphic rocks, and how they relate to the grains around them. As such, it is impossible to describe a metamorphic rock without referring to its texture, and some of the most commonly used terms were introduced in Chapter 1. Sometimes, textures of metamorphic rocks are inherited from the original protolith; for example, relic igneous pyroxenes may survive in low-grade metaba-sites (Chapter 5, Figure 5.7b), but most arise during metamorphism and reflect the distinct challenges for mineral grains (crystals) to grow in a solid rock with very limited porosity and tight grain boundaries.

Well-shaped crystals with planar faces (**euhedral** or **idioblastic** grains), common in igneous rocks, are relatively rare in metamorphic rocks. Most grains are irregular (**anhedral**) or have only weakly developed crystal faces (**subhedral**) but a few minerals do tend to form grains with a clearly developed crystal form, even in metamorphism (Figure 7.1). Garnet is probably the best known but tourmaline is also a widespread example and most of the pelite zone index minerals can form grains with a reasonably well-developed crystal shape when they occur as porphyroblasts (Chapter 1, Figure 1.3; Figure 7.1). In metabasites, amphiboles often develop crystal faces but grains tend to be elongate or **acicular** (Figure 5.7c), except at the highest metamorphic grades. Often, metamorphic rocks are made up of interlocking grains of similar size (e.g. Figure 5.7d, g). If the grains are more or less equidimensional and randomly oriented, as in many quartzites, this is known as a **granoblastic** texture (Figure 5.7d, f).

Of course, in many metamorphic rocks there is a strong alignment of the constituent mineral grains, resulting in schistosity. Most commonly, this is defined by phyllosilicate grains (e.g. Figures 1.11c, 4.2 and 4.7). Other minerals can also form aligned grains, giving rise to a schistose rock with a tendency to split in a particular direction. Amphiboles are commonly elongate, and can align to give rise to

(a)

(b)

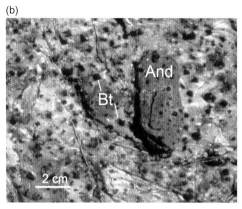

Figure 7.1 Examples of metamorphic minerals with an approximately euhedral habit. (a) Staurolite twins in graphitic garnet schist. (b) Coarse andalusite in muscovite schist. Note the abundant smaller, subhedral black porphyroblasts of biotite.

schistose metabasites (e.g. Figure 5.7c). If they align with their long axes parallel, the rock will also have a visible **lineation**.

It is unusual for metamorphic minerals to grow into open porosity, but this can happen in skarns (e.g. Figures 6.15 and 6.16a) and in low-grade meta-igneous rocks (Figure 5.7b). In such situations, well-developed crystal faces typically grow into the open space, but the side of the grain attached to the surrounding rock is usually irregular.

Metamorphic mineral grains are seldom pure; poikiloblasts, and most porphyroblasts, contain inclusions of other minerals that are usually present in the matrix also. Sometimes, inclusions are relics of minerals that were formerly present but have broken down elsewhere; an example has been illustrated earlier in Figure 2.10b. The presence of matrix inclusions in a porphyroblast means that the grain as a whole has a composition intermediate between that of the pure porphyroblast mineral and the rock in which it has grown. As a result, less movement and redistribution of chemical components was required to form it than would have been necessary to form a pure mineral grain.

In any one regional metamorphic suite and lithology, the grain size normally increases with metamorphic grade, but there are sometimes large differences in grain size between rocks of the same grade from different metamorphic belts. For example, the staurolite–garnet schists of Connemara, Ireland, generally have garnet and staurolite porphyroblasts less than 2 mm across, with many less than 1 mm. In contrast, rocks of this grade in many other areas have grain sizes that are much coarser, as is the case for the staurolite grains illustrated in Figures 1.1a or 7.1a.

Grain size and inclusion density are features of the rock that are likely to reflect the rate and duration of metamorphism, and this will be explored further in Section 7.5.

7.2   How Crystals Grow

Textures tell us about the history of metamorphic rocks, and they can also provide clues about whether mineral assemblages truly formed at equilibrium, but to use them effectively it helps to understand something about how crystals are able to grow. The study of equilibrium mineral assemblages is based on the thermodynamic properties of minerals treated as infinite perfect crystals, but real crystals are not perfect and, as a result, they have additional free energy. In extreme cases, this affects their stability. In early experimental studies, finely-ground mineral grains were found to react more rapidly than coarser grains, making experiments easier to run, but unfortunately it also transpired that they broke down under different conditions from those required in nature, because they were made less stable by the grinding. To understand the growth and breakdown of crystals during metamorphism it is necessary to take into account how textural features of grains affect their Gibbs free energy (Chapter 3), and to consider precisely how atoms may move through the bulk of a rock, or through its constituent mineral grains. These concepts were first developed in solid-state physics and have proved invaluable to geologists. Further information is provided in Boxes 7.1 and 7.2.

BOX 7.1 Surfaces and defects – their influence on grain energy

Real mineral grains do not have a perfect arrangement of atoms extending to infinity. Where there are flaws in the structure, or where atoms are situated at the edge of a grain, there will be an additional energy contribution to the overall energy of the grain reflecting the distortion of the bonds around these imperfections. The additional energy may affect the stability of specific grains and how they react relative to other grains of the same mineral nearby.

Energies of Crystal Surfaces

Atoms at or near the surfaces of a crystal are not bonded in such a stable way as those in the interior; they have unsatisfied bonds and, as a result, there is an excess energy associated with them, known as **surface energy**. Surface energy can be defined as the energy needed to increase a surface by unit area, and is therefore measured in units of J/m^2. The presence of this extra energy results in a tendency for foreign atoms or molecules to be attracted to mineral surfaces and loosely bound to them or **adsorbed**.

The magnitude of the energy associated with a surface depends on the nature of the substance on each side. For example, a surface between a mineral and air may have a higher surface energy than one between the same mineral and water. Surface energies usually make only a very small contribution to the total Gibbs free energy of a mineral in a rock, and so they are unlikely to affect the P–T conditions for equilibrium between mineral assemblages. However, when grains are extremely small, a high proportion of their constituent atoms occur near surfaces, and the surface free energy then becomes significant. A very fine-grained mineral assemblage will have a larger free energy than a coarse-grained assemblage, and this could affect the temperature and pressure at which a reaction takes place. For a sphere of quartz of 1 cm radius the surface free energy $\sigma \cong 6.3 \times 10^{-4}$ J; if the same mass of quartz is broken up into spheres of radius 10^{-3} cm, then $\sigma \cong 0.63$ J, while if it is reduced to spheres of 10^{-7} cm radius, $\sigma \cong 6280$ J and surface energy will make a significant contribution to the total free energy of the quartz. The effect of surface energy is particularly important when new mineral grains first begin to grow, because it is inevitable that at first most atoms will be near a surface.

Defects in Crystals

Mineral lattices inevitably contain imperfections within individual crystals as well as at their surfaces. These imperfections are termed **defects**, and can be grouped into three classes: point defects, line defects and surface imperfections.

Point defects are usually centred around a single lattice site or a pair of sites and may be due to: (a) an empty site or **vacancy**; (b) **substitution** of an impurity atom in place of the species that would normally occupy the site; or (c) **insertion** of an extra atom into a hole in the structure that would not normally be occupied.

Line defects or **dislocations** are one of the most important types of defect and play a major role in facilitating the deformation of grains. Dislocations can be visualised as resulting from the slip of one block of the lattice past another along a plane. Imagine a prism made up of a single crystal of a mineral with a simple cubic lattice as shown in Figure 7.2a. A cut is made along its length from the outside to the centre and represents the **slip plane**. If the lattice on one side of the slip plane is pushed up by exactly one unit cell dimension there will be no disruption across the slip plane; however, the ends of the prism are now spiral ramps instead of

(a) (b)

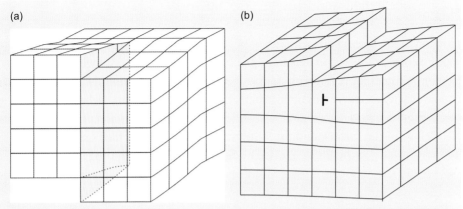

Figure 7.2 Schematic representations of dislocations in a crystal with an idealised cubic lattice. (a) Screw dislocation emerging from the upper surface, the plane on which the lattice has slipped is shown in yellow. (b) Edge dislocation with an extra plane of atoms in the right half of the block. The core of the dislocation is marked with "⊦" where it emerges at the front face.

planar surfaces. Near the centre of the prism, where the slip plane comes to an end, the lattice must be somewhat distorted and this zone of distortion extends along the entire length of the prism, hence the classification as a line defect.

The type of dislocation shown in Figure 7.2a is known as a **screw dislocation**. An alternative type of dislocation is illustrated in Figure 7.2b, whereby an extra plane of atoms is inserted in the right-hand half of the block which is absent from the left. The lattice is distorted along the edge of this extra 'half plane' and this constitutes an **edge dislocation**. Real dislocations can be intermediate between these two end-members and can change character and orientation along their length.

If the bonds around a dislocation break and reform, the dislocation effectively moves through the crystal. Progressive creation and movement of dislocations through a crystal allow it to deform while breaking only a small number of atomic bonds at a time, and this is why real crystals that contain dislocations are very much weaker than would be the case for perfect crystals.

Surface imperfections include any change in the orientation or composition of atom planes across a planar boundary. The most extreme type is the **grain boundary**, which separates crystals with different lattice orientations, and often of different mineral species. The ordinary grain boundaries in rocks have very little match across them but some boundaries between grains of the same mineral species involve only a slight atomic mismatch and are known as **low-angle** or **tilt boundaries**. They are most often seen in quartz and olivine in thin section, and appear when a large and apparently uniform grain is viewed in cross-polars and rotated to extinction. The large grain may then prove to be made up of a mosaic of smaller grains, or **sub-grains**, each of uniform but slightly different extinction position.

Much more detailed information about crystalline materials is available in the texts included as Further Reading.

7.2.1   Nucleation and Growth of Grains

Just because a particular mineral offers the lowest energy configuration that a group of atoms can assemble itself into, does not mean that crystals of it will form spontaneously. Supersaturated solutions, including undercooled melts (melts cooled below their melting temperature), exist because crystals do not always grow under conditions where thermodynamics tells us they should. This is because, when a small number of atoms first comes together to make a new crystal, most of them are at or close to the surface and are a source of additional **surface energy** (Box 7.1). Before a grain can begin to grow, enough atoms must come together to make a sufficiently large starting crystal for the energy saved by the reaction to compensate for the extra surface energy. This starting crystal is known as a **stable nucleus**.

Consider a simple reaction with a single mineral product, initiating at some temperature above the equilibrium temperature. We can define the drop in the **bulk Gibbs free energy** of this system, ΔG_V, as the difference between the Gibbs free energy of the reactants and products. The value of ΔG_V depends on the amount by which the equilibrium temperature has been **overstepped**, and is zero at the equilibrium temperature. This is illustrated in Figure 7.3a. The further the temperature of reaction is from the equilibrium temperature, the bigger the drop in bulk free energy accompanying any reaction. Figure 7.3b shows the bulk energy saved by a simple reaction at two temperatures, T_1 and T_2, as a function of the size of the nucleus of the product mineral. Gibbs free energy is an extensive parameter, so the bigger the crystal, the more energy saved. However, there is also an increase in surface energy from growing the new crystals, and this is also represented schematically on Figure 7.3b. The net free energy change from the reaction is the sum of the changes in surface energy and bulk energy, and varies with the size of the newly-formed nucleus. Once this is larger than a critical radius, an increase in its size results in a decrease in the net free energy. At this point it is stable and will continue to grow. Comparing the curves for T_1 and T_2 on Figure 7.3b, the further from equilibrium that the reaction is taking place, the smaller the size of the stable crystal nucleus, and therefore the fewer atoms that must come together to make it.

Reactions that involve the formation of a new crystalline phase (and hence require it to nucleate) do not take place until the equilibrium temperature has been passed (or overstepped), whereas reactions that add to existing grains or produce only phases with no definite crystal structure, such as melts, can take place at almost exactly the equilibrium temperature because no nucleation step is needed.

It will also be easier to form a stable nucleus if the surface energy of the product mineral can be reduced, i.e. the surface energy curve on Figure 7.3b is lowered. This is the case if the nucleus forms on a substrate to which some aspects of its structure have a close match, and is known as **heterogeneous nucleation**. In metamorphism, new minerals often nucleate on pre-existing ones in a particular orientation so that there is the best possible match between the two lattices. Growth of a crystal over a

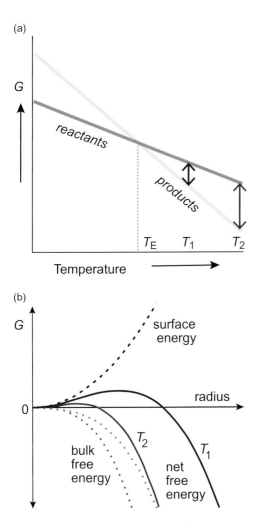

Figure 7.3 Schematic representation of the competing energy changes that lead to the formation of a stable nucleus from a simple reaction with one mineral product. (a) G–T plot to illustrate the variation in the bulk Gibbs free energy of reactants and products. The energy difference, ΔG_V, increases away from equilibrium (T_E) and T_1 and T_2 are two arbitrary temperatures used in part (b). (b) Changes in free energy with the radius of the nucleus of the product mineral. New surface energy is shown by the dashed black line and is assumed independent of temperature. Dotted lines show the reduction in the bulk Gibbs free energy as a result of the reaction and solid lines show the net free energy change from the addition of the other two curves. Colours are coded for T_1 and T_2 from part (a).

pre-existing one in a specific orientation is known as **epitaxy**, and arises as a result of **epitaxial nucleation**. The tendency for sillimanite to grow in specific orientations on biotite (Figure 2.10d), is an example of this effect. It is likely that epitaxy is a more common phenomenon than is usually recognised. For example, epitaxial nucleation of garnet on biotite has been documented in Italy (Moore *et al.* 2015). In extreme cases, a product mineral shares a significant part of the lattice with the

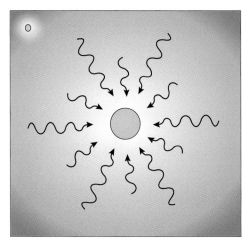

Figure 7.4 Schematic representation of the balance between transport, growth and further nucleation during reaction. Colour intensity reflects the notional concentration of a mineral component. Arrows denote diffusion of material towards an initial grain, outlined in black. In this scenario, the rock is depleted in the mineral component around the growing grain, but not at a distance. As a result, a new nucleus has formed in the top left corner.

mineral it is replacing and so even less disruption of the lattice is necessary. This is known as **topotaxy**, and common examples include the replacement of intermediate plagioclase by albite and the retrograde replacement of biotite by chlorite.

Once a stable nucleus has formed, there is a chemical potential gradient between the reactant grains and the product nucleus, and this will drive transport of material towards the newly formed grains of the product minerals, allowing **growth**. In regions of the rock that are too far away from the initial nuclei for material to migrate to them, additional nuclei of the product grains are likely to form (Figure 7.4). If growth is relatively slow while nucleation is rapid, the result is a fine-grained rock; more growth on the initial nuclei leads to a coarser grain size. What determines the relative rates of nucleation and growth depends on imposed factors, such as heating rate, the mineralogy of the rock itself, and on the presence and properties of a fluid phase, a major factor controlling movement of atoms in the rock.

The shapes in which grains grow are influenced by how new atoms attach to their surface, and how easily atoms can move through the rock (see below, Section 7.2.3). Once the product minerals begin to grow, different factors influence their shape. Some grow to keep surface area to a minimum, especially if surface energy is large and the degree of overstepping is small, and this leads to crystals with a more nearly euhedral shape and few inclusions. Often the shape of a grain is dictated by the ease with which atoms can attach to the growing crystal. For example, if atoms attach to one crystal face more easily than to others, that face will grow faster and the grain will become elongated. Imagine a small square, representing a crystal nucleus. If all the faces grow at the same rate, the crystal remains square, but if opposite faces grow at different rates, one pair will still be close to the initial nucleus while the other pair are distant; in other words, the crystal will now be elongated, with the small faces at the ends being the fastest growing ones since they are now furthest from the starting nucleus. Poikiloblastic textures imply that reaction was sufficiently overstepped to compensate for the energy of the additional surface area, while the included material means that growth was not constrained by removal of superfluous material by diffusion.

7.2.2 Grain Boundaries

The boundaries between grains play an important role in metamorphic transformations, because they allow relatively rapid movement of material through the rock and they are the surfaces to which atoms are added or removed as grains grow or break down. Any fluid phase forms a film within the three-dimensional grain-boundary network, and greatly enhances transport, but even without a fluid, atoms along grain boundaries are potentially more mobile because they are less strongly bonded than those within grains. Despite their obvious importance, it has proved very difficult to get information about the precise nature of grain boundaries during metamorphism. Scanning Electron Microscopy (SEM) and Transmission Electron Microscopy (TEM) are two instrumental techniques with the power to image grain boundaries, but the development of a Focussed Ion Beam (FIB) technique to prepare exactly the right part of a sample for electron microscopy has been an important complementary development (Wirth 2009).

SEM imaging studies of quartz grains in water-saturated amphibolite-facies mylonites show grain boundaries pitted by depressions (Figure 7.5) (Mancktelow & Pennacchioni 2004). In contrast, quartz grain boundaries in mylonites formed

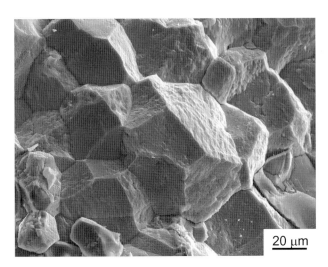

Figure 7.5 SEM image of a foam-textured quartzite mylonite. The quartz surfaces are covered in small pits, inferred to represent the sites of fluid-filled pores. Mylonites formed under dry conditions generally lacked such pitting. From Mancktelow & Pennacchioni (2004).

20 μm

under notionally dry conditions generally lack pores except where other minerals were also present. In view of this, the pits were interpreted as representing interconnected fluid-filled pores, formed during metamorphism.

In some rocks, the grain boundaries are likely modified after the peak of metamorphism and the features that can be observed today mainly developed during cooling and exhumation. For example, one combined TEM and FIB/SEM study showed significant porosity along quartz grain boundaries in both contact and regional rocks (Kruhl *et al.* 2013). The data suggested that porosity developed as rocks cooled and exhumed due to differential volume changes of the adjacent minerals, which allowed water to enter.

These types of observations provide an invaluable insight into the nature of grain boundaries in crystalline rocks, but still leave open the issue of what the grain boundaries were like under peak metamorphic conditions, particularly when the grains themselves were participating in reactions.

Irrespective of their structure, grain boundaries are a source of energy associated with every mineral grain, and while the energy differences between grains of different shapes and sizes are usually small, surface energy can be a driving force for recrystallisation. In very high-grade rocks, and in marbles, grains have often recrystallised to shapes which minimise surface area. They are often of similar size, and more or less equidimensional, while the grain boundaries themselves are planar rather than irregular. This is known as **granoblastic polygonal** or **foam** texture (Figures 5.7f and 7.5) and contrasts markedly with the complex grain shapes with large surface areas that are often present at lower grades.

7.2.3   Transport

Most reactions leading to the growth of new metamorphic minerals require the transport of atoms from the sites where reactant minerals are breaking down to those where product minerals are growing. Provided the different phases in a rock (the fluids, melts and minerals) are not moving relative to one another, then atoms are transported through the rock by **diffusion**. Diffusion arises from the random movements of atoms in the presence of a chemical potential gradient, of which the simplest type is a concentration gradient in a solution. As a result of the gradient, there is a net movement of atoms in a particular direction. In contrast, if there is a fluid or a melt moving relative to the minerals, and it carries atoms with it, these are said to move by **advection**.

In broad terms, **volume diffusion,** through solid mineral grains, may be distinguished from **grain-boundary diffusion**, which includes both diffusion through tightly-fitting grain boundaries as well as through the **grain-edge fluid** (Watson & Baxter 2007) (Figure 7.6). These pathways for diffusion differ in how tightly the moving atom is held in place: where an atom is more-securely

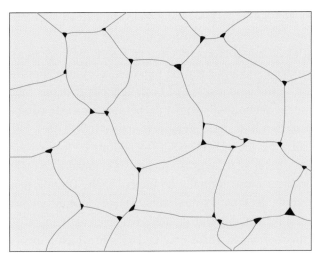

Figure 7.6 Schematic illustration of diffusion pathways through a crystalline rock. The bulk of the rock is made up of minerals through which volume diffusion is possible, but likely to be very sluggish (light blue). Grain boundaries offer somewhat faster diffusion, but make up only a very small proportion of most rocks. The fastest diffusion pathways are through a grain-edge fluid (dark blue), and these dominate the transport properties of rocks provided fluid is present. Based on Watson & Baxter (2007).

bonded, it requires more energy to allow it to break free and move than if it is only loosely attached. The stronger an atom is held, the higher the **activation energy** for diffusion. Volume diffusion has the highest activation energy, and in most minerals it is not an important process until high temperatures are attained (Box 7.2). Under all but the highest temperature metamorphic conditions however, the most important factor enhancing diffusion is the presence of an aqueous fluid in pores along grain boundaries (Figure 7.5) and especially along grain edges (Figure 7.6). Not only is it easier for atoms to detach from mineral surfaces into water, but atoms present in solution can readily diffuse through the pore water network. This mechanism is not available if the rock does not contain an aqueous fluid phase.

Diffusion accompanying metamorphism of a homogeneous rock layer typically allows the growth of progressively coarser metamorphic minerals, so that material is redistributed within the layer but the composition of the layer overall does not change. As the grain size gets coarser, progressively larger samples are needed to be representative of the composition of the layer as a whole. Sometimes, porphyroblasts are surrounded by a zone of distinct composition, depleted in the constituents of the porphyroblast. This is likely to be an indication that diffusive transport through the rock mass was relatively sluggish and restricted porphyroblast growth (as represented in Figure 7.4). Diffusion is also important where two rocks of contrasting composition are in contact. For example, diffusion between limestone and pelite under medium- to high-grade conditions results in reactions that produce calc-silicate minerals at the interface (Figure 6.1c). The limits on how far material can be transported by diffusion during metamorphism are not known with any certainty. However, experimental data suggest that diffusion through water-saturated crystalline rocks can extend over tens of metres on a million-year timescale if there is a sufficient driving force.

Advection has the potential to be much more effective at modifying rock composition than diffusion, and we have already seen the importance of advection in geothermal systems for modifying the composition of basic igneous rocks as they undergo metamorphism (Chapter 5, Section 5.6). Under deeper metamorphic conditions, where fluid pressure approaches lithostatic pressure (Chapter 1, Section 1.4.3), water cannot recirculate; it can only move irreversibly to regions of lower fluid pressure. The high fluid pressures (relative to hydrostatic pressure) mean that rocks undergoing prograde metamorphism must normally be impermeable. Despite this, examples of large-scale advection, leading to metasomatism, do occur but are associated with specific structural or metamorphic changes which create temporary permeable pathways through the rock mass. These may be zones in which porosity has been increased as a result of metamorphic reactions driving a decrease in solid volume, such as those which break down carbonate minerals (Chapter 6, Section 6.2.2). Alternatively, they may be faults or shear zones.

BOX 7.2 Diffusion in solids

Diffusion is the process by which atoms, ions or molecules are transported through matter. Even in a crystalline solid, where atoms are strongly bonded, the continuous thermal vibrations mean that individual atoms are in motion, exchanging positions within the crystal. These random motions within a chemically homogeneous crystal may be detected by using a distinctive isotope of one of the constituent elements as a tracer, and are known as **self-diffusion**. Where compositional gradients are present, for example in a zoned grain of a solid-solution mineral, there will be a tendency for volume diffusion to occur to make the grain homogeneous. In the case of a zoned olivine crystal for example, Fe will tend to diffuse away from the fayalite-rich portions towards the parts enriched in forsterite, while Mg will tend to diffuse in the opposite direction. The opposing movements must be exactly balanced to prevent the local development of charge imbalances.

Volume diffusion attempts to eliminate concentration gradients within the crystal. The rate at which diffusion takes place is measured in terms of the **flux** of matter, J. This is the mass of material diffusing across a unit area of an imaginary surface within the crystal in unit time. The flux is proportional to the concentration gradient that is driving the diffusion, so if we represent concentration by C and the direction in which C changes is designated x we have:

$$J \propto dC/dx \qquad\qquad [7.1]$$

If we designate a constant of proportionality, D, then:

$$J = -D \cdot (dC/dx) \qquad\qquad [7.2]$$

The constant D is known as the **diffusion coefficient** and Equation [7.2] represents Fick's first law of diffusion. In many instances where diffusion occurs, the concentration gradient dC/dx changes as diffusion proceeds, and many equations have been developed to express this and measure changes in composition with time in bodies of different shapes.

The amount of material that can be moved by diffusion depends on both the time available and the size of the diffusion coefficient D. In minerals, D is very strongly temperature dependent and, as a result, it is possible for diffusion within a crystal to be negligible at one temperature, but very rapid, in geological terms, if the temperature is raised by only 50 or 100 degrees. For example, garnet crystals in pelites from the garnet and staurolite zones typically display strong chemical zonation in the major elements which developed as they grew over a range of conditions, whereas those from migmatites are more or less unzoned (Section 7.4.1). Diffusion of the major elements in garnet commonly becomes effective at the temperatures of high-grade metamorphism (Carlson 2006).

7.3   Crystallisation Textures

The mineral textures that develop as a metamorphic reaction progresses are dependent on a number of factors, including the way in which new grains nucleate, the ease of diffusion through the rock matrix and the degree of overstepping of the equilibrium conditions. Here we explore briefly how different factors can control the size and shape of metamorphic minerals.

7.3.1   Influence of Nucleation Characteristics

The way in which mineral products nucleate has an important impact on the rock's final texture. In the simplest cases, reaction products need not form a new nucleus on which to grow because the same mineral is already present in the rock. The new material may form a distinct overgrowth or rim to the pre-existing grains of the same phase (Figure 7.7), or new and old material may recrystallise together to form new grains. In some instances overgrowths can be readily identified, but in others their presence is much more equivocal.

We have seen that nuclei form most readily on a substrate where there is a close match to the structure of the new mineral. Where favourable nucleation sites are

Figure 7.7 Example of a zoned amphibole crystal, As Sifah, Oman. The zoned blue core is of glaucophane (the zoning reflects Fe^{3+} concentrations). The darker, bluish-green rim is of an Na–Ca amphibole, intermediate between actinolite and glaucophane, which has formed an overgrowth on the earlier amphibole.

common, a mineral product may form large numbers of small grains rather than a few large ones. This style of nucleation gives rise to schists in which the sillimanite occurs as numerous long, thin fibres, known as fibrolite, rather than the distinct prismatic crystals formed by kyanite or andalusite (Figure 2.10b and d). In detail, there are specific orientations in the biotite structure which have a particularly close match to sillimanite, giving rise to the alignments seen in Figure 2.10d. Minerals that do not nucleate readily form isolated large grains: porphyroblasts or poikiloblasts. On this basis, andalusite appears to be an example of a mineral that is difficult to nucleate (e.g. Figure 7.1b). A wide range of minerals form porphyroblasts in some circumstances but not others. Biotite and plagioclase can be either porphyroblast or matrix minerals in metapelites, and likewise hornblende and lawsonite may occur in either manner. In many high-grade rocks, minerals such as garnet that we think of as forming porphyroblasts, actually form grains similar in size to matrix minerals.

7.3.2   Growth and Dissolution Characteristics

Minerals grow by adding atoms to their surfaces, and so grains with large surface areas can grow more quickly than equidimensional grains, even though they are less stable. Where a reaction is sufficiently overstepped, the small energy differences between different grain shapes are insignificant, and minerals may grow in forms that are dictated by convenience and kinetics. Examples include growth as dendritic crystals or as poikiloblasts with a network of inclusions, both of which have a large surface area for atoms to attach to and so increase the growth rate. As already noted, less movement of material is required for minerals to grow with abundant matrix inclusions rather than as pure crystals, and so poikiloblastic textures reflect a balance between growth and transport, providing a large surface area for attachment of atoms, while ensuring that the product mineral grows through the rock mass, minimising transport distances.

Metamorphic mineral growth requires complementary dissolution of other minerals, and similar considerations can apply. Mineral surfaces typically become pitted as the mineral breaks down and dissolves into pore fluid (e.g. Figure 7.5), with the pits sometimes related to defects in the mineral structure. With further breakdown, mineral surfaces often become distinctly corroded. It is generally the case that growing surfaces of grains have a 'convex outward' form, often with planar surfaces, whereas reactant grains become irregularly corroded as they break down, thereby increasing the surface area from which atoms can be released.

Some crystals grow more rapidly on some faces than on others, and this influences the final shape of the grain, as described earlier. Anisotropic growth is especially important in determining the textures of metamorphic rocks at low grades, where minerals such as amphiboles or pumpellyite are frequently acicular and may occur as bundles of radiating crystals (Figure 5.7b). At higher grades most minerals form more or less equidimensional grains (see Figure 5.7d).

7.4   Disequilibrium Textures

The textures of many metamorphic rocks preserve evidence of deviations from equilibrium that are of great importance in interpreting the metamorphic history of the rock. These were introduced in Chapter 2, Section 2.9, and may take the form of the local preservation of minerals or mineral compositions that were once present throughout the rock, but broke down subsequently. Alternatively, textures may preserve evidence of incomplete reaction between minerals, providing an insight into the reactions that took place. Note, however, that incomplete reaction does not always indicate disequilibrium, because many reactions are continuous and will progress over a range of P–T conditions.

7.4.1   Chemical Zonation in Minerals

Solid-solution minerals often form grains that vary in composition between the core and rim. For some minerals, such as tourmaline or amphibole, chemical zonation is accompanied by colour differences that are readily observed under the microscope (e.g. Figure 7.7). In others, zoning can only be detected by chemical analysis of a series of points across the crystal using an electron microprobe analyser, or another imaging technique that is sensitive to chemical composition (e.g. Figure 7.8). In recent years, improved technology has allowed the zoning of trace elements to be studied alongside that of the major elements in a mineral, and this can provide new insights into how mineral assemblages have evolved (e.g. Konrad-Schmolke *et al.* 2008; Regis *et al.* 2016).

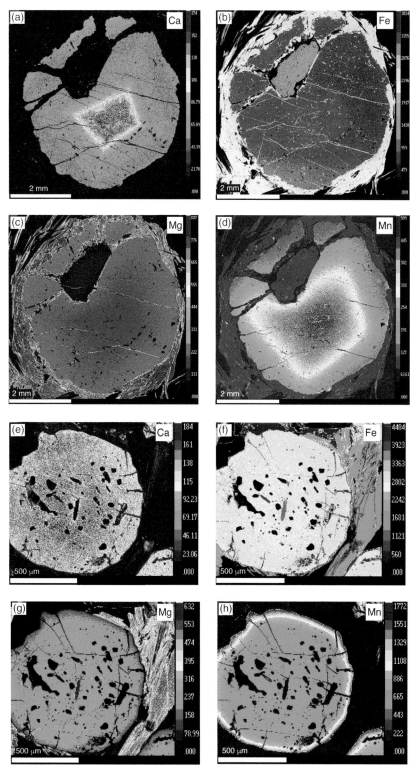

Figure 7.8 False colour images of contrasting styles of chemical zoning in two garnets from pelitic gneisses in Bhutan (Mottram *et al*. 2014, 2015). Each part is a map of the distribution of a single element,

Zoning may form in a variety of ways, but the fact that it is present means that, at best, the zoned grain has only partially equilibrated with the other minerals in the rock. Material in the centres of zoned grains is effectively removed from the rock and is not available to react with the matrix.

Growth Zoning This is zoning that develops as a crystal grows, most typically during progressive metamorphism, and hence if it is preserved today we can conclude that, even at the peak of metamorphism, volume diffusion in the particular mineral was too sluggish to eliminate the original compositional gradients. The zoned amphibole shown in Figure 7.7 preserves early-formed amphibole compositions beneath later overgrowths.

Chemical zoning in garnet provides a record of an extended period of metamorphism. Zoning in garnet is not apparent optically; element concentration data are instead collected by electron microprobe. Garnets in medium-grade metapelites often exhibit growth zoning, and an example is illustrated in Figure 7.8a–d. Cores are characteristically rich in Mn, which is fractionated into garnet when it first grows. The distribution coefficient (K_D) (Chapter 2, Section 2.4.3) for partitioning of Mn between garnet and the other Fe–Mg–Mn minerals is invariably large, which means that Mn is preferentially concentrated into garnet once it starts growing. As a result, the small amounts of garnet that appear first scavenge Mn from the rest of the rock and are Mn-rich. As garnet growth continues, the Mn-content of the reactant phases remaining in the matrix becomes much lower, and therefore the Mn-content of the later-formed parts of the garnets is also lower than that of first-formed garnet. This model for Mn-rich garnet cores is sometimes termed Rayleigh Fractionation, based on the model for distillation of liquids.

Although Rayleigh Fractionation can explain the formation of zoned garnets at constant temperature, in nature garnet is produced over a range of temperatures, leading to further, subtle growth zoning, because the K_D values for Fe–Mg and Fe–Mn exchange between garnet and other minerals are temperature sensitive (Chapter 3, Section 3.3.5). For example, the Mg/Fe ratio may increase in the parts of garnet formed at higher temperatures (compare Figure 7.8, parts (c) and (d)).

Not all growth zoning requires such extensive investigation before it can be interpreted. We have already come across examples of amphibole grains in which a core of sodic amphibole is rimmed by a margin of actinolite or hornblende (Figure 5.3). This implies that metamorphism progressed from blueschist-facies conditions to greenschist-facies conditions.

Figure 7.8 (*cont.*) coloured to show relative concentrations. The warmest colours are the highest concentrations in each image. (a)–(d) Growth-zoning patterns. Note that the core is distinctly rich in both Ca and Mn (parts (a) and (d)), and retains the original euhedral form. A subtle increase in X_{Mg} is apparent from comparison of parts (b) and (c). (e)–(h) Retrograde-zoning patterns. The distribution of all the illustrated elements is very uniform across most of the grain, but the rim is strongly enriched in Mn and depleted in Mg (parts (h) and (g) respectively).

Retrograde Diffusion and Mineral Zoning Zoning patterns may also result from volume diffusion in minerals after they first grew. For example, garnets from very high grades (upper sillimanite zone and above) are typically more or less homogeneous because they form at temperatures at which volume diffusion in garnet appears to be effective (Figure 7.8e–h). However, the rims of high-grade garnets often exhibit a certain amount of zoning which is believed to form as a result of diffusive exchange with matrix minerals during cooling, and is known as **retrograde zoning**. As the rock cooled, diffusion in garnet became less effective and only a progressively narrowing outer rim was able to equilibrate with the rock matrix (Figure 7.8g, h). The homogeneous core is unlikely to preserve the composition at the peak of metamorphism because if diffusion was rapid when the grain began to cool, the composition of the entire garnet would have been modified by exchange with the matrix.

7.4.2 Relic Minerals

Relic minerals are remnants from an earlier stage in the metamorphism, preserved within later porphyroblasts after they have completely broken down in the rest of the rock. For example, a relic staurolite inclusion in plagioclase in Figure 2.10b is from an upper sillimanite zone schist. Staurolite is no longer present in the rock matrix because metamorphic conditions exceeded those at which staurolite can co-exist with quartz (Chapter 4, reaction [4.12]), but inside the plagioclase there is no quartz to react with it and it remains stable. Other common examples include inclusions of chloritoid in garnet or plagioclase from staurolite schists, or inclusions of glaucophane or actinolite in plagioclase from metabasites with hornblende in the rock matrix. Most relic minerals owe their preservation to the sluggishness of volume diffusion through the host porphyroblast, which serves to isolate them from phases in the rock matrix with which they would otherwise react. An exception is the preservation of relic coesite in garnets formed during *UHP* metamorphism discussed in Chapter 5, Figure 5.12, since no chemical change is involved in the coesite–quartz reaction.

7.4.3 Determining Metamorphic *P–T* History for Rocks with Disequilibrium Textures

It has been an assumption throughout this book that, after careful petrographic observation, we can identify metamorphic mineral assemblages, and reach conclusions about the conditions under which they formed, by treating them as having formed at equilibrium. But if metamorphic rocks actually contain textural evidence for disequilibrium, to what extent is this assumption really valid? The textures just discussed have been examples of disequilibrium in which early-formed minerals do not completely react during the later metamorphism in which the surrounding minerals grew. Their existence means that careful observation is needed to interpret the rocks, but does not question the validity of treating the later assemblage as an equilibrium one.

If a rock's textures can be used to identify equilibrium mineral associations, some of the methods used to determine the *P–T* conditions of metamorphism, introduced in Chapter 3, can still be applied to rocks with disequilibrium textures in order to throw light on their metamorphic history. This is most straightforward where distinctive minerals present as relic inclusions clearly formed under different conditions from the assemblage of interconnected minerals. The inclusions of coesite or diamond as markers of *UHP* metamorphism (Chapter 5) are an excellent example. It may also be possible to use geothermometers and geobarometers to calculate specific temperatures and pressures for different stages in the metamorphic history, provided it is possible to recognise compositional zones in different minerals that formed together in equilibrium and have not been modified subsequently by diffusion. This is not always a straightforward requirement. Garnet may preserve compositional zones from early in its growth history, but other minerals, such as micas, experience diffusion down to much lower temperatures. A biotite inclusion in garnet is unlikely to preserve the composition it had when first trapped because of the likelihood of diffusive exchange of Fe and Mg with the immediately adjacent garnet. Nevertheless, *P–T* histories have been deduced from garnet zoning in a number of now-classic studies (Spear *et al.* 1984; St-Onge 1987), and the methods have been used to underpin many further regional investigations.

Under some circumstances, it is possible that the assumption that metamorphic rocks contain mineral assemblages that grew in stable equilibrium may not be true. A mineral or assemblage grown under conditions far-removed from equilibrium need not be the lowest free-energy combination of atoms predicted by equilibrium calculations, but simply one that is of lower energy than the reactants and easy to assemble. A common phenomenon in geothermal fields is the growth of chalcedony, a locally-disordered form of silica. It grows instead of quartz because the less-ordered structure makes it easier to nucleate and grow. It appears that there are metamorphic minerals that are relatively hard to nucleate and sometimes appear at higher temperatures than indicated by equilibrium calculations. A number of studies now report convincing examples of non-equilibrium growth (Castro & Spear 2017; Spear 2017; Waters & Lovegrove 2002) and explore the possibility that some assemblages do not form at equilibrium (Spear & Pattison 2017). The implication of these studies is that some of our methods for estimating conditions of metamorphism may be less accurate than we currently believe.

7.5 Rates and Mechanisms of Metamorphic Mineral Growth

We have already discussed how crystals grow and the factors that influence their textures, but for a reaction to proceed, the growth and dissolution of the participating minerals are coupled. Metamorphic textures can throw light on how a reaction proceeds overall and, by indicating which factors controlled the processes that took place, can also provide information about rates of reaction.

7.5.1   Solution-Reprecipitation

Metamorphic transformation is achieved by grains dissolving in some parts of the rock while others grow nearby (Ruiz-Agudo *et al.* 2014). Material moves between sites of dissolution and growth by diffusion in the pore fluid, even though the fluid phase makes up only a very small proportion of the rock mass and its composition probably remains rather constant. It is the mechanism of solution-reprecipitation, coupled to diffusion through the grain-boundary network, that allows grains to react with other minerals that are nearby but not in contact, and to take part in reactions under conditions where diffusion within grains is negligible. We have seen already that diffusion rates depend on temperature, and only at very high metamorphic grades does volume diffusion in minerals become extensive. Despite this, some reactions at very high grades are incomplete and involve partial local reaction between contrasting pre-existing grains (see Figure 5.1c, and Section 7.5.3 below). This is widely considered to be a hallmark of rocks from which a fluid phase was absent for much of the time, so that solution-reprecipitation was not very effective.

7.5.2   Coupled Reaction Cycles

The specific mechanisms by which particular reactions take place are often rather complex, involving multiple reactants and products. Although we can write metamorphic reactions indicating the growth of some specific mineral from a particular group of reactants, there is often no textural evidence to confirm that the reaction products have indeed grown from the supposed reactants. However, local replacements identified in one part of a thin section can in many cases be balanced by other replacements in adjacent parts of the sample so that the overall change corresponded to a simple metamorphic reaction (Carmichael 1969). The polymorphic transition from kyanite to sillimanite is an excellent example of this because sillimanite rarely grows directly from kyanite. Typically, kyanite is rimmed by muscovite as it breaks down, and this local transformation, supposing it takes place at a contact between kyanite and quartz, can be described by the reaction:

$$3\,\text{kyanite} + 6\,\text{quartz} + 2\,K^+ + 9\,H_2O \rightarrow 2\,\text{muscovite} + 2\,H^+ + 3\,Si(OH)_4 \qquad [7.3]$$

In this reaction K^+, H^+, and $Si(OH)_4$ are general ways to denote the occurrence of these components in solution in the pore-fluid phase. Since the reaction is not balanced in terms of conventional solid or fluid components and requires the participation of these ionic species, it is known as an **ionic reaction**. Note that the reaction has been balanced so that Al does not appear as an ionic component; this is an arbitrary choice, although it reflects its low solubility.

Once kyanite has been completely rimmed by muscovite so that it is no longer in contact with quartz, it can continue to break down by the reaction:

$$3 \text{ kyanite} + 3 \text{ Si(OH)}_4 + 2 \text{ K}^+ \rightarrow 2 \text{ muscovite} + 2 \text{ H}^+ + 3 \text{ H}_2\text{O} \qquad [7.4]$$

which is essentially the same as reaction [7.3], except that Si must now gain access to the reacting surface of the kyanite through the fluid phase. Suppose, however, that the rim of muscovite around the kyanite becomes sufficiently thick that the diffusion of K and Si through it is inhibited. The reaction at the kyanite surface might then be:

$$4 \text{ kyanite} + 3 \text{ Si(OH)}_4 + 2 \text{ K}^+ \rightarrow 2 \text{ muscovite} + 1 \text{ sillimanite} + 2 \text{ H}^+ + 3 \text{ H}_2\text{O} \qquad [7.5]$$

and this would produce rims of muscovite containing crystals of sillimanite. This is a common texture, known from the sillimanite zone in the south-eastern Highlands of Scotland (outlined in Chapter 1 and Chapter 4) and many other localities. These reactions require complementary ionic reactions in adjacent parts of the rock to act as the sources for the dissolved reactant components such as K^+ and as the sinks for products such as H^+. For example, if muscovite breaks down nearby according to the reaction:

$$2 \text{ muscovite} + 2 \text{ H}^+ \rightarrow 3 \text{ sillimanite} + 3 \text{ quartz} + 2 \text{ K}^+ + 3 \text{ H}_2\text{O} \qquad [7.6]$$

this leads to the replacement of muscovite by sillimanite + quartz, which is again a common texture in thin section, while releasing K^+ and consuming H^+. The occurrence of reactions [7.5] and [7.6] in nearby parts of a rock, coupled with diffusive exchange between these two domains and the solution and reprecipitation of quartz according to:

$$\text{Si(OH)}_4 = 1 \text{ quartz} + 2 \text{ H}_2\text{O} \qquad [7.7]$$

can account for the growth of sillimanite in different parts of a rock from those where kyanite is breaking down. The coupled reactions are summarised in Figure 7.9. Despite the complexity of the coupled reaction cycle, the overall reaction, obtained by adding reactions [7.3], [7.6] and $3 \times$ [7.7], is simply:

$$3 \text{ kyanite} \rightarrow 3 \text{ sillimanite} \qquad [7.8]$$

Other, comparable reaction cycles have since been described, usually from pelitic rocks. A local change in composition in one part of a rock is balanced by changes taking place simultaneously nearby. The precise reason why a reaction should proceed in such a complex fashion is not always clear, but it is probably related to the heterogeneous nucleation of the reaction products on particular substrates. For example, sillimanite does not nucleate on kyanite as readily as it does on micas, and this selective nucleation may drive the complex cycle of mass transfer shown on Figure 7.9.

(a)

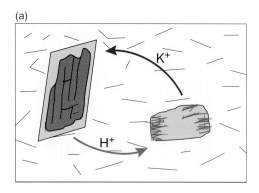

(b)

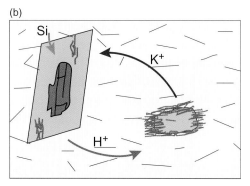

Figure 7.9 Schematic representation of an ionic reaction cycle linking breakdown of kyanite (blue) to growth of sillimanite (brown) in a quartz-bearing matrix (pale yellow). (a) Corroded kyanite mantled by muscovite (green) due to reaction [7.3] is complemented by the breakdown of a muscovite grain to sillimanite, reaction [7.6]. (b) More-advanced reaction; kyanite is now separated from quartz by muscovite but continues to breakdown by reaction [7.5], so that some sillimanite is present in the muscovite mantle.

7.5.3   Transport-Controlled Growth

The pattern of sillimanite nucleation illustrated in Figure 7.9 leads to much more chemical redistribution than would be the case if sillimanite simply replaced kyanite. Components can readily diffuse through the rock matrix between reactants and products, suggesting that the reaction is not limited by constraints of transport. This is not always the case for other metamorphic reactions.

In some instances garnet-forming reactions are constrained by the ease of diffusion of the reacting elements through the rock matrix (Carlson 2011). If diffusion through the rock matrix is relatively rapid, then once the first garnet nuclei have formed, components can migrate to them and no further nuclei are likely to form (cf. Figure 7.4). Because garnet grows from continuous reactions and its composition changes with time, then if all the garnet grains in a rock nucleated at the same time, they would all have the same core composition even if they ended up being of different sizes. In contrast, if diffusion through the rock matrix is sluggish enough to limit the growth of the first-formed garnets, new garnets will continue to nucleate as the reaction progresses, and the later-formed ones will have cores whose compositions are the same as intermediate zones of the first-formed grains; they are also likely to be smaller. Pioneering studies to look for such effects were carried out using painstaking mechanical approaches to determining the distribution of garnet grains of different sizes in a rock and separate them out for analysis (e.g. Kretz 1974). More recently, the application of micro-X-ray tomography to imaging rocks has greatly speeded up the process (Denison *et al.* 1997). Most studies of garnet report that small grains do indeed lack the first-formed zones encountered in larger crystals, implying that nucleation continued for much of the period of garnet growth, but there are also a few examples where all garnet grains appear to have nucleated at the same time.

Reaction Rims and Corona Textures   Reaction rims develop around a potentially reactive mineral grain which has survived metamorphism despite partially breaking down. The extent of reaction has been limited by the supply of reactants. Corona textures are a particularly well-developed type of reaction rim texture in which metamorphic minerals such as garnet develop at the interface between relic igneous minerals during later high-grade metamorphism (Figures 5.1c and 7.10a). Typically, rocks with corona textures preserve little or no evidence of their history of cooling, burial and metamorphism between the original emplacement of the pluton and the high-grade metamorphic event in which the corona formed. Because the reaction has been inhibited, it is possible to see the reactant phases and often to determine their original textural relationships. **Symplectite** is distinctive texture often associated with the coronas and generally linked to decompression at high temperatures. In a symplectite, two or more minerals are intimately grown together; they make up a single shell of a reaction rim or corona (Figure 7.10b).

Corona textures arise when reaction is incomplete, despite the close proximity of the reactants and the high temperature, which should favour diffusion. They probably form in rocks that lack an aqueous fluid phase; grain-boundary diffusion is much less effective if water is not present.

Reaction rims also develop during retrograde metamorphism, where the development of low-grade minerals is limited by the transport of water to the site of reaction. The development of chlorite rims around garnet is a very common example.

7.5.4   Interface-Controlled Growth

Reactions that take place when the equilibrium conditions have been overstepped may show evidence that their progress was constrained by the attachment of material to the growing grains, rather than by its transport to the site of growth. For example, a marble close to a granite contact in California contains many small forsterite crystals close to the contact, but fewer, larger, grains further away. We

(a)

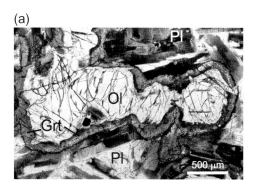

(b)

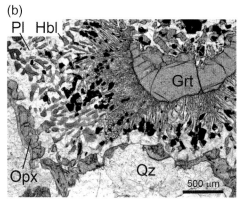

7.10 Reaction rims. (a) Corona of fine-grained garnet around a relic olivine, formed at the interface with plagioclase. West Norway. (b) Symplectite of intergrown orthopyroxene, plagioclase and oxides around a corroded garnet core. The outer part of the symplectite is of green hornblende intergrown with plagioclase and this is surrounded by a shell of orthopyroxene. The mineral outside this complex structure is quartz. Bhutan.

have already seen that the effect of increased overstepping is to facilitate nucleation and in this case it appears that there is a rapid increase in nucleation rate with increased overstepping. In contrast, diffusion rates increase linearly with overstepping (Figure 7.11; Roselle *et al.* 1997). The observed grain-size distribution in the aureole demonstrates proportionately more nucleation relative to growth close to the contact, and this reflects greater overstepping of the equilibrium conditions towards the contact. Growth rates are also enhanced near the contact, reflecting the greater surface area available for growth where there are many, smaller, grains, and so are controlled by processes at the grain surfaces rather than by transport. This is known as **interface-controlled growth.**

7.5.5   Heat-Flow-Controlled Growth

An important overall constraint on most metamorphic reactions is the supply or removal of heat. We noted in Chapter 2 that reactions which release fluid are strongly endothermic. It isn't enough to 'just' raise a rock to the temperature for such a reaction to proceed, more heat must be supplied to allow it to progress, just as

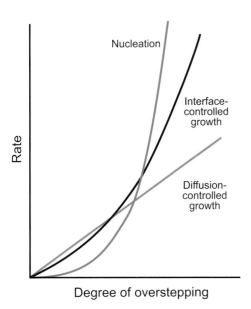

Figure 7.11 Schematic representation of the way in which the rates of crystal growth processes vary with the degree of overstepping of the reaction. Nucleation is strongly dependent on overstepping (Figure 7.3) whereas the rates of diffusion to supply growth increase linearly. The rate of interface-controlled growth also increases with overstepping. After Roselle *et al.* (1997).

it is necessary to keep heating a pan of boiling water for it to continue turning to steam. Just as in the kitchen, the faster heat is added, the faster the metamorphic dehydration reaction.

The supply of heat influences the progress of a metamorphic reaction in several ways. First, the equilibrium temperature must be overstepped so that product phases can nucleate and begin to grow. If growth is slow relative to the heating rate, the rock will continue to get hotter, the reaction will become more over-stepped and so reaction rate will increase (Figure 7.11) until there is a balance between reaction rate and heating rate. In the case of the ionic reaction cycle described in Section 7.5.2, there was widespread but very selective nucleation of the product sillimanite, so reaction was probably not greatly overstepped. Since transport was extensive, it is unlikely to have constrained the progress of the reaction, and it is possible that the overall reaction rate was determined by the supply of heat.

If a reaction proceeds only after it has been strongly overstepped, it can draw down the temperature of the rock back towards the equilibrium temperature, by consuming heat faster than heat is being added. As the temperature approaches equilibrium, the reaction rate will slow and come into balance with the heating rate. At a large scale therefore, the rates of endothermic reactions must be coupled to the rate of heat supply, although there are circumstances where they need not be in balance.

SUMMARY

In this chapter we have seen that the textures of metamorphic rocks reflect the way in which they developed and preserve information about their history. The clues that metamorphic textures provide about the mechanisms by which reactions have proceeded yield further insights into metamorphic rates and processes. Mineral growth involves separate steps of nucleation and growth of the solid products, and also requires transport of the constituent atoms from the places where they reside in reactant minerals to the sites where the products are growing. A range of factors control how reactions are able to proceed and these are reflected in rock textures. They include the pre-existing mineralogy of the rock as well as characteristics of the product minerals. Transport of material through the grain-boundary network of the rock and potentially through minerals themselves is essential for almost all metamorphic reactions and is strongly influenced by the presence or absence of water. Even in rocks whose mineral assemblages largely developed at equilibrium under specific conditions of temperature and pressure, the textures often retain evidence of earlier stages. Disequilibrium features like these provide an opening into understanding the metamorphic history of a rock as well as the peak P–T conditions that it attained.

EXERCISES

1. What are the two stages involved in the production of visible crystals of a new metamorphic index mineral?
2. List the mechanisms by which atoms can move through a metamorphic rock (a) in the absence of water and (b) if it is water-saturated.
3. What is the difference between growth zoning and retrograde zoning in porphyroblasts? A kyanite-zone rock contains garnets with small inclusions of chloritoid. Which type of zoning would you expect it to display?
4. Olivine commonly occurs with plagioclase in gabbros crystallised in the upper crust, but these minerals react together at relatively high pressures (Figure 7.10a). How is it possible for both these minerals to be present in a metagabbro that has experienced high-pressure granulite-facies metamorphism?

FURTHER READING

Cai, W. & Nix, W.D. (2016). *Imperfections in Crystalline Solids*. Cambridge University Press, 519 pages.

Heinrich, W. & Abart, R. (Eds.) (2017). Mineral reaction kinetics: microstructures, textures, chemical and isotope signatures. *EMU Notes in Mineralogy, volume 16.* The Mineralogical Society, 651 pages.

Putnis, C. & Ruiz-Agudo, E. (Eds.) (2013). The mineral–water interface. *Elements,* 9(3), 177–216.

Vernon, R. H. (2018). *A Practical Guide to Rock Microstructures.* Cambridge University Press, 432 pages.

8 Metamorphism Linked to Deformation

Deformation of some sort often accompanies metamorphism and, when it does, it is possible to learn more about the metamorphic processes by also considering the accompanying deformation. Detailed analysis of the deformation of metamorphic rocks is beyond the scope of this book, but further reading is suggested at the end of the chapter. Here, we will look at how rocks can deform during metamorphism in general terms, and then focus on three aspects of the relationships between metamorphism and deformation.

The textures of a regionally-metamorphosed rock often document a history of mineral growth and the development of fabrics. These textures do not necessarily imply any causative links between metamorphism and deformation, but make it possible to determine the sequence of mineral growth and metamorphic grade at which successive deformation episodes took place.

Deformation can also be a direct *cause* of metamorphism. Highly-deformed rocks such as mylonites owe their character as much to the strain that they have undergone as to the mineral reactions that may have taken place, but the link may be more subtle. Fracturing, which allows water to penetrate, is often necessary for retrograde metamorphism to take place. Under low-grade conditions, where volume diffusion within grains is limited, deformation can break down original grains, allowing their interiors to react with other grains nearby.

Finally, metamorphism can also act as a *driver* for deformation, because the release of fluid at high pressures, and the growth of new minerals which initially are fine grained, can reduce rock strength, so that rocks begin to deform in response to the existing tectonic stresses, without additional stress being applied.

In this chapter, you will learn to interpret metamorphic textures to unravel the history of regionally-metamorphosed rocks, and will be introduced to the, sometimes controversial, issues around the influence that metamorphism and deformation exert on each other.

8.1   How Do Rocks and Minerals Deform?

During most types of metamorphism, ductile or plastic deformation is usually dominant. In **ductile deformation** the rock body as a whole changes shape, and often the individual grains are also deformed, whereas **brittle deformation** is localised along discrete fracture planes. Brittle deformation can occur in metamorphism, even at high temperatures, under conditions where rocks are strong, or where fluid pressure is high. Ductile deformation involves the movement of individual grains past one another (**intercrystalline deformation**) and together with changes to the shape of individual grains by **intracrystalline deformation**. It also involves the breakdown of existing grains and the growth of new ones (recrystallisation).

There are several ways by which crystals can deform, but two mechanisms dominate in metamorphism. Deformation of grains by movement on dislocations is known as **dislocation creep** and has been studied extensively for synthetic materials in materials science, and for a small number of rock-forming minerals. Dislocation creep results in distorted or strained grains, and leads to the crystallographic axes of different grains of the same mineral species becoming aligned. This is known as a **crystallographic-preferred orientation**. The other dominant deformation mechanism is **pressure solution**, or more generally **dissolution–precipitation creep**. In pressure solution, parts of the grain remain unchanged, but surfaces that are under stress dissolve, with the material being re-precipitated on surfaces in different orientations. Transfer of material is not always simply a matter of moving atoms from one part of a grain to another; some entire grains may tend to dissolve while others grow. The distinction between the distortion of grains by dislocation creep and changing shape by pressure solution is illustrated in cartoon form in Figure 8.1.

As with so many aspects of metamorphism, temperature and the availability of fluids are the key influences on how rocks deform. Dissolution–precipitation creep is only possible in the presence of a water-rich fluid into which minerals can dissolve. Dislocation creep requires volume diffusion to enable the movement of dislocations through the crystal and allow them to move around other imperfections in the structure, and this is favoured by high temperatures and by the presence of water. In the case of quartz, the effect of water is particularly marked and is commonly referred to as **water weakening**. Dry quartz is strong, but the presence of water in

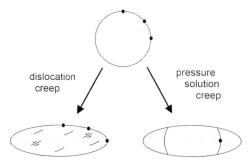

Figure 8.1 Alternative mechanisms for changing the shape of grains. An initially circular grain distorted by dislocation creep retains its original surface and the markers on it (black spots) remain in the same relative positions. If the grain is deformed by pressure solution, it is partially dissolved at sites of high stress, but this material is reprecipitated in low-stress regions. Markers on the original surface are either removed by dissolution or grown over so that they become incorporated into the grain.

dislocations makes them mobile and allows quartz to deform readily under all but the lowest-grade metamorphic conditions. Irrespective of the dominant deformation mechanisms, the strength of crustal rocks is strongly dependent on whether or not an aqueous fluid is present.

In metamorphism, deformation by diffusive mass transfer occurs most commonly by pressure solution. Except perhaps at the highest temperatures, this mechanism requires a pore-fluid phase to be present for the stressed grain to dissolve into, and this is likely to be the case in many lithologies during prograde metamorphism, because of the release of water by dehydration reactions. In general, pressure solution is most obvious at low metamorphic grades, where specific textures may document the redistribution of mineral material, but it probably continues to play a role at higher grades where dissolution–precipitation creep involves multiple mineral grains.

The net effect of large numbers of grains each distorting independently means that the rock must have additional deformation options available to accommodate them. To some extent, grains may move past one another by grain-boundary sliding, but some minerals are also likely to recrystallise, with material from multiple original grains becoming incorporated into a range of new grains to accommodate strain.

Different deformation mechanisms predominate in different metamorphic environments, with the nature of the rock types and their mineralogy both having an important influence in addition to temperature and the availability of water. The deformation mechanism that overall predominates in different parts of the crust is important for determining the way the crust deforms and flows under tectonic forces. For example, dissolution–precipitation creep (pressure solution) appears to be a major deformation mechanism over much of the range of crustal metamorphic conditions (Wassmann & Stöckhert 2013). It is also important in polymineralic rocks; for example, there is more effective dissolution of quartz at boundaries with phyllosilicates rather than where it is in contact with other quartz grains (Wheeler 1992). The existence of lattice-preferred orientations, especially for quartz and omphacite, suggests that dislocation creep is important in some crustal rocks at medium to high grades.

8.2   Textures Produced by Deformation During Metamorphism

The effects of deformation may vary greatly according to the mineralogy and grain size of the rocks concerned, as well as temperature, availability of fluid, strain rate and the pre-existing texture. They include the production of tectonic fabrics such as foliations and lineations, and the recrystallisation of minerals in response to strain. In addition, metamorphic reactions and deformation processes may interact with one another. This interaction may be purely passive, as for example when a poikiloblast grows while deformation is proceeding, so that the inclusion trails within it record the evolving schistosity of the rock, or there may be dynamic feedbacks whereby deformation can enhance reactivity, and vice versa.

8.2.1   The Development of Tectonic Fabrics and Mineral Alignments

One of the most characteristic features of deformed metamorphic rocks is the presence of fabrics such as cleavage, schistosity or mineral lineations. Metamorphic fabrics commonly reflect the alignment of some or all of the constituent mineral grains in particular orientations. Unlike primary layering, such as bedding, meta-morphic fabrics are usually **penetrative**, i.e. they are fabrics that are developed throughout the rock mass. Platy minerals such as micas, or elongate minerals such as amphibole, are especially likely to become aligned during deformation, but intracrystalline deformation can also give rise to crystallographic-preferred orientations of other minerals such as quartz or calcite. These may be detectable only in the laboratory by measurements of the orientations of crystallographic axes, but can also involve the formation of aligned elongate grains.

Preferred orientations may develop as a result of the following.

1. Mechanical rotation of originally asymmetric grains (e.g. mica flakes) into new orientations, usually accompanied by slip along grain boundaries or pressure solution to accommodate their changing orientations.
2. Recrystallisation or crystallisation of minerals to form new grains growing directly in the preferred orientation.
3. Deformation of pre-existing grains to a new, aligned, asymmetric shape.

Mechanical rotation is the simplest of these mechanisms, and provides the classical explanation of the origin of cleavage. To rotate all the platy grains in a rock into parallelism requires very large strains. However, the removal of some of the intervening material can facilitate their rotation. This seems intuitively improbable, but was documented for a series of very-low-grade slaty rocks, where distortion of fossil graptolites provided evidence for a 50% volume loss during formation of the first cleavage (Wright & Platt 1982). This study implies that the first fabric to form in a metamorphic rock can sometimes date back to a very early stage of burial and the expulsion of pore water.

Growth of new grains is now recognised as the dominant factor in cleavage formation in slates, as well as in the formation of schistosity and other fabrics at higher grades. As metamorphic reactions take place, the reaction products crystallise preferentially in an orientation that reflects the imposed stress field and the strain that is taking place. As the new phyllosilicate grains continue to grow, and the old ones break down, so the new cleavage becomes pervasive and completely replaces the earlier one. In low-grade rocks, new and old grains may be distinguished by differing chemical compositions. In higher-grade rocks, any such distinctions are more subtle, but modern analytical techniques are generally sufficiently sensitive to be able to distinguish between different textural generations of matrix minerals even in schists.

Changes in the shape of individual grains are especially important for the development of preferred orientation in minerals such as quartz, olivine and calcite. These minerals readily deform by dislocation creep, resulting in lattice-preferred orientations, although shape changes may also be produced by pressure solution.

8.2.2 Metamorphic Segregation Layering

Some regionally metamorphosed rocks exhibit compositional layering, usually parallel to the schistosity, that is caused by mineral segregation. For example, layers rich in quartz and feldspar may alternate with layers rich in micas. In some cases such **segregation layering** may reflect original sedimentary banding, but in others it is of metamorphic origin. Metamorphic segregation layering is best known from high-grade rocks; it is a characteristic of gneisses and a common feature of migmatites. However, the same phenomenon is also known at lower grades, especially if deformation is intense. In extreme cases, metamorphic segregation layer features become indistinguishable from veins concordant with the foliation.

The thermodynamic rationale for segregation layering is probably that grain boundaries between crystals of the same, or a structurally-similar phase, are likely to have a lower surface energy than those between wholly unlike phases. By segregating the rock into layers, each dominated by one mineral type, the total free energy of the system is reduced. However, this rationalisation provides no information about the way in which the layering originates. At low to medium metamorphic grades, crenulation of the schistosity can lead to the segregation of mica-rich and quartz-rich layers, corresponding to the attenuated limbs of the crenulations and their hinge zones respectively (Figure 8.2). These are known as P (phyllosilicate) and Q (quartz)

Figure 8.2 Crenulated phyllite in which the limbs of the crenulations concentrate dark phyllosilicates (P domains) while the hinges concentrate quartz (Q domains). As a result, the outcrop displays segregation layering parallel to the axial surfaces of the crenulations. South Island, New Zealand.

domains. In other cases, segregation of a melt phase or movement of pore fluid may be important for the development of distinct compositional layering.

8.2.3 Highly-Deformed Rocks – Mylonites

Intense deformation results in the reduction of grain size and the formation of rocks that are strongly foliated. Under metamorphic conditions, deformation is predominantly ductile and the deformed rocks are termed mylonites (Chapter 1, Section 1.6.3, Figure 1.14a, Figure 8.3). Mylonites most commonly occur in distinct **shear zones**. These may correspond to major faults which juxtapose rocks that are clearly very different, or may cut through gneissic basement rocks which appear similar on each side. Mylonites are characterised by the development of grains that are strained and reduced in size, although different minerals may have responded very differently to deformation, even within a single hand specimen. The development of a mylonite fabric is a dynamic process. As original grains are deformed they acquire extra free energy associated with dislocations and this destabilises them. Smaller but undeformed grains grow around them to produce a **mortar texture**, but as these grow larger (reducing surface energy) they inevitably become strained and are themselves replaced by new, undeformed grains in a continuing cycle known as **syn-tectonic recrystallisation**.

Mylonites are usually classified according to the extent to which the original texture has been modified. The least-deformed category of mylonites is **protomylonite**. These are rocks which contain many strained grains but they have not recrystallised extensively and the original grains can often be identified (Figure 8.3a). Often, they display mortar texture (Figure 8.3b). The amount of recrystallised fine-grained matrix is in the range 10–50%. For a rock to be rigorously classified as mylonite, the recrystallised matrix makes up more than 50% of the rock (Figure 8.3c) and the relic original grains are rounded or fractured.

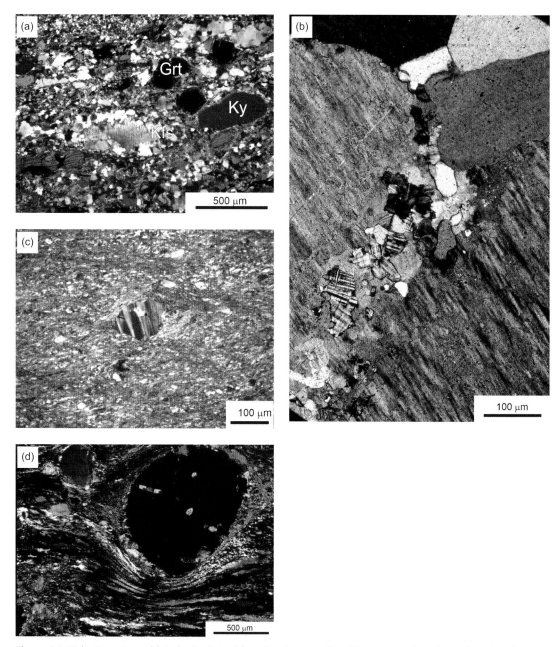

Figure 8.3 Mylonite textures. (a) Mylonite derived from kyanite granulite with many porphyroclasts. Note that the K-feldspar porphyroclasts are perthitic whereas the mortar of smaller K-feldspar grains around them is much closer to the K-feldspar end-member in composition. Sligo, Ireland. XPL. (b) Mortar texture composed of small microcline grains marking a deformation zone across a coarse perthitic K-feldspar grain. XPL. (c) Intensely deformed mylonite with only sparse feldspar porphyroclasts remaining. Glencoul, Scotland. XPL. (d) Ribbon texture developed in quartz in a high-strain zone around a garnet porphyroclast. Skagit Valley, Washington, USA. XPL.

Since they are now coarser than the deformed matrix, the coarse relics are termed **porphyroclasts**. In many mylonites, the matrix is dominated by quartz which has undergone syn-tectonic recrystallisation. If the strain is high, larger quartz may have recrystallised into ribbons of small polygonal sub-grains, hence the name **ribbon texture** (Figure 8.3d).

8.3 Determining the Relative Timing of Metamorphism and Deformation

It is not unusual for a regionally metamorphosed rock to exhibit more than one tectonic fabric. Often, it will be apparent that one is younger and is associated with deformation that affects the earlier fabric. We can also use fabrics to unravel the relative age of metamorphic minerals and deformation episodes, making it possible to assign depths and temperature to the deformation.

8.3.1 Relative Timing of Porphyroblast and Poikiloblast Growth

Most porphyroblasts, and all poikiloblasts, contain inclusions of matrix minerals, and these often reflect the fabric of the rock at the time when the porphyroblast grew. Multiple inclusions may occur in distinct planes, corresponding to the foliation of the rock matrix at the time the porphyroblast grew, or individual elongate inclusions may likewise preserve the alignment of the original foliation. Further deformation taking place after the growth of the porphyroblast may completely destroy the foliation in the rock matrix outside the porphyroblast, and replace it with a new one, but the inclusions inside the porphyroblast are preserved, and with them information about the earlier fabric.

The major textural criteria for determining the relative ages of deformation and metamorphic mineral growth are based on the relationship between the schistosity preserved by the pattern of inclusions in a porphyroblast (known as the **internal schistosity**, S_i), and the dominant foliation in the rest of the rock (known as the **external schistosity**, S_e). Variations on these relationships are illustrated in Figure 8.4.

The **post-tectonic growth** of porphyroblasts leads to simple patterns in which the inclusion trails define a fabric that is parallel to, and continuous with, the external schistosity. This type of pattern is particularly common in thermal aureoles where the aureole rocks experienced an earlier regional metamorphism (Figure 8.4a), but it is also found in regional metamorphism, especially at relatively low pressures. Sometimes, the external schistosity that is overgrown had been previously crenulated, so that there are small folds in the inclusion trails within the porphyroblast. In this case, the porphyroblast is said to exhibit **helicitic texture** (Figure 8.4b).

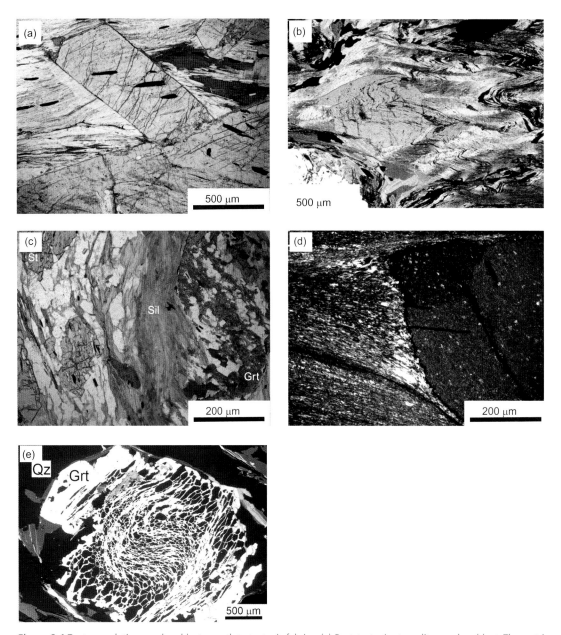

Figure 8.4 Textures relating porphyroblast growth to tectonic fabrics. (a) Post-tectonic staurolite porphyroblast. The matrix muscovite fabric includes trails of ilmenite and graphite dust that are preserved in the staurolite and continue through unbroken. Some distortion of the fabric around the porphyroblast records minor later deformation. (b) Crenulated mica schist with a post-tectonic albite porphyroblast (grey) containing helicitic trails of inclusions recording the folded fabric. XPL. (c) Sillimanite garnet staurolite schist with a fabric defined by biotite. Pre-tectonic garnet is wrapped by later sillimanite (fibrolite). Smaller staurolite grains overgrow the matrix fabric. (d) Phyllite with pre-tectonic porphyroblast of ankerite, mantled by aligned white micas and flanked by a pressure shadow of coarser and less-strongly-aligned white mica with quartz. XPL. (e) Back-scattered electron image of a 'snowball' garnet in metachert, Saas-Fee, Switzerland. Garnet appears white, quartz is dark grey. (Photo: Dave Prior.)

In contrast, if the porphyroblast grew before the deformation (**pre-tectonic growth**), then the internal fabric, S_i, is likely to be discordant with S_e (Figure 8.4c). In addition, the foliation in the matrix often wraps around the porphyroblast because it remained rigid as the matrix deformed (Figure 8.3d). Minor bending of the schistosity around porphyroblasts may, however, arise from relatively minor, later deformation. An angular discordance between S_i and S_e is the best indication of pre-tectonic porphyroblast growth, but in the absence of a suitable internal fabric in the porphyroblast other indications include cracking or straining of the large crystal, or the development of a **pressure shadow** beside it (Figures 8.3d and 8.4d). Pressure shadows are areas of coarsely-recrystallised matrix material next to a porphyroblast. They develop in the region that was shielded from the maximum compressive stress during deformation by its proximity to the rigid porphyroblast. Material probably migrates here by pressure solution.

Some porphyroblasts grew in whole or in part as their host rocks were deforming (**syn-tectonic growth**). The classic syn-tectonic texture consists of a spiral pattern of inclusions within a porphyroblast (Figure 8.4e), and is known as a 'snowball texture'. Where the apparent rotation is less than $90°$, the curved inclusion trails result from recrystallisation of the schistosity into a new orientation during growth. Inclusion trails in pre- or syn-tectonic porphyroblasts are typically almost parallel throughout all the examples in an outcrop, even across folds.

Porphyroblast growth may outlast the deformation event that produced the schistosity. Growth can include pre- to syn-tectonic stages or syn- to post-tectonic stages. Most grains exhibiting extended growth are garnets, and this likely reflects the fact that garnet can be produced by continuous reactions over a large temperature interval and an extended period of time (the value of such crystals for unravelling metamorphic history is explored further in Chapter 9).

8.3.2 Fabric Minerals and Timing of Metamorphism

The aligned minerals that define metamorphic fabrics are usually abundant rock-forming minerals, but also include oxides and accessory minerals. Often these are minerals that have been stable for much of the metamorphic episode, but sometimes fabric-forming minerals also provide an indication of metamorphic conditions during the deformation event in which the fabric formed. Metabasites commonly have a fabric defined by amphiboles whose composition is sensitive to temperature and pressure. For example, the blueschist illustrated in Figure 5.7g has a fabric defined by glaucophane, indicating that the deformation occurred under high-P, low-T conditions. This type of relationship makes it possible to determine the depth and temperature at which specific structural features formed.

At relatively high grades, metapelites develop sillimanite which often occurs as bundles of fibres or fibrolite which can be aligned with tectonic fabrics. Figure 8.5 is

Figure 8.5 Fold in interbedded metapelites and metapsammites. Aligned bundles of sillimanite fibres with intergrown quartz (examples arrowed) parallel the axial plane of the fold, demonstrating syn-tectonic growth. Connemara, Ireland.

an example of a fold in upper-sillimanite-zone metapelite from Connemara, Ireland, in which fibrolite bundles are aligned parallel to the axial plane of the fold. In this case, the sillimanite is clearly syn-tectonic and so this episode of folding took place at the peak of metamorphism.

8.4 Feedbacks Between Deformation and Metamorphism

The relationship between metamorphism and deformation is complex because deformation may facilitate metamorphism and vica versa. To understand how regionally-metamorphosed rocks have evolved, it is important to look at structural and metamorphic aspects in an integrated manner, not in isolation.

Many metamorphic belts have revealed complex histories of fabric formation and mineral growth, using the techniques just outlined, but it is almost always the case that the final episode of *pervasive* deformation took place close to peak metamorphic conditions. Medium- to high-grade rocks often preserve evidence of earlier pervasive deformation and folding during prograde metamorphism. This is certainly the case in the Connemara region, Ireland, where the folds with axial planar sillimanite (Figure 8.5) themselves deform earlier folds and fabrics associated with lower metamorphic grades. Once the region began to cool, however, deformation was limited to the development of broad, open folds and localised shears and faults. This is a pattern that is repeated in orogenic belts worldwide: prograde metamorphism is accompanied by pervasive folding, which comes to an end as soon as the metamorphic peak is passed. The change in behaviour could conceivably be the result of a change in stress regime, but it is more likely to reflect the way in which metamorphic reactions determine the availability of water and hence influence rock strength.

During prograde metamorphism, water is generally present in all lithologies and fluid pressure is generally close to lithostatic pressure (Chapter 1, Section 1.4.2). Once metamorphic rocks have begun to cool, however, most metamorphic rocks 'dry out'. The small amounts of remaining pore water are consumed by retrograde reactions once peak metamorphic conditions are passed, so that the high-grade minerals co-exist with the products of incipient retrogression. This mineral combination provides a buffer (Chapter 2, Section 2.8) that controls the fugacity of water. Only small amounts of cooling from peak conditions are required for water fugacity to be buffered at such low values that the rock is effectively dry (Yardley & Valley 1997). This is true whatever the peak metamorphic grade. We have already seen that rock strength is linked to the availability of water, with dry rocks very much stronger than wet rocks at all but the highest temperatures, and so this effect means that rocks must become much stronger after the peak of metamorphism. Figure 8.6 is a schematic representation of the variation of rock strength with depth for 'wet' and 'dry' conditions, and shows a large difference in rock strength between prograde and retrograde conditions in the mid to lower crust. Cooling rocks are periodically infiltrated by water and it can be seen from Figure 8.6 that this will make them weak, at least until the water has been consumed by further retrograde reactions. We can conclude that the ubiquitous association of pervasive regional folding with

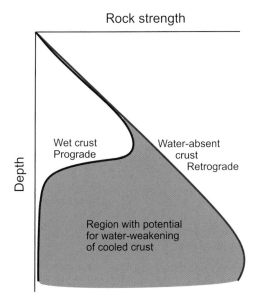

Figure 8.6 Schematic representation of the variation of the strength of typical crustal rocks with depth, according to whether they are wet or dry. The blue and red curves correspond to bulk rocks experiencing heating (prograde conditions) and cooling (retrograde conditions) respectively. Cooled crust is likely to lie in the shaded region, where its strength can vary between the bounding curves as water is introduced but then consumed by reactions.

prograde metamorphism, but not with cooling and exhumation, simply reflects the way in which metamorphic mineral interactions control water in the deep crust.

8.4.1   Metamorphic Reactions Enhanced by Deformation

Irrespective of the overall metamorphic cycle, both ductile and brittle deformation can accelerate reactions in a number of ways, including the following.

1. Juxtaposing reactant mineral grains as they move past one another.
2. Fracturing zoned grains and so exposing their interiors to the grain-boundary fluid.
3. Increasing the free energy of reactant grains by straining them or breaking them into smaller pieces with larger surface area.
4. Providing sites for preferential nucleation on newly-created fracture surfaces.
5. Allowing fluid to access rocks, thereby triggering hydration reactions and enhancing reaction rates in general.

These processes can be summarised in terms of the parameters discussed in Chapter 7 (Section 7.5) as enhancement of transport of material between reactants, enhancement of the nucleation of product minerals and providing access for aqueous fluid. An example of a reaction apparently triggered by deformation was shown in regional chlorite-zone metapelites around an intrusion in the Spanish Pyrenees. There, the first appearance of biotite was closely linked to the development of a new tectonic fabric that is present only in the aureole (Dempster & Tanner 1997).

An example of preferential nucleation of new metamorphic minerals on syn-metamorphic fractures is provided by andalusite. In eastern Connemara, Ireland, the staurolite zone passes through a staurolite–andalusite zone before it reaches the staurolite–sillimanite zone. Andalusite invariably forms large porphyroblasts, much coarser than any other mineral in these rocks (cf. Figure 7.1b), and this suggests that nucleation is difficult. The first appearance of andalusite in the field is in veins rather than in the rock matrix (Figure 8.7), suggesting that these mark fractures which provided favoured nucleation sites. Nucleation in the rock matrix required further overstepping and so occurred at slightly higher temperatures (Chapter 7, Section 7.2.1).

High-grade basement terranes, characterised by gneisses, migmatites and granulites, are commonly cut by shear zones, faults and fractures in which the high-grade assemblages are overprinted by lower-grade ones. The overprinting assemblages are typically the result of hydration reactions and their formation sometimes also involve metasomatism. In north-west Scotland, amphibolite-facies shear zones cut granulites, and the transformation involved the addition of Na and K with removal of Ca (Beach 1976). Similar transformations have been described from many other areas, but the chemical changes are not always the same. In the eastern Alps,

Figure 8.7 Vein composed almost entirely of stubby pink prismatic andalusite cutting fine-grained staurolite schist, Connemara, Ireland. Andalusite does not occur in the schist matrix at this outcrop, but is present at slightly higher grades. Note that a quartz vein is visible in the lower left corner but most of the white material on the outcrop is lichen.

Austria, granodiorite was converted to aluminous schists in shear zones with loss of Na and Si while some mafic elements increased in concentration (Selverstone *et al.* 1991). These changes appear to arise from the removal of part of the rock by dissolution during the alteration. More recent work dating the changes in these rocks are discussed in Chapter 9, Section 9.3.1, Figure 9.7.

A particularly distinctive type of metamorphism is associated with shear zones in deep crustal rocks that have experienced a high-pressure metamorphic overprint. Precambrian granulite-facies basement rocks in west Norway were locally converted to eclogite during Caledonian (Palaeozoic) continental collision (Austrheim & Griffin 1985), especially in shear zones (Figure. 8.8). During burial, the intervening rocks between the shear zones were also subject to eclogite facies *P*–*T* conditions but did not react, so on the face of it this is a simple case of deformation allowing the reaction to take place. However, the eclogite assemblages invariably contain hydrous minerals such as epidote and phengite, and are hydrated relative to the original granulites. The infiltration of water along the deformation zone was at least as important as the deformation in allowing reaction to take place (Austrheim 1987). There has been much debate over the years about whether the fluid infiltration facilitated the deformation or whether the deformation allowed the fluid

Figure 8.8 Precambrian anorthosite with garnet coronas mantling remnant pyroxene (lower right), cut by a fine-grained eclogite shear zone of Palaeozoic age (below the hammer). Bergen arcs, Norway.

infiltration. Careful and detailed petrographic study on a number of deep crustal shear zones has shown that the strong, dry lower crust first fractured by brittle faulting (Giuntoli *et al.* 2018; Jamtveit *et al.* 2018). These fractures allowed water to penetrate, leading to two complementary effects. Firstly, metamorphic hydration reactions consumed the water and took place under conditions that were far from equilibrium. Secondly, the introduction of water greatly reduced rock strength, leading to extensive ductile deformation and the formation of a shear zone focussed on the original fault.

8.4.2 Metamorphism as a Trigger for Deformation

Because metamorphic reactions commonly involve a change in mineral shape, size, type and location, they may also cause deformation by weakening the rock so that it yields in response to smaller deviatoric stresses than would otherwise be required. Many metamorphic reactions either release water, facilitating most deformation mechanisms, or consume water, which precludes deformation by mechanisms that involve water. Rocks have a finite strength which the applied stresses must exceed before plastic deformation can occur, even under high-grade metamorphic conditions. However, reactions can reduce rock strength in several ways.

1. Maintaining an aqueous pore-fluid phase at high pressure, which enables deformation by dissolution creep (pressure solution) and facilitates grain-boundary sliding and hydrofracturing.
2. Producing new minerals that are weaker than those originally present, for example micas produced by retrograde breakdown of feldspars.
3. Producing large numbers of small grains, and so making it easier for the rock to deform by grain-boundary sliding.
4. Setting up local stresses as a result of the volume change of the reaction, which can then act in combination with the tectonic stresses to permit deformation in response to a smaller applied stress.

In addition, at very high metamorphic grades the formation of melts has a fundamental impact on rock strength. Rosenberg and Handy (2005) reported that rock strength decreases linearly with melt fraction up to about 7% melt. At this stage it is likely that the grain-boundary network has become saturated with melt. Further loss of strength takes place when there is about 25% melt, sufficient for the migmatite to begin to break up.

Extensive deformation can sometimes take place in response to an anomalously small applied stress. This is known as **superplastic flow**, and examples have been documented from metamorphic rocks at all grades. At low grades, the growth of fine-grained phyllosilicates in a fault zone weakens the rock so that further movement is concentrated in the vicinity of the initial fault (Wintsch *et al.* 1995). Under very different conditions, rocks metamorphosed under eclogite-facies conditions from the Sesia-Lanzo zone of northern Italy locally developed jadeite + quartz (from sodic feldspar) and were intensely deformed with mylonitic fabrics (Rubie 1983). However, it was the deformation that became concentrated in those parts of the rock in which feldspar began to break down, rather than the other way round. This reaction is an ideal one for causing weakening because the products form fine-grained intergrowths, quartz is more readily deformed than feldspar under most conditions, and the reaction involves a significant volume reduction.

SUMMARY

In this chapter we have introduced the basic terminology for describing the textures of deformed metamorphic rocks, and shown how it is possible to use them to unravel the relative timing of the development of penetrative fabrics, such as schistosity, and the growth of metamorphic minerals. This allows deformation to be correlated with the temperature and pressure of metamorphism (Chapter 3) and to be dated in absolute terms (Chapter 9).

The relationship between deformation and metamorphism is complex. Prograde metamorphism liberates water-rich fluid, which is required for effective rock deformation by mechanisms such as pressure solution. As a result, most rocks are weak during heating, and deform pervasively with the development of fabrics such as schistosity. However, deformation can itself facilitate reaction, either directly or by allowing water to access rocks, and is particularly important for enabling retrograde metamorphism. Once an initial fracture has formed, allowing water to access cooled rocks, retrograde reactions are concentrated around the fracture. They will often reduce rock strength so that further deformation affects the same general volume of rock.

EXERCISES

1. Regionally-metamorphosed metapelites are characterised by the sub-parallel alignment of phyllosilicate grains (i.e. cleavage or schistosity). How might this mineral alignment be brought about?
2. Describe two characteristics of pre-tectonic porphyroblasts in regionally-metamorphosed rocks.
3. Why does the presence of a water-rich fluid reduce the strength of rocks under medium- to high-grade metamorphic conditions?

FURTHER READING

Jamieson, R. A., Unsworth, M. J., Harris, N. B., Rosenberg, C. L. & Schulmann, K. (2011). Crustal melting and the flow of mountains. *Elements*, **7**(4), 253–60.

Jamtveit, B. & Austrheim, H. (2010). Metamorphism: the role of fluids. *Elements*, **6**(3), 153–8.

Passchier, C. W. & Trouw, R. A. (2005). *Microtectonics*. Springer Science & Business Media.

Vernon, R. H. (2018). *A Practical Guide to Rock Microstructures*. Cambridge University Press, 432 pages.

9 The Duration of Metamorphism

Much of the material in this book so far has skirted around the problem of how quickly metamorphic rocks form and how long metamorphism lasts. Regional metamorphism can be thought of as taking place in a **metamorphic cycle**, involving burial, heating, exhumation and cooling. The question of 'how long' a metamorphic cycle takes from start to finish may be determined either directly or indirectly. Indirect methods involve calculations of how long it takes for rocks to heat up, cool down, or be buried or exhumed. Based on the thermal properties of rocks, indirect approaches have been used for many years to calculate the rates of heating and cooling associated with igneous intrusions.

Direct methods involve radiometric dating of metamorphic minerals to determine when they grew. Early results of the application of radiometric techniques to dating metamorphic events were often very uncertain and required large samples. Since the late 1990s, the advent of much more sensitive analytical instrumentation, and improved spatial-resolution sampling by lasers and ion beams, have provided significant improvements in accuracy and precision (Box 9.1) and allow dating of much smaller volumes of material, including *in-situ* measurements of individual mineral grains or separate chemical zones within them.

This chapter introduces indirect means of calculating heating, cooling, burial and exhumation rates to provide estimates of the duration of metamorphic cycles. We then look at the minerals that make good geological clocks (**geochronometers**) and discuss the different ways in which we can link the ages of these minerals (or different chemical zones within them) to different parts of the metamorphic history of the rock that hosts them. We end with a summary of what we currently know about the duration of metamorphic episodes of different types. By the end of this chapter you should be able to estimate the timescale and spatial scale of heating caused by an igneous intrusion and calculate an age based on the radioactive decay from U to Pb. You will also have an understanding of different dating techniques and the advantages and drawbacks of dating different minerals in different circumstances.

BOX 9.1 **Accuracy and precision**

Geochronologists strive to date rocks both precisely and accurately. 'Precision' indicates the degree to which measurements are repeatable and reproducible, whereas 'accuracy' expresses how close a measurement comes to the true value (Figure 9.1).

The yellow dots in Figure 9.1a cluster with both high accuracy and high precision. All the dots are in the bull's-eye (they are accurate) and they are not scattered. In contrast the dots in part (d) are neither accurate nor precise – the dots do not hit the bull's-eye and they are scattered. In part (b), the dots are accurate (in that on average they cluster around the bull's-eye) but are imprecise, because they scatter. In part (c), the dots show high precision (the dots are all clustered) but low accuracy (they don't hit the bull's-eye).

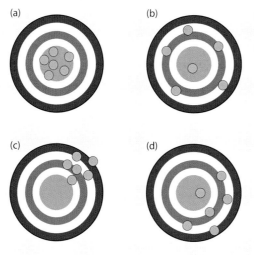

Figure 9.1 A series of bull's-eye targets to illustrate the difference between accuracy and precision. See text for details.

9.1   Indirect Estimation of Metamorphic Timescales

Throughout this book we have seen that most metamorphic reactions happen as a consequence of changes in temperature and/or pressure. Whilst changes in pressure can generally only be achieved by physically moving the rocks with respect to the surface (the ways in which rocks may be buried and brought back to the surface again are detailed in Chapter 10), changes in temperature can also be achieved by heat moving through rocks without the rocks needing to move anywhere (Box 9.2).

BOX 9.2 **Conduction and convection**

Heat passes through rock by both conduction and convection. **Conduction** is the process by which heat is transferred directly from a hotter region to a colder region by the increased motion of atoms and molecules. The amount of conduction depends on the temperature difference between the two regions and the thermal properties of the materials through which heat is being transferred. Most of the heat that is transferred from the mantle to the crust is transferred by conduction.

Rocks are good insulators and therefore conduction is slow, even on geological timescales. When rocks are buried, they therefore heat slowly as a result, but, given sufficient time, conduction allows deeply-buried, surface-derived rocks to reach high metamorphic temperatures at depth. Conversely, rocks are slow to cool as they are exhumed, and in actively-rising (and rapidly-eroding) young mountain ranges, hot rocks occur much nearer the surface than in stable cratons, even where there is no igneous activity taking place.

Convection is the process by which heat is transferred through the movement of hot materials. As hot fluids, especially melts, move through the crust, they carry heat which can then be transferred to the surrounding rocks, much as a tap heats up when hot water runs through it. Magmas derived from the mantle carry large amounts of heat into the crust by convection; this heat is then distributed to the surrounding rocks. At depth, this distribution happens slowly by conduction and, in the middle and upper crust where surrounding rocks are relatively cool, conduction gives rise to contact metamorphism in the form of a thermal aureole (e.g. Chapters 1 and 4). At shallow levels (within a few kilometres of the surface), if water is available and the rocks are permeable, heat can be redistributed very efficiently by convection of heated groundwater in a geothermal system (Chapter 5, Section 5.6.1). This process produces hydrothermally metamorphosed rocks in the area surrounding the intrusion but the maximum temperature is limited by the boiling curve of water (Chapter 1, Figure 1.10). There may be little or no associated contact metamorphism.

There are a number of heat sources which may contribute to metamorphic changes in the crust. The mantle supplies the most heat, but because this heat supply is constant, mantle heating will only cause metamorphism when some aspect of the crust changes (such as crustal thickness during rifting, for example, Chapter 10). Mantle heat may also be rapidly transferred directly into the crust via the intrusion of mantle-derived melts.

The crust itself has an internal heat supply, generated by the radioactive decay of unstable elements. Each individual decay releases heat and, over geological timescales, this heat builds up. Heat may also be produced locally by frictional heating caused by movement on faults and shear zones. Finally, due to the insulating properties of the crust (Box 9.2), movement of crust along faults and shear zones may juxtapose hotter crust against colder crust. The hotter crust will therefore act as a local heat source for the colder crust.

We saw in Chapter 7 that overall reaction rates of endothermic metamorphic reactions are normally controlled by the rate at which heat is added to the crust. The duration and geographical extent of a metamorphic event is therefore intricately linked to the amount of heat supplied to the system, the spatial extent and longevity of the heat source and the thermal properties of the rocks involved.

9.1.1   Rates of Heating During Contact Metamorphism

Contact metamorphism relies on cold country rock being heated by hot magma intruding into it (Chapter 1). A simple scenario is one involving the intrusion of magma into the crust in a single, short-timescale pulse, followed by cooling. In such a case, a simple mathematical relationship can be formulated between the size and shape (e.g. sheet-like body, sphere or cylinder) of the intrusion and the size and timescale of heating of the thermal aureole. Mathematical equations have been formulated for a variety of cases; the more complex the scenario, the more complex the resulting calculation. Box 9.3 outlines a simple example that describes the patterns and timescales of heating in the vicinity of a sheet-like intrusion that results in contact metamorphism.

In reality, detailed investigations of a number of intrusions have shown that plutons are built incrementally over time (e.g. Glazner *et al.* 2004). Heat is therefore added to the crust incrementally, resulting in longer timescales of metamorphism than the simple scenarios suggest. The simple cases are still useful, however, as they provide an order-of-magnitude calculation of heating rates against which the more-complex natural examples can be compared.

As well as assuming instantaneous injection of magma, the calculations outlined in Box 9.3 involve many further simplifications, such as assuming that the heat of

BOX 9.3 Heating in the vicinity of a sheet-like intrusion

The quantities that we need to define in order to calculate the development of this ideal type of thermal aureole are outlined in Table 9.1.

The quantity α has been determined for a wide range of rocks and only varies by a factor of about three between extremes. It is therefore convenient to fix α at a value of $31.5 \, m^2/a$, which is a good approximation for many rocks. The calculations are performed using two ratios:

$$d = \frac{x}{y} \tag{9.1}$$

$$r = \frac{\alpha t}{y^2} \tag{9.2}$$

In both cases the units cancel out and the ratios are dimensionless. The temperature at any point in the aureole specified by the distance ratio d, after a period of time t from the assumed instantaneous intrusion of the magma, is given by a mathematical function of d and r:

$$\frac{T}{T_0} = \phi(d, r) \tag{9.3}$$

Values of the function ϕ are given in Table 9.2 for a limited range of values of d and r (Chapman & Furlong 1992; Jaeger 1964).

Values of T_{max} (°C), the maximum temperature reached, can be calculated by using Equation [9.3] with the values of ϕ_{max} given by values of r_{max} and d in Table 9.3.

At the immediate contact of the intrusion ($d = 1$), the maximum temperature reached is half-way between the magma temperature and the initial temperature of the country rock. As a result, aureole temperatures will only exceptionally approach those of the magma, even for short periods of time. The fact that

Table 9.1 Quantities required for calculations of heating around sheet-like intrusions

Symbol	Unit	Quantity
y	m	half-thickness of intrusion
x	m	distance from centre of sheet to the point of interest
T	°C	temperature at the point and time of interest
T_0	°C	initial temperature of magma
α	m^2/a	thermal diffusivity (a measure of the rate of transfer of heat)
t	a	time in years

Table 9.2 Values of ϕ as a function of d and r, after Jaeger (1964)

d	r					
	0.01	0.04	0.2	1.0	4.0	10.0
0	1.000	1.000	0.886	0.520	0.276	0.177
0.4	1.000	0.983	0.815	0.503	0.274	0.176
0.8	0.921	0.760	0.622	0.455	0.266	0.174
1.2	0.079	0.240	0.376	0.384	0.253	0.171
1.6		0.017	0.171	0.303	0.237	0.166
2.0			0.057	0.222	0.217	0.160
4.0				0.017	0.106	0.119
6.0					0.032	0.073

Table 9.3 Values of r_{max} and ϕ_{max} for country rocks hosting an intrusive sheet for different values of d, after Jaeger (1964)

d	1.0	1.2	1.5	2.0	3.0	4.0	5.0
r_{max}	0	0.500	0.932	1.820	4.33	7.83	12.33
ϕ_{max}	0.500	0.407	0.324	0.242	0.161	0.121	0.097

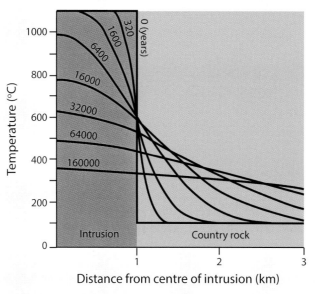

Figure 9.2 Temperature profiles showing the thermal evolution of a 2-km-thick sheet-like intrusion and its aureole. At the time of the intrusion (time t_0) the intrusion temperature was 1100 °C, emplaced into country rock at 100 °C. Successive thermal profiles are labelled with time in years after the timing of the intrusion, assuming conductive cooling only. Data from Jaeger (1964).

very-high-temperature rocks do occur in the innermost parts of many aureoles suggests the model is oversimplified. In particular, if magma flows past the contact for a time before solidifying, additional heat will be introduced by convection.

Figure 9.2 illustrates a series of successive temperature profiles across the contact of a vertical, sheet-like intrusion. Note how country rocks progressively further from the contact reach their maximum temperatures progressively later.

Worked Example

(a) A 400-m-wide diorite dyke with a temperature of 1000 °C intrudes into country rock at a temperature of 0 °C. What is the temperature of the country rocks 200 m from the contact after 1250 years to the nearest 10 degrees?

x	y	T_0	α
400 m	200 m	1000 °C	31.5 m^2/a

From [9.1], $d = x/y = 400/200 = 2$.
From [9.2], $r = \alpha t / y^2 = 31.5 \cdot 1250 / 200^2$
$= 0.984 \approx 1$.
From Table 9.2, $\phi = 0.222$.
Therefore, from [9.3], $T = \phi T_0 = 0.222 \cdot 1000 = 220$ °C.

(b) What is the maximum temperature that a point 200 m from the contact will reach?

From Table 9.3, $\phi_{max} = 0.242$. Since T_0 was 1000 °C, this means a maximum temperature of c. 240 °C.

crystallisation given off by the magma as it solidifies is balanced by the heat of reaction consumed by the endothermic reactions in the aureole. In detail, parameters such as thermal diffusivity are dependent on temperature rather than being fixed values, and the transfer of heat via convection is also ignored. Nevertheless, simple calculations such as these provide useful first-order insights into how long it takes for the country rocks to heat up following magma intrusion. They suggest that the timescale for contact metamorphism around a simple intrusion is likely to be in the order of tens to hundreds of thousands rather than millions of years. We will compare these predictions with results from field examples in Section 9.4.1.

9.1.2 Rates of Heating During Regional Metamorphism

Regional metamorphism occurs when rocks are heated at depth on a regional scale. The sources of heat for regional metamorphism include the mantle, mantle-derived melts intruding the crust on a regional scale, radioactive decay, frictional heating and local juxtaposition of crustal blocks of different temperatures along faults and shear zones.

Simple one-dimensional thermal models of instantaneously-overthrust continental crust (that emplace hotter crust on top of colder crust) suggest that it takes tens of millions of years to supply the heat required for regional metamorphism in this way. This long timescale is due to the insulating properties of the crust and the slow supply of heat (England & Thompson 1984). Figure 9.3 shows the evolving geothermal gradient in continental crust that is instantaneously overthrust by a 35-km-thick slice of the same crust (grey line marked 0 Ma). The figure also shows the evolving pressure–temperature–time path recorded by a rock buried to a depth of 40 km by the thrust. In a case with no erosion (Figure 9.3a), a rock 5 km beneath the thrust would stay buried at 40 km depth and would continue to heat up until thermal equilibrium – and a maximum temperature of >900 °C – were achieved after c. 200 Ma. However, if the surface is eroded at 1 mm/a for a period of 35 Ma, until the overthrust slice has been completely removed, the same rock is exhumed more quickly than the isotherms can evolve (Figure 9.3b). The rock would reach a maximum temperature of only c. 550 °C before starting to cool, with this temperature reached about 20 Ma after peak pressure. The initial rate of heating is the same in both cases (c. 10 °C/Ma). Note that the thermal evolution of the crust in Figure 9.3b is more complicated than in

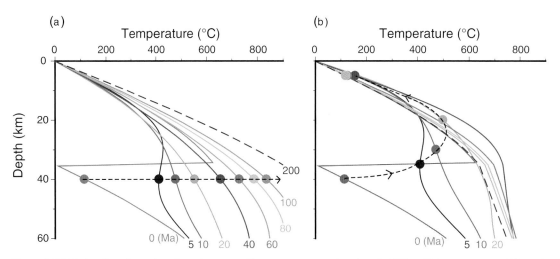

Figure 9.3 Calculated geothermal gradients for a case of instantaneous overthrusting of a 35-km-thick upper crust at 0 Ma and subsequent thermal relaxation. The coloured dots and black dotted line mark the thermal evolution of a rock initially buried at 40 km depth. The time stamps on the isotherms are the same in both panels. The model assumes a crustal thickness of 35 km, a heating rate of 1 μW/m² and a 200-km-thick lithosphere which is initially in thermal steady state. (a) No exhumation, or erosion. (b) Erosion at a rate of 1 mm/a until crust is restored to its original thickness. Figure from Alex Copley.

Figure 9.3a because of the upwards movement of crustal material. The final geothermal gradient eventually returns to where it had been before thrusting.

These thermal calculations suggest that regional metamorphism in response to burial is a slow process relative to the plate movements that drive it. The calculations also demonstrate that metamorphic pressure–temperature changes are a function of heat supply, erosion rates and tectonic movement.

9.1.3 Rates of Burial and Exhumation

The possible rates of burial during metamorphism may be estimated from rates of plate movement today. Modern plate-motion rates can be measured directly by long-term global positioning systems (GPS). For example, India and Asia are still colliding at a rate of 4 cm/a and oceanic crust is being subducted into the Mariana Trench at a rate of 5–8 cm/a. These rates of motion, along with estimates for the angle at which one plate is buried beneath another, provide us with estimates for the rates of crustal burial in different tectonic settings.

GPS measurements also provide information about modern uplift rates. For example, measurements suggest that due to the ongoing India–Asia collision, Mount Everest is still rising by ~0.5 cm/a. In general, tectonic uplift rates in mountainous regions at plate boundaries are of the order of 1–5 cm/a, a similar order of magnitude to tectonic burial rates. As we saw in Chapter 1, however, uplift involves the movement of rocks upwards compared to a fixed datum, e.g. the centre of the Earth or mean sea level (Figure 9.4a). Uplift on its own does not result in the rocks of the uplifting crust exhuming or experiencing a decrease in pressure.

Tectonic uplift is mainly driven by movement of rock along faults and shear zones beneath the surface. Rocks are both buried and transported towards the surface in this way but won't be exhumed completely without assistance from erosion. Erosion acts in tandem with tectonic uplift to remove overlying rocks and reduce lithostatic pressure (Figure 9.4c). It acts fastest in areas of steep topography, so mountainous areas will erode more rapidly than flatter ones (Figure 9.4b). A feedback loop is therefore created: mountainous regions form in areas of tectonic uplift, which itself maintains steep topography and fosters high erosion rates.

Estimates for rates of erosion vary from ~1 cm/a, or even higher, in areas of steep topography to $\ll 0.01$ cm/a in stable areas of muted topography (Figure 9.4c). It is also difficult to estimate long-term averages of erosion rates: modern estimates appear to be faster than long-term averaged estimates in the geological past (Kirchner *et al.* 2001).

9.1.4 Summary of the Physical Constraints on Metamorphic Cycles

During metamorphic cycles, rocks heat up (sometimes with and sometimes without associated burial), transform into new stable mineral assemblages and eventually

(a)

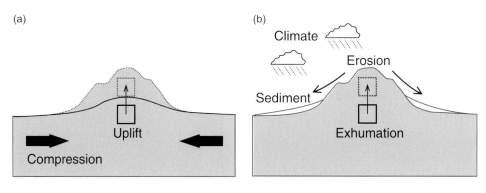

(c)

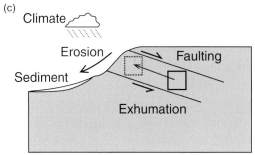

Figure 9.4 Cartoons showing the difference between uplift (a) and exhumation (b) and (c). (a) Uplift only. (b) Erosion-driven exhumation only. (c) Tectonically-driven exhumation assisted by erosion.

cool and decompress as they travel towards the surface. Geologists have developed a number of mathematical models based on physical properties such as conductivity and rates of tectonic processes to provide estimates of the timescales of metamorphism of different types and the way in which temperature and pressure vary relative to one another. The resulting predictions need testing against observations of natural rocks. As we have seen throughout this book, different minerals and mineral assemblages record different parts of this history. Later on in this chapter we will see how different minerals provide ages for different parts of the rock's evolution making it possible to estimate the rates at which pressure and temperature change through the metamorphic cycle and providing further tests for the predictions of how metamorphic belts evolve.

Modern measurements and estimates of tectonic movement and erosion rates are always discussed in units of distance/time, e.g. cm/a or km/Ma. However, metamorphic rocks record their pressure history (in MPa or GPa) not their depth history. We make a number of assumptions to convert pressure to depth, including assuming that lithostatic pressure is effectively equal to the total pressure experienced by the rock, and then convert to depth via a simple calculation (Chapter 1, Box 1.3). It is important to allow for the uncertainties involved in metamorphic pressure and depth estimates when comparing modern (GPS-derived) plate-tectonic rates with those estimated from the study of metamorphic rocks.

9.2   Minerals as Geological Clocks

In the past 25 years, developments in geochronology have made it possible to directly date minerals formed at different stages of a metamorphic cycle, and so test the rates of heating, cooling, burial and exhumation suggested by the theoretical models. The absolute timing of events in the geological record is determined by using the constant rate at which unstable elements break down via **radioactive decay** (Box 9.4). As radioactive decay always happens at a rate that is constant for the isotope that is decaying, minerals that incorporate radioactive elements can therefore act as geological clocks, or geochronometers.

Many of the most robust geochronometers are accessory phases; remember from Chapter 1 that these are minerals that might be fairly common in the sample, but are generally small in size and make up only a small proportion of the rock (typically a few percent or less). Many accessory phases (for example zircon, monazite, rutile) contain high concentrations of U. Because U decays to Pb through two different

BOX 9.4 **Radioactive decay**

Many elements occur as a number of different **isotopes**: atoms with the same number of protons and electrons but different numbers of neutrons in the nucleus. Some of these isotopes are unstable, leading to the eventual decay of the nucleus. Radioactive decay is the process by which the unstable atomic nucleus of the **parent isotope** breaks down to produce a new **daughter isotope**, releasing energy in the process. For example, uranium (U) has three naturally-occurring isotopes, ^{234}U (which makes up 0.006% of the total U), ^{235}U (0.72%) and ^{238}U (99.27%), all of which are radioactive and decay to ^{230}Th, ^{207}Pb, and ^{206}Pb respectively (Table 9.4). Potassium (K) also has three naturally-occurring isotopes, ^{39}K (93.3% of the total K), ^{40}K (0.012%) and ^{41}K (6.7%). Only ^{40}K is radioactive and it is one of a few isotopes that decays into two daughter isotopes: ^{40}Ca (89% of the time) and ^{40}Ar (11% of the time).

Radioactive decay is a random process, in that it is impossible to predict which atom will decay in any point of time. However the proportion of atoms in a population that will break down or decay over a fixed time interval is predictable. The amount of time that it takes half the atoms in a population to break down is known as the **half-life**. Different radioactive isotopes have different half-lives and geologists exploit these to provide absolute ages for different geological events.

Some of the most common radioactive decay schemes that are used in metamorphic geochronology are detailed in Table 9.4.

Table 9.4 Common radioactive decay schemes used in metamorphic geochronology

Parent isotope	Daughter isotope	Half-life (Ga)	Common mineral(s)
^{238}U	^{206}Pb	4.468×10^3	Zr, Mnz, Rut, Aln, Ttn, Xen
^{235}U	^{207}Pb	7.04×10^2	Zr, Mnz, Rut, Aln, Ttn, Xen
^{232}Th	^{208}Pb	1.40×10^4	Mnz, Aln
^{176}Lu	^{176}Hf	3.76×10^4	Grt
^{147}Sm	^{143}Nd	1.06×10^5	Grt, Ep
^{87}Rb	^{87}Sr	4.97×10^4	Mus, Bt, Fsp
^{40}K	^{40}Ar	1.251×10^3	Mus, Bt, Fsp, Amp

Half-lives from the *International Atomic Energy Agency* Table of Nuclides.

The age of the rock or mineral is determined by measuring the amount of the radioactive isotope present today and the amount of daughter isotope that has been produced and applying a simple equation:

$$t = \ln \frac{\left[\frac{(D-D_0)}{N} + 1 \right]}{\lambda} \qquad [9.4]$$

where t = time in millions of years,

D = number of atoms of daughter isotope in the sample now,

D_0 = number of atoms of daughter isotope in the sample when it first formed,

N = number of atoms of parent isotope in the sample now,

λ = the decay constant = (ln 2)/(half-life of the parent isotope)

The longer the half-life, the slower the radioactive decay, and the more useful the system is for determining ages of older geological events, because there is relatively more of the parent isotope left in the sample to measure. Equally, it may not be possible to use isotopes with very slow decay rates to date more-recent events because not enough daughter isotope has formed to be measured precisely. For example the decay of ^{147}Sm, with a half-life of 1.06×10^{11} a, is used to date solar system history, but ^{14}C, with a half-life of only 5730 a, is better for dating archaeological remains.

In practice, we can measure D/N in a mass spectrometer. In many cases we can safely assume that D_0 is zero (see Section 9.2.1). Given D/N we can therefore calculate an age.

Example Age Calculation

The ^{206}Pb/^{238}U in a zircon that crystallised in a migmatite is 0.0015. When did the zircon crystallise?

Answer

$(D-D_0)/N = 0.0015$

$t = \ln\left[(D-D_0)/N + 1\right]/\lambda$

$\quad = \ln\left[0.0015 + 1\right]/(\ln 2/4.468 \times 10^9)$

$\quad = 0.001499/1.55 \times 10^{10}$

$\quad = 3.03 \times 10^7\,\mathrm{a}$

$\quad = 30.3\ \mathrm{Ma}.$

This migmatite therefore crystallised in the Oligocene.

decay schemes (Table 9.4), a comparison of the ages provided by the two provides an in-built accuracy cross-check. If the ages from both decay schemes match, neither isotope system has been disturbed.

Common metamorphic rock-forming geochronometers include micas and feldspars (containing radioactive K and Rb) and garnet (Sm). These tend to provide less-reliable ages than the U-bearing accessory phases because there is no second decaying isotope to provide a cross-check that the system has not been disturbed. Micas and feldspars in particular are more prone to alteration than many other geochronometers.

Except in very high-grade metamorphic rocks, where residence at high temperatures may have erased chemical information by diffusion (Chapter 7), accessory mineral grains are often composites of material that has crystallised at different times – much like tree rings. For example, ancient zircon grains that have survived weathering, erosion and sedimentary deposition (known as detrital grains) may become overgrown with new zircon formed when the host sediment was metamorphosed during a later orogenic episode (Figure 9.5). Zircon is a very refractory mineral, which means that it does not get destroyed easily during igneous, metamorphic or sedimentary processing. Instead, each magmatic or metamorphic episode it experiences leaves a new layer of zircon wrapping the older core. Monazite similarly records multiple episodes of growth but is less robust; it only tends to record growth events experienced during the latest orogenic episode.

The development of microbeam techniques has allowed the elemental and isotopic composition of these individual growth zones to be analysed precisely. These *in-situ* techniques have fueled the recent revolution in accessory phase geochronology.

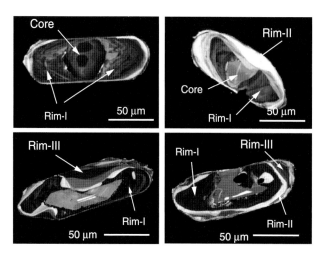

Figure 9.5 Cathodoluminescence images of polished zircon cross-sections showing cores and multiple episodes of rim growth (see also Figure 6.15). Here the cores are detrital remnants of a much older (and temporally unconnected) phase of zircon growth. The three later rims that have grown on these zircons are related to the most recent orogenic episode: here during the Himalayan orogeny. Figure from Regis *et al.* (2016).

9.2.1   Geochronometers

Early attempts to date rocks by radioactive decay involved measuring the total amount of U and Pb in mineral concentrates extracted from a U–rich Devonian rock (Holmes 1911). Holmes noted that in order for the concentration of Pb to provide a reliable age in any mineral, three assumptions needed to be justified:

1. That negligible Pb was present in the mineral at the time of its formation.
2. That all the Pb in the mineral formed by radioactive decay and not by any other mechanism.
3. That no Pb or U was added to or removed from the mineral after it formed.

These assumptions, generalised to all parent and daughter isotopes, still underpin geochronology today, although modern geochronology is based on measurements of individual isotope abundances and not simply the total amount of the element. Provided it is possible to analyse just a single growth stage, accessory minerals such as zircon and monazite make very reliable geochronometers because they fulfil all three of Holmes's assumptions. Other radioactive decay schemes, including many of those applied to common rock-forming minerals such as micas, are more difficult to interpret because significant amounts of the daughter isotope may have been incorporated into the mineral at the time of growth, or lost by diffusion since the mineral grew (see Chapter 7, Box 7.2). Corrections and further assumptions can be made to counteract these issues, but they do increase uncertainty in the result. Such approaches are detailed in geochronology textbooks (see Further Reading). Geo-chronometers that record an age that is younger than the time that mineral crystallised, due to daughter isotope loss by diffusion, are known as **thermochronometers** (see Section 9.2.2 below).

Metamorphic minerals used regularly as geochronometers include garnet, micas and a number of accessory minerals. Garnet commonly incorporates Sm (in metapelites) and Lu (in metabasites) into its crystal structure and can be dated using the Sm–Nd and Lu–Hf decay systems. As we have seen in earlier chapters, garnet is widespread in medium- to high-grade metabasites and metapelites. Although diffusion of both parent and daughter isotopes is possible at high grades, garnet Sm–Nd and Lu–Hf geochronometers are considered to provide robust crystallisation ages under most common metamorphic conditions (Smit *et al.* 2013).

Micas are commonly dated by the K–Ar (analysed as $^{40}Ar/^{39}Ar$ in the laboratory) and Rb–Sr decay systems, and are useful as geochronometers in low-grade rocks such as slates (Sherlock *et al.* 2003). In higher-grade rocks, micas are more commonly used as thermochronometers (see below).

Rutile incorporates U into its crystal structure, as well as a small amount of Pb, therefore it is most commonly dated by using the U–Pb system. Rutile is a common accessory phase in medium- to high-grade and high-pressure metabasites and metapelites. It is used as a geochronometer under most common metamorphic conditions, but in high-grade rocks rutile may also be used as a thermochronometer.

Zircon incorporates U but shuns Pb when it crystallises. It is most commonly dated by using the U–Pb system. Zircon is an accessory phase that is relatively common in medium- to high-grade metabasic and high-grade pelitic rocks, especially when they have reached conditions that allow partial melting. Zircon is very refractory, which means that in metasedimentary rocks zircon nearly always retains a core that is inherited from one or more previous tectonic episodes.

Monazite incorporates U and Th into its crystal structure, as well as a small amount of Pb. It is most commonly dated by using the U–Pb or Th–Pb systems. Monazite is a common accessory phase in medium- to high-grade metapelites. It only rarely retains inherited (detrital) cores except at very low grades.

Zircon and monazite commonly occur as small inclusions in biotite, and in thin section they are surrounded by dark haloes of radiation-damaged mica (Figure 9.6). These haloes are best developed in very old rocks or around inclusions with particularly high uranium concentrations. Where they occur elsewhere in the rock, they may be harder to spot.

Other accessory phases that are used as geochronometers include xenotime, apatite, titanite and allanite. These minerals are all dated using the U–Pb system, and further information can be found in the Further Reading at the end of this chapter.

9.2.2 Thermochronometers

We saw above that reliable metamorphic geochronometers should not lose any daughter elements by diffusion or other processes after they first grew. However,

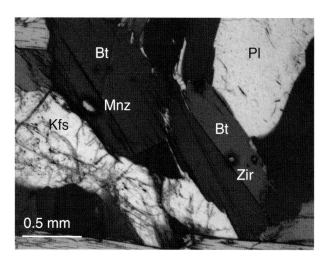

Figure 9.6 Photomicrograph of a granulite-facies pelite, Ivrea Zone, Alps, showing radiation damage 'haloes' around zircon and monazite inclusions in biotite. (Photo: Barbara Kunz.)

the loss of daughter isotopes can be turned to our advantage for dating these processes. Loss of daughter isotopes is most commonly associated with thermally-activated diffusion. The temperature below which a daughter isotope is retained in the mineral structure rather than escaping by diffusion is known as the **closure temperature** (Dodson 1973). Strictly speaking, a range of temperatures exists over which diffusion changes from slow to fast on geological timescales, and Dodson's simple empirical formulation, which provides a closure temperature somewhere in the middle of that range, is defined as *the temperature of a mineral at the time of its apparent age.*

Argon, the daughter isotope of K–Ar decay, is a noble gas which does not bind into the crystal structure and is therefore particularly susceptible to diffusive loss at high temperatures. The closure temperatures of K-bearing minerals such as muscovite and biotite are low, in the range of 300–450 °C for muscovite and 300–350 °C for biotite, but depend in detail on composition, grain size and cooling rate (Dodson 1973; Harrison *et al.* 1985, 2009). Measurement of the accumulation of Ar in micas therefore allows the timing of exhumation through the mid-crust to be estimated. Other processes such as deformation, dissolution and recrystallisation may also remove ^{40}Ar from micas, so the precise interpretation of ^{40}Ar/^{39}Ar mica ages is currently debated (Villa 2016).

Decay of U and Th generates alpha particles, which are equivalent to He nuclei. As a result, helium accumulates in U- or Th-bearing accessory minerals such as apatite or zircon. Helium, also a noble gas, is even more mobile than argon and is easily lost from mineral structures at low temperatures. Closure temperatures are around 60–80 °C for apatite and 180–190 °C for zircon, although these may be lowered still further if the crystal structure is imperfect (Farley 2000; Guenthner *et al.* 2013). Measurement of the accumulation of He in these minerals allows the timing of the final stages of exhumation to be determined.

The decay of U and Th in apatite and zircon can also be used as a thermochronometer in a different way. As U and Th decay, alpha particles are violently ejected from the parent nucleus, a bit like bullets. These ejections leave behind damage trails, known as fission tracks, in the crystal structure. At high temperatures, the crystal repairs itself, annealing the track. At low temperatures the tracks remain, providing a record of the number of decays and thus the time that has elapsed since the mineral passed into the 'track retention zone'. Both the number of tracks and the track lengths provide information about thermal history (Tagami & O'Sullivan 2005).

In high-grade (granulite-facies) rocks, minerals that are used as geochronometers in lower-grade rocks become useful thermochronometers for constraining the early stages of cooling. Such phases include garnet (Lu–Hf and Sm–Nd systems) and rutile (U–Pb system). Both of these have closure temperatures in excess of ~600 °C (Cherniak 2000; Smit *et al.* 2013).

9.3   Linking Age to Metamorphic History

Many metamorphic rocks preserve a history of mineral growth, through the presence of inclusions, chemically-zoned minerals and grains formed at different times relative to deformation (Chapter 7). As analytical techniques have improved it has become possible to date these distinct **metamorphic stages** of the metamorphic cycle. Geologists used to talk about metamorphic 'events' as if they happened instantaneously: e.g. 'The eclogite facies rocks in Kaghan, Pakistan, formed 46 Ma ago'. Increasingly, the ages of distinct episodes of mineral growth can be differentiated: e.g. 'Allanite in the Kaghan eclogites crystallised at 46.5 ± 1.0 Ma on the prograde path at high pressures, zircon with inclusions of coesite grew at 46.4 ± 0.1 Ma at peak temperatures of 700 °C, titanite cooled through 600 °C by ~44 Ma and rutile cooled through 500 °C shortly after' (Parrish *et al.* 2006); see also Figure 9.13b. Metamorphic geologists therefore now have lots of different tools available to them to find out not only *when* the burial and exhumation cycle took place, but also *how quickly* metamorphic changes progressed and *how long* the metamorphic cycle took overall.

To understand how metamorphic terranes form, it is necessary to piece together multiple pieces of information from different rocks exposed within them, including identification of different equilibrium assemblages, their relative timing of crystallisation, their connection to episodes of deformation, and the *P–T* conditions of their formation, as outlined in Chapter 3. Absolute dating of minerals that form part of those equilibrium assemblages allows geologists to add absolute time to metamorphic *P–T* history and therefore to piece together absolute **metamorphic pressure–temperature–time (*P–T–t*) paths**. These paths allow us to constrain absolute rates and timescales of natural metamorphic cycles and allow rate and timescale predictions from laboratory-derived experimental rates and heat flow calculations to be tested.

Linking 'age' to 'metamorphic stage', 'assemblage' or 'process' is key to constraining absolute metamorphic P–T–t paths. Unless you only want to determine whether a metamorphic terrane formed during, for example, the Cretaceous or Miocene, a mineral date is not much use for understanding the metamorphic history without knowing the stage of the metamorphic process (e.g. prograde, peak or retrograde) to which that age refers. Metamorphic geologists therefore now commonly distinguish between '**date**', which is the number in millions of years that derives from the ratio of parent to daughter isotope in a sample, and '**age**', which is a date with a geological interpretation. Accessory phases such as zircon and monazite can yield highly-precise dates; however, the P–T conditions of their crystallisation reactions are seldom well constrained, and thus interpreting dates as ages can be difficult. Furthermore, zircon rarely crystallises during reactions that involve the main rock-forming minerals. This means that it is difficult to link the timing of zircon crystallisation into the P–T path determined from the major rock-forming minerals. In order to use accessory phase chronometers to constrain rates and timescales, therefore, it's often necessary to use indirect techniques to link them to the evolution of the main mineral assemblage.

9.3.1   Direct Links: Dating Index Minerals

Garnet has long been recognised as a particularly useful mineral for metamorphic geologists because it crystallises over a wide portion of the prograde metamorphic path, in a wide compositional range of rocks (Chapters 4 and 5), and the reactions responsible for its formation are generally well-constrained (e.g. Baxter & Scherer (2013). This means that the links between P–T conditions and garnet major-element composition are well known for a wide variety of bulk rock compositions. The timing of garnet growth therefore provides a direct link to the timing of parts of the metamorphic cycle.

However, the very fact that garnet can grow over such a protracted period of metamorphic history means that a whole garnet crystal date determined from Sm–Nd or Lu–Hf analysis will provide only an *average* date for the entire growth span. This is fine if the aim is to determine whether, for example, the metamorphic terrane is Cretaceous or Jurassic in age, but not so useful if the aim is to constrain the timescale of garnet growth during that particular metamorphic cycle.

In practice, the bulk of the parent isotope is concentrated into a single part of the garnet. For example lutetium is strongly partitioned into garnet cores (Section 2.4.3) such that the concentration in the core may be up to 100 times that of the rim. A garnet Lu–Hf date is therefore more likely to be biased towards the timing of garnet core growth. Samarium concentrations in garnet, on the other hand, are commonly not so strongly zoned, or may be higher in the garnet rim. A whole-grain Sm–Nd date is therefore likely to be younger than a Lu–Hf date determined from the same grain. In practice, it is uncommon to find garnets with high enough Sm, Nd, Lu *and* Hf concentrations to be able to date them by both methods: Lu enrichment is more common in mafic garnets, whereas Sm enrichment is more common in metasedimentary (pelitic) garnets.

Recent technical developments such as sampling by laser ablation or micro-drilling coupled with high-precision mass spectrometry allow particular parts of individual grains of garnet to be dated accurately and precisely. Figure 9.7 shows Sm–Nd dates of different zones, sampled by micro-milling, in a 6 cm diameter garnet porphyroblast from the Austrian Alps. These dates show that this garnet crystal grew over 7.5 Ma, between 27.5 Ma and 20 Ma, in two periods of rapid growth bracketed by periods of slower growth (Pollington & Baxter 2010).

This particular garnet was sampled from a shear zone in which a granodioritic rock was transformed into a chlorite–garnet schist during metasomatism. Independent thermodynamic modelling work for this particular area suggested that the garnet grew during decompression from 1.1–0.7 GPa at $T < 550$ °C (Selverstone *et al.* 1991). Assuming a crustal density of 2550 kg/m^3, this implies that the rock travelled 16 vertical km in 7.5 Ma. The garnet-zone ages therefore constrain a vertical exhumation rate for this area of the Alps of c. 2 mm/a averaged over that time (Pollington & Baxter 2010).

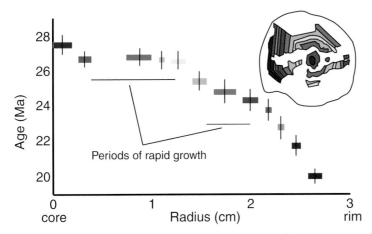

Figure 9.7 Radial Sm–Nd ages across a 6 cm diameter garnet from the Austrian Alps (shown schematically in the top of the figure). The white areas in the top figure represent areas that were cracked or otherwise not suitable for sampling. In the bottom chart, the width of the coloured bar represents the thickness of the garnet 'ring' that was sampled. The vertical black bars are the 2σ uncertainties on the ages. Figure modified from Baxter & Scherer (2013); data from Pollington & Baxter (2010).

9.3.2   Textural Correlation

The textural relationships between minerals in a metamorphic rock provide significant clues about relative timing of metamorphic change in general, and equilibrium relationships in particular. As noted in Chapter 7, inclusions can be particularly useful, as in many cases they provide insight into the equilibrium mineral assemblage present at the time when the host mineral(s) was/were growing. Dating inclusions of geochronometer phases in index minerals such as garnet, staurolite or kyanite constrains the timing of growth of the index mineral. This presupposes that the index and inclusion mineral grew at the same time and were therefore in equilibrium (Mottram *et al.* 2014). Inclusions of index minerals inside geochronometer minerals likewise constrain the timing of mineral growth along specific parts of the *P*–*T* path under the assumption that they grew in equilibrium. Although minerals such as zircon and rutile are commonly very small compared to minerals such as garnet and biotite, they may trap inclusions indicative of particular metamorphic conditions as they grow.

Garnet is particularly useful because it commonly hosts significant quantities of inclusions of datable accessory minerals such as rutile, zircon and monazite, which can themselves provide further constraints on the timing of garnet growth. For example, one detailed study showed how the ages of monazite inclusions trapped in a particularly inclusion-rich garnet porphyroblast from the Grouse Creek Mountains, Utah, changed from core to rim (Figure 9.8; Hoisch *et al.* 2008). The monazite ages ranged from 72 Ma in the garnet cores to 37 Ma in the garnet rims, suggesting not only that the garnets grew over a timespan of 35 Ma, but also that the monazites were forming via one or more continuous reactions such that each monazite trapped by the growing garnet formed at the same time as the engulfing garnet. If the monazites had formed only once, by a single (discontinuous) reaction, then all the monazite inclusions would yield the same age, regardless of where they were in the garnet. The conclusion of the study was that the monazites were co-crystallising with the garnet via a muscovite-breakdown reaction, thus allowing them to be locally formed and easily engulfed by the growing garnet.

As another example, zircon can trap and preserve inclusions of coesite, the *HP* polymorph of quartz (Chapter 5, Section 5.5.3), during metamorphism at ultra-high pressures. Remember that coesite forms at pressures >2.6 GPa and rapidly (in a matter of minutes) transforms back to quartz at lower pressures unless trapped in a strong mineral that prevents its expansion. The presence of inclusions of coesite in zircon shows that the host zircon grew around (and trapped) coesite at pressures >2.6 GPa. The age of the zircon in the immediate vicinity of the inclusion can therefore constrain the timing of *UHP* metamorphism. Tiny inclusions of other pressure-indicator minerals such as diamond, omphacite, albite and kyanite in zircon may also provide indications of pressure in different settings.

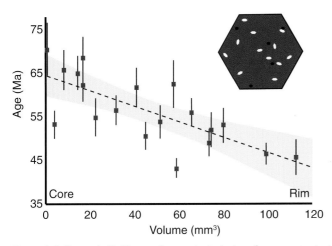

Figure 9.8 Change in U–Pb age of monazite inclusions from core to rim in a garnet from the Grouse Creek Mountains, north-west Utah. Errors are 1σ on each inclusion age. The 95% confidence interval is shown in yellow with the best fit shown as a dotted line. The insert shows a schematic garnet crystal with representative distributions of monazite (white) and other mineral (black) inclusions. The relative inclusion size is illustrative. Figure modified from Kohn & Penniston-Dorland (2017); data from Hoisch *et al.* (2008).

9.3.3   Chemical Correlation

As metamorphic minerals grow, they incorporate trace elements into their crystal structures. These trace elements, not essential to the mineral crystal structure, may act as passive 'tape recorders' of other mineral reactions taking place in the rock at the same time as the host mineral is growing. Figure 9.9 shows the changes in concentration of a number of different elements measured from the core to rim of a garnet porphyroblast in a migmatite. These changes provide tantalising clues about the crystallisation and dissolution of the accessory geochronometer phases in the rock, allowing their crystallisation age to be linked to the timing of garnet growth.

We first need to look at the rock as a whole, to provide the framework within which the trace-element concentration changes can be interpreted. As we saw in Chapter 4, garnet is likely to start growing in a pelite long before temperatures become high enough for the rock to start melting. So, in this case, it is likely that the garnet core grew in the solid state. Garnet can also crystallise in the presence of melt, especially where biotite is involved in the melting reactions (see reactions [4.17]–[4.19] in Chapter 4, Section 4.3.1). As this rock is a migmatite, we need to leave open the possibility that part of the garnet, for example the rim or outer rim, therefore grew in the presence of melt.

As we saw above, Lu is strongly partitioned into garnet when it crystallises – high Lu concentrations are shown in the garnet core in Figure 9.9. Lu concentrations are considerably decreased in the garnet rim, and this suggests that another phase is competing with the garnet for Lu at this point. This could be the co-crystallisation of

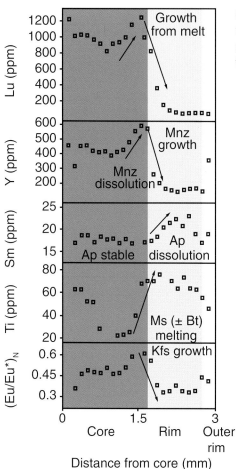

Figure 9.9 Changes in concentration of different trace elements and trace element ratios across a garnet from a migmatite, Bhutan. Figure modified from Regis *et al.* (2014).

a mineral such as zircon, which, unlike monazite, has a high preference for Lu. Zircon most easily crystallises in metapelites in the presence of melt, so the Lu concentrations provide one piece of evidence that melt may also have been present when the garnet rim grew.

Similarly, Y is strongly partitioned into monazite but not into zircon. When monazite co-crystallises with garnet, there is less Y available to be incorporated into the garnet. Figure 9.9 shows a pronounced decrease in Y concentrations in the garnet rim compared to the core, which provides a strong hint that monazite was co-crystallising at that time. Monazite easily crystallises from melt, so this evidence still fits with our working hypothesis that melt was present.

In contrast, Sm concentrations in the garnet rim in Figure 9.9 are higher, rather than lower, than in the garnet core. Minerals such as apatite are rich in Sm, and apatite is also highly soluble in melts.

Ti is strongly fractionated into micas, especially biotite. The higher Ti concentrations in the garnet rim support the hypothesis that biotite was melting at the same time as the garnet rim was crystallising.

Finally, Eu partitions strongly into feldspars, which form as a product in many mica melting reactions (Chapter 4, Section 4.3.1). As a result, when garnet and feldspar grow together, the garnet contains less of the Eu released from reactants than garnet grown independently. Geochemists reference the concentrations of Rare Earth Elements, such as Eu and Sm, to their concentrations in primitive, chondrite meteorites and often find that the actual concentration of Eu is different to what we would predict, based on patterns in chondrites. This deviation is known as a 'Europium Anomaly', represented as $(Eu/Eu^*)N$. The drop in the Europium Anomaly at the garnet rim in Figure 9.9 suggests that the rim grew at the same time as K-feldspar.

In summary, there are a number of trace elements that track the production or destruction of different minerals and melt during metamorphism. Minerals such as garnet that may grow over a range of temperatures during metamorphism, though multiple mineral reactions, may record evidence for specific reactions in their trace element concentrations. Unlike the major elements, whose concentrations may be affected by diffusion at high temperatures (Chapter 3), trace elements have lower diffusivities and are thus able to retain the record of metamorphic histories to higher temperatures.

9.4   Timescales of Metamorphism

As we saw in Section 9.1, thermal modelling suggests that pulses of heat causing contact metamorphism last on the order of tens to hundreds of thousands of years, while regional metamorphic cycles last on the order of tens of millions of years. As analytical equipment has become more sensitive and precise over the past couple of decades, metamorphic geologists are now able to interrogate the rock record to determine the accuracy of their theoretical predictions.

9.4.1   Contact Metamorphism

Theoretical predictions suggest that contact metamorphic episodes last in the order of tens to hundreds of thousands of years (Section 9.1.1). For example, the duration of metamorphism of the contact aureole surrounding the South Mountain granite in Nova Scotia, Canada – which had a similar metamorphic evolution to the Bugaboo aureole example described in Chapters 1 and 4 – was determined by modelling the thermal conductive history that best fit the distribution of peak metamorphic temperatures away from the contact (Hilchie & Jamieson 2014). Their calculations suggested that peak temperatures within 1 km of the intrusive contact were reached within 50 ka after granite emplacement and that the outer part of the intrusion likely reached peak temperatures 200–450 ka later.

In South Africa, the intrusion of an 8-km-thick sheet of ultra-mafic rocks (the Bushveld intrusion) led to the contact metamorphism of a 4-km-thick, 25-km-wide aureole in pelitic rocks (Waters & Lovegrove 2002). Following a similar logic to that outlined in Box 9.3, these researchers suggested that the timescale of metamorphism of this huge aureole was around 2 Ma.

These timescales are too short to be currently resolvable by isotopic techniques, when dealing with geological events that took place more than a few million years ago. This is because analytical uncertainties on geological ages, even on the most precise currently-available instruments, are between 0.5% and 1%. The further back in time, the larger the absolute magnitude of the age uncertainty: at an age of 1 Ma, 0.5% uncertainty is 5 ka but at 1 Ga, 0.5% uncertainty is 5 Ma. The analytical uncertainty on the age therefore rapidly becomes greater than the duration of the geological event. The Bushveld aureole, for example, is ~2 Ga old (Scoates & Friedman 2008). Confirmation of the predicted 2 Ma timescale for contact meta-morphism by direct dating methods is therefore outside the currently-available analytical precision.

9.4.2   Regional Metamorphism

We saw in Section 9.1.2 that simple models of the thermal consequences of burying continental crust during orogenesis (such as by thrusting) predict that regional metamorphic cycles should proceed over tens of millions of years (Figure 9.3). This prediction has now been confirmed in many regionally metamorphosed terranes by using direct dating techniques and the trace-element 'detective' methods outlined in Section 9.3.3.

The Songgpan-Garzê fold belt in the eastern Tibetan Plateau provides a good example (Figure 9.10). The area was originally a sedimentary basin containing 6–8 km of turbidite sequences, and was metamorphosed and folded in the late Triassic. In the region around the town of Danba, an antiformal dome exposes a typical moderate-temperature, moderate-pressure (*MT–MP*) prograde sequence, with chlorite-grade rocks exposed on the outer limbs and an increase in grade towards migmatites in the core.

Monazite grains in the staurolite-, kyanite- and sillimanite-grade samples pro-vide constraints on the timing of peak metamorphism. These minerals suggest that peak conditions were reached in staurolite-grade rocks at 192 Ma, in kyanite-grade rocks at 184 Ma and in sillimanite-grade rocks at 179 Ma (Figure 9.11). These data imply relatively continuous heating at a rate of ~7 °C/Ma over at least ten million years, compatible with theoretical predictions of conductive heating models during slow burial during continental collision.

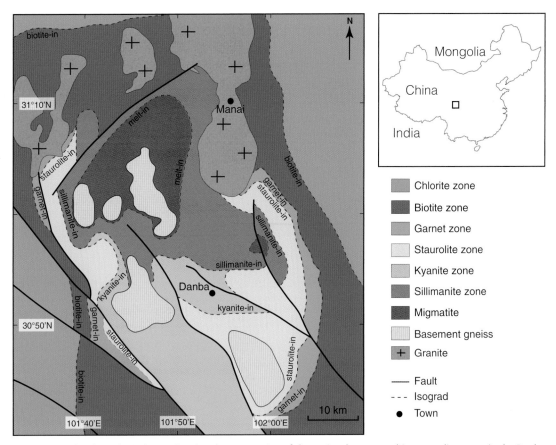

Figure 9.10 Sketch geological map showing the progression of the regional metamorphic metapelite zones in the Danba dome, Songgpan-Garzê fold belt, China. The index mineral zones have the same colour scheme as the Scottish Highlands regional metamorphism example in Figure 1.2. Modified after Weller *et al.* (2013).

Metamorphic Field Gradients We can use the *P–T–t* information recorded by metamorphic rocks to provide information about the local heat flow at the time of metamorphism. The *P–T* line defined by the peak *P–T* conditions of a series of rocks from the same metamorphic region is termed a **metamorphic field gradient,** or **apparent geotherm** (see Section 1.4.1 and the grey lines on Figure 9.11). In actively-deforming areas, the crustal geotherm (Chapter 1, Section 1.4.1) is generally **transient,** or constantly changing, as colder and hotter rocks are juxtaposed next to each other in different parts of the system at different times. Peak metamorphic conditions are therefore also reached by different parts of the system at different times, as shown in Figure 9.11.

Metamorphic field gradients are useful for unravelling different thermal events that affect different parts of a metamorphic region at different times. For example, a compilation of 116 *P–T* paths for an area of the Central European Alps (the area

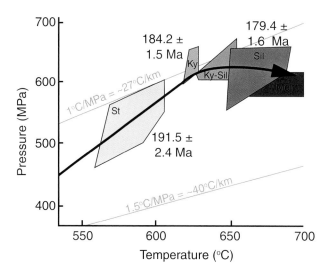

Figure 9.11 Pressure–temperature–time (*P–T–t*) path of the Danba region, China, moderate-pressure, moderate-temperature (*MP–MT*) metamorphism. Grey lines are geotherms, coloured fields are conditions for the matching zones in Figure 9.10.. Modified from Weller *et al.* (2013).

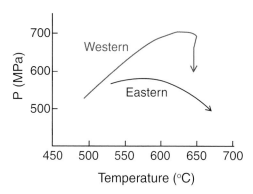

Figure 9.12 Metamorphic field gradients for the western and eastern parts of the Central Alps. The lines are drawn through the peak *P–T* conditions of a number of samples collected along two transects that run from north to south across the studied area. Figure modified after Todd & Engi (1997).

shown in Figure 6.6) which reached amphibolite-facies conditions, showed that two different thermal events had affected the area at different times (Todd & Engi 1997). In the western part of the area, an earlier higher-pressure event was preserved (Figure 9.12); in the eastern part of the area, a younger thermal event had metamorphosed the region under lower-pressure conditions. This study resolved a long-standing discussion about the link between metamorphism and tectonics in this region.

High-Pressure (*HP*) Metamorphism High-pressure (*HP*) metamorphic rocks, including blueschists and eclogites, tend to yield their age and timescales of formation begrudgingly. Accessory-phase geochronometer minerals such as zircon are relatively rare and tend to be rather small, making them difficult to analyse precisely (Rubatto & Hermann 2007). Additionally, the lack of tightly constrained geobarometers and the non-equilibration of mineral assemblages due to the relatively cool temperatures of metamorphism make linking 'age' to 'stage' rather difficult. Inclusions of *(U)HP* mineral indicators such as kyanite, omphacite, coesite or diamond, although rare, provide some of the necessary compelling evidence that the rocks crystallised at high-pressure conditions.

A number of recent detailed *P–T–t* studies have shown that *HP* rocks – and especially those that have reached *UHP* conditions – form and exhume much faster than lower pressure, regionally metamorphosed rocks (Chapter 10). The *(U)HP* rocks exposed in the Dora Maira Massif, Western Alps, provide a good example (Figure 4.16). This Massif contains Permian–Carboniferous-aged metapelites with mafic and granitoid intrusions. The entire unit was metamorphosed to ultra-high pressures during the Tertiary Alpine orogeny. The calc-silicate units in this Massif contain multiple generations of titanite that were dated by using the U–Pb decay scheme (Rubatto & Hermann 2001). The first titanite generation documents crystal-lisation during pre-Alpine metamorphism at c. 250 Ma, the second documents formation at *UHP* conditions during the Alpine orogeny (at c. 35 Ma), and the last two generations document crystallisation at lower pressures during exhumation (at c. 33 Ma and 32 Ma). Petrological and chemical links between the titanite generations and the host rock assemblages constrained the *P–T* conditions that existed at the time the titanite generations crystallised. Together these data suggest that the Dora Maira Massif was buried to and metamorphosed at (mantle) depths of 110–120 km 35 Ma ago, but was transported back to (mid-crustal) depths of c. 20 km by 32 Ma ago (Figure 9.13a). This equates to vertical speeds of 2–5 cm/a, similar to horizontal plate tectonics rates.

Overall the data suggest burial and exhumation cycles for *UHP* rocks from the surface, down to mantle depths and back to near the surface again, last on the order of 5–10 Ma. When first proposed for the Dora Maira, these rates seemed astonishingly quick. However, since then, similarly-rapid exhumation rates have been suggested for *UHP* eclogites exposed in eastern Papua New Guinea (Baldwin *et al.* 2004), and in Kaghan, western Himalaya (Parrish *et al.* 2006); Figure 9.13. The evidence for rapid exhumation rates from *UHP* conditions from three separate regions suggests that these rates must relate to the exhumation process itself (covered in Chapter 10).

High-Temperature (*HT*) Metamorphism Accessory mineral ages from *HT* (>700 °C) and *UHT* (>900 °C) terranes most commonly yield spreads of ages across tens of Ma, suggesting that high-temperature events last considerably longer than high- and ultra-high-pressure metamorphic cycles (Kelsey & Hand 2015). For example, the Reynolds Range in Australia records a complete regional metamorphic gradient in basement and metasedimentary rocks from greenschist facies through to granulite facies. Zircon and monazite that crystallised in partially-melted granulite-facies rocks record ages spanning 30 Ma, suggesting both that the high temperatures were experienced over that

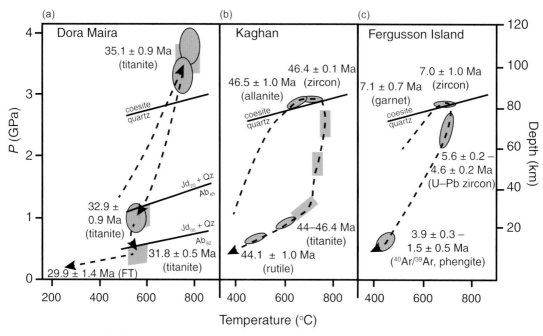

Figure 9.13 *P–T–t* paths for *UHP* rock exhumation in (a) Dora Maira (modified from Rubatto & Hermann (2001)), (b) Kaghan (modified from Parrish *et al.* (2006)) and (c) Fergusson Island (modified from Baldwin & Das (2015)). FT= fission track data.

time and also that zircon and monazite were continuously crystallising or recrystallising throughout (Rubatto *et al.* 2001).

At the more extreme end of the scale, metasedimentary gneisses in the Southern Granulite Terrane in India, which reached temperatures of >900 °C, record zircon and monazite ages spanning 100 Ma (Taylor *et al.* 2016). Careful investigation has shown that this long time frame is separable into multiple prograde and retrograde zircon and monazite crystallisation events, driven by multiple episodes of heating and fluid infiltration events affecting rocks of different bulk composition (e.g. Johnson *et al.* 2015; Taylor *et al.* 2014).

The reason why accessory mineral age data from *HT* terranes do not generally yield single age populations is because crystallisation is driven by different processes in different chronometers and in different composition rocks. Another complicating factor for *UHT* terranes in particular is that the combination of high temperatures and long timescales mean that daughter elements such as Pb can mobilise and diffuse both within, and beyond, host grains or the chemical zones within them. The overall effect of this mobilisation is dependent both on grain size and temperature, with the greatest effect being seen in smallest grain sizes in the hottest rocks. Recent advances in analytical resolution have allowed these to be distinguished, especially in cases where the high temperatures have been experienced for a sufficient length of time.

Ultimately the timescales of *(U)HT* metamorphic cycles will depend on the longevity of the heat source. High temperatures may be sustained in deep cratonic crust for tens of millions of years due to the longevity of a mantle heat source and slow removal of the overburden by erosion (at rates of <1 mm/a). However, as we have seen throughout this book, minerals will only crystallise when there is a chemical driving force for them to do so, so chemical and thermal perturbations are required to make the geological clock start ticking.

SUMMARY

In this chapter we have shown how the duration of metamorphic episodes can be estimated theoretically by thermal modelling, and confirmed or disputed independently by geochronological methods based on radioactive decay. In many cases the results of both approaches appear to be consistent, except in instances where metamorphism was not controlled predominantly by conductive heat transfer.

Direct dating of metamorphic minerals relies on the process of radioactive decay. Different radioactive elements with different decay rates are partitioned into a variety of mineral structures. Common accessory-phase geochronometers include zircon, monazite, rutile and titanite. Garnet is a useful rock-forming mineral chronometer. The best geochronometers incorporate the parent element but exclude the daughter element when they first crystallise. They have slow diffusion rates for the daughter element at the metamorphic temperatures and timescales of interest. When geochronometers lose their daughter elements by diffusion at high temperatures, and therefore yield an age that is younger than the time of mineral crystallisation, they are called thermochronometers. Mica is a commonly-used example. Other processes such as deformation and fluid flow may also affect isotope ratios and hence mineral age.

Rocks may be heated by both conduction and convection. The rate of heating depends on the temperature gradient, the rate of heat supply and lithology. Thermal modelling predicts that rocks in the vicinity of a local intrusion will heat up quickly, and contact metamorphism will take place within tens to hundreds of thousands of years depending on the size, shape and temperature of the intrusion. These timescales are currently too short to be verified by isotopic dating. However, given the rapid pace of improvement in analytical precision, this may change soon. In contrast, because heat is supplied over longer timescales during regional metamorphism, these cycles may last millions to tens of millions of years. These estimates have been shown to be within the right order of magnitude by absolute geochronology.

EXERCISES

1. What is the difference between a geochronological dataset that is accurate and one that is precise?
2. Using the terms 'conduct' or 'conduction', explain:
   (a) why geotherms in subduction zones are depressed (refer to Figure 1.6 if required), and
   (b) why peak pressures are commonly reached before peak temperatures in many metamorphic rocks.
3. A basaltic sheet 200 m thick and at a temperature of 1200 °C is emplaced into the upper levels of the continental crust, which you may assume to be at 0 °C.
   (a) What is the maximum metamorphic temperature reached at the contact, according to the simple thermal model set out by Jaeger (1964)?
   (b) How far away from the intrusion does the temperature reach 300 °C after 1270 years?
   (c) At what time are peak temperatures reached 50 m from the contact?
4. Some minerals, like zircon, incorporate negligible concentrations of the important geochronological daughter isotope (in this case Pb) into their structures during growth. Other minerals incorporate measurable quantities of the relevant daughter isotope into their structures when they crystallise. If uncorrected, would the effect of this 'excess' daughter isotope be to make the mineral appear older or younger than it really is? Why?
5. A metapelitic gneiss contains an equilibrated assemblage of garnet, biotite, muscovite, kyanite, plagioclase and quartz. There are inclusions of monazite in the garnet core and the garnet rim.
   (a) Which minerals might constrain the time of prograde, peak and retrograde metamorphism?
   (b) Which decay schemes would you use to date those minerals?
6. Monazite grains occurring as inclusions in a garnet core, mantle and rim yield the following U/Pb ratios.

Position	$^{238}U/^{206}Pb$	$^{235}U/^{207}Pb$
Core	0.0339	0.236
Mantle	0.0334	0.232
Rim	0.0326	0.226

   (a) Use the half-lives in Table 9.4 to calculate values of λ^{238} and λ^{235} to three significant figures, remembering that λ (decay constant) = ln(2)/half-life of the parent element. (Be careful with units! Remember the values in Table 9.4 are provided in Ga.)

(b) Calculate the $^{238}U/^{206}Pb$ and $^{235}U/^{207}Pb$ ages of the monazite inclusions in the garnet (in Ma to three significant figures). Are the ages for both decay systems the same? What can you conclude from this?

(c) Estimate the length of time the garnet took to grow (to the nearest Ma). What assumption(s) have you had to make in this calculation?

FURTHER READING

Baxter, E. F., Caddick, M. J. & Dragovic, B. (2017). Garnet: A rock-forming mineral petrochronometer. *Reviews in Mineralogy and Geochemistry*, **83**(1), 469–533.

Dickin, A. P. (2018). *Radiogenic Isotope Geology*. Cambridge University Press.

Engi, M. (2017). Petrochronology based on REE-minerals: monazite, allanite, xenotime, apatite. *Reviews in Mineralogy and Geochemistry*, **83**(1), 365–418.

Fowler, C. M. R. & Fowler, M. (2005). *The Solid Earth: an Introduction to Global Geophysics*, 2nd edition. Cambridge University Press. (*Especially Chapter 6.*)

Kohn, M. J. (2016). Metamorphic chronology – a tool for all ages: Past achievements and future prospects. *American Mineralogist*, **101**(1), 25–42.

McClelland, W. C. & Lapen, T. J. (2013). Linking time to the pressure–temperature path for ultrahigh-pressure rocks. *Elements*, **9**(4), 273–9.

Rubatto, D. (2017). Zircon: the metamorphic mineral. *Reviews in Mineralogy and Geochemistry*, **83**(1), 261–95.

10 Metamorphism and Tectonics

Some of the most important outcomes that emerge from the study of metamorphic rocks are the insights they provide into the past thermal structure and tectonic behaviour of the Earth. In order for metamorphic rocks to form, their protoliths must have become buried and heated. In order for us to be able to study them at the surface today, they must have been brought back to the surface in such a way that they preserve the mineralogical evidence for their history. Together, the evidence for these events document how the Earth's crust has operated at different periods of geological time.

This chapter highlights the relationships between different tectonic environments and the types of metamorphic rocks and facies series that are likely forming within them. Throughout this book we have seen that metamorphic rocks record mineral assemblages, textures and chemical compositions acquired over a history of changing physical (and sometimes chemical) conditions. These features in turn provide quantitative information about depths of burial, prevailing temperatures, styles of deformation and the rates and timescales of changes in different tectonic environments. Because there are so many different inputs into the evolution of any one tectonic system, not all metamorphic rocks that have formed in similar tectonic settings are the same. Equally, some types of metamorphic rocks may form in a number of different tectonic environments. By studying similar metamorphic rocks that formed at different times in the Earth's past we can see how the pressure and temperature conditions inside different tectonic environments have changed over the past 4600 million years of Earth history.

This chapter first describes modern plate-tectonic settings from a metamorphic perspective, and defines their distinctive physical conditions. We then explore the different metamorphic associations that form in these environments and take apart the metamorphic histories of two relatively young orogens, the Himalaya and the European Alps, to investigate how the juxtaposition of rocks of different grade and history can help to provide insight into how an orogen forms. We address the tectonic mechanisms by which different types of metamorphic rock exhume and preserve evidence for their metamorphic history. Finally, we look at the insights that ancient metamorphic rocks provide us about changes in tectonic processes during Earth history. By the end of this chapter you should understand how to link mineralogical, textural and contextual clues together, to determine the tectonic setting(s) in which any particular metamorphic rock formed.

10.1 Modern Tectonic Settings and their Metamorphic Implications

We saw in Chapter 1 that thermal gradients through the crust today vary considerably between different places on Earth (Figure 1.6). These variations generally relate to the type and age of crust, the movement and juxtaposition of the crustal plates, and magmatic activity. Temperatures change rapidly with depth, and surface heat flow is higher in the places where magmas are being emplaced or crust is being thinned. Temperatures change more slowly with depth, and surface heat flow is lower, where old continental crust is thick and stable (e.g. the cratons) or where a thick blanket of sediments covers the underlying crust. The lowest geothermal gradients exist above sites of subduction, where relatively cold crustal rocks are carried down into the mantle. These different geothermal gradients lead to the formation of different types and associations of metamorphic rocks in different tectonic settings (Figure 10.1).

The tectonic settings that dominate the formation of metamorphic rocks on Earth today are outlined below; note that we'll be looking only at the *general* points that are pertinent to metamorphism in each setting. In detail, there will always be subtle variations in protolith(s), pressure, temperature, stress field and tectonic history that make the metamorphic history of each location unique. Further details on the metamorphic rocks that form in each setting are outlined in Section 10.2.

10.1.1 Mid-Ocean Ridges

As the plates separate at mid-ocean ridges, mantle rocks move towards the surface and melt, and it is this melt that creates new oceanic crust as it cools and solidifies.

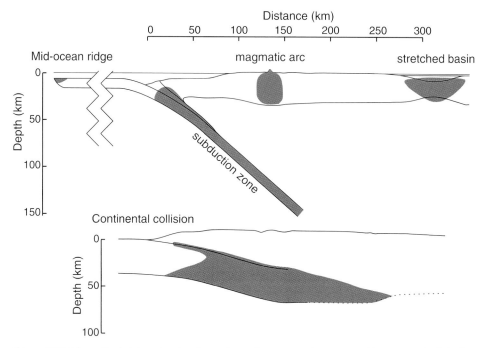

Figure 10.1 Schematic sketches showing the regions where prograde metamorphism occurs (red) in the different tectonic settings discussed in this chapter. See also Figure 1.6, which shows the typical geothermal gradients for the same tectonic settings.

The geothermal gradient at mid-ocean ridges is therefore extremely high – the mantle, at a temperature of 1000 °C or more, is only a few kilometres beneath the surface (Figure 1.6a).

The metamorphic and metasomatic processes occurring at mid-ocean ridges were described in Chapter 5, Section 5.6.2. The extent of hydrothermal alteration at modern mid-ocean ridges varies considerably today, depending on the thermal regime and the spreading rate (see review by Poli & Schmidt 2002). At slow-spreading ridges, extensive faulting and tilting of crustal blocks allows water to penetrate fractures through the entire crustal section. At fast-spreading ridges, continuous magmatic activity supplies enough heat to maintain extensive and vigorous hydrothermal systems.

The overall percentage of altered oceanic crust is relatively low: most of the Earth's upper oceanic crust preserves the igneous characteristics that allows it to retain important tectonic information, for example remnant magnetism. However, hydrothermally-altered oceanic crust and upper mantle play a critical role in subduction zone tectonics. During subduction, they release volatiles, which can trigger melting and volcanism in the over-riding plate.

10.1.2   Subduction Zones

In contrast to mid-ocean ridges, where oceanic crust is created, subduction zones are locations where oceanic crust is recycled back into the mantle and mostly destroyed. Subduction is occurring on Earth today at the margins of the Pacific Ocean and in the eastern Caribbean, for example. The subducting oceanic crust and sedimentary cover are cold, and heating by thermal conduction cannot keep pace with the rate of burial. Rocks buried to depths of 30–40 km in subduction zones – equivalent to a pressure of ~1.0–1.5 GPa – are therefore significantly colder (300–500 °C) than rocks at similar depths in other tectonic settings (Figure 1.6a). In detail, the thermal structure of any individual subduction zone depends mainly on the age of the subducting plate, the amount and type of sedimentary cover, the convergence rate and the angle of subduction. These different factors will therefore affect the P–T–t paths of the metamorphic rocks that form in subduction zones. Their metamorphic evolution provides useful information about the thermal structure of different subduction zones in the geological past.

10.1.3   Regions of High Magmatic Activity

Chains of volcanoes commonly form magmatic arcs above subduction zones: the Andes in South America is a modern example of a continental arc and there are numerous examples in the Pacific of modern oceanic arcs, including the Izu, Bonin and Mariana Arc. Areas of high volcanism also occur away from subduction zones, for example above hotspots (Yellowstone, USA, or Iceland), and in areas of rifting (along the Rift Valley, eastern Africa). The crust in volcanic regions is characterised by high surface heat flows, and thus high temperature gradients. Melt that has stalled in the crust, or passed through it and erupted at the surface, contributes to the heating of the country rocks.

Metamorphism can take place at a number of levels in the crust in regions of volcanic and magmatic activity, but all are characterised by relatively high temperatures at relatively low pressures, with evidence for heating at roughly constant pressure. In detail, the thermal structure of these regions depends on a number of factors including the thickness of the crust, the type and volume of magma, the longevity of the magmatic activity and the location in the crust where the magmatism pools.

10.1.4   Regions of Extension

There are a number of regions on Earth today where the crust is rifting apart. The most obvious example is along mid-ocean ridges, discussed in Section 10.1.1, but continental crust may also rift, for example in back-arc basins such as the Basin and Range Province in the USA and the Aegean region of Greece, and in places where the continents are pulling apart to form incipient ocean basins such as in the Middle

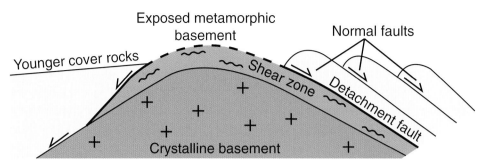

Figure 10.2 Schematic cross-section of a metamorphic core complex.

East and eastern Africa. Regional association with magmatic and/or volcanic activity is common.

As the lithosphere extends, the upper crust breaks along brittle normal faults, and the middle and lower crust rifts along ductile shear zones. This stretching can exhume deeply buried crust in the footwalls of these structures, and result in a region of lower-grade rocks cored by higher-grade rocks: a **core complex** (Figure 10.2) (Whitney *et al.* 2013). Core complexes may, somewhat paradoxically, also form in regions of overall compression (the formation of back-arc basins in the hanging wall of a subduction zone, such as form in the Aegean, is a classic example).

In extensional environments, the mantle is closer to the surface, and its high temperature causes high surface heat flows, high geothermal gradients and accompanying *HT* metamorphism at relatively low pressures. The detailed thermal structure of a rift zone will depend on the thickness of the crust, the rate of thinning, the thermal structure of the underlying mantle and the extent of magmatic activity.

10.1.5   Orogenic Belts

Ocean basins do not last forever, at least on geological timescales. When the subduction of oceanic crust outpaces its production, ocean basins will eventually close, their bounding magmatic arcs or continents will collide and mountain belts (**orogens**) will form. Collisional orogens form in different ways depending on the type of tectonic plates that are colliding (Cawood *et al.* 2009; Jamieson & Beaumont 2013). They form either during a collision of one continent with another (a continental collision orogen) or during the collision of smaller fragments of continental crust, arcs or oceanic crust with a continent (an accretionary orogen).

The Himalaya and Alpine orogens provide two modern examples of continental collision orogens. The Himalaya are an example of a 'hot', magmatic orogen, which formed when the relatively thick sedimentary margin of Greater India collided with Asia and caused significant crustal thickening and mid-crustal melting. At the other end of the scale, the Alps are an example of a 'cold', non-magmatic orogen, which formed when relatively thin crustal fragments buckled against a stronger piece of continental crust before collision ceased. The metamorphic evolution of both of these orogens will be discussed in more detail in Section 10.3.

The south-east Asian region surrounding Sundaland and stretching from Sumatra into eastern Indonesia and the Philippines provides a modern example of an accretionary orogen. It contains a core of older continental crust, onto which other continental and oceanic fragments have sutured during episodic collision since the Mesozoic. Abundant magmatic activity is provided by oceanic-crust-related arc volcanism along much of the length of the orogen. Collision between crustal fragments occurs when neighbouring oceanic basins or mini-basins close.

Metamorphism is an integral part of orogenesis because the colliding fragments are buried, heated and deformed. As in subduction zones, heating by thermal conduction cannot keep pace with the rate of burial. Rocks caught up in collision zones also become buried more rapidly than they can heat up. In detail, an orogen's thermal structure evolves according to the thickness and composition of the colliding crust, and the rate of collision.

10.2   Linking Facies Series to Tectonic Settings

As we have already seen in Chapters 4–6, rocks of different metamorphic facies do not generally form in isolation. Instead, they develop in facies series (Chapter 3, Section 3.2.1) within **metamorphic terranes** that have mostly behaved as a distinct entity during metamorphism. These terranes usually include a range of metamorphic grades and facies (which may vary considerably). Different facies series record different pressure–temperature regimes which reflect different tectonic settings.

10.2.1   High-Pressure, Low-Temperature (*HP–LT*) Metamorphism

HP–LT metamorphic rocks are found in terranes that have a number of common features, including being elongate and narrow, and being bounded by major faults. The mineral assemblages include glaucophane, lawsonite, kyanite, garnet, omphacite or talc (depending on bulk composition and grade, see Chapters 4 and 5). Minerals such as sillimanite and orthopyroxene are never formed. Peak blueschist and eclogite-facies metamorphic mineral assemblages are commonly overprinted by lower-grade (greenschist-, blueschist- or amphibolite-facies) assemblages, but may be evidenced by relics such as inclusions in garnet or zoned minerals (Chapter 7). Some rocks may be so overprinted by moderate-pressure, moderate-temperature (*MP–MT*) assemblages that no evidence of their *HP* history remains. The *HP* history can then only be inferred from the rock and mineral textures and field associations.

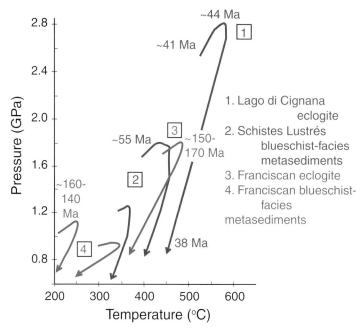

Figure 10.3 Typical *P–T–t* paths of rocks with well-preserved subduction zone assemblages. Alpine data from: Amato *et al.* (1999), Rubatto *et al.* (1998) and Agard *et al.* (2009). Californian data from: Anczkiewicz *et al.* (2004) and Wakabayashi *et al.* (1999). See also Figure 9.13.

Where *HP–LT* metamorphic rocks are well-preserved, their *P–T* paths generally show a 'clockwise' sequence of steeply increasing pressures at relatively low temperatures, followed either by isothermal (fixed-temperature) decompression or decompression associated with minor cooling (Figure 10.3).

As we have seen above, subduction zones promote metamorphism under *HP–LT* conditions and the *P–T-t* history of rocks metamorphosed along a high-*P* facies series provides us with information about the thermal and physical structure of subduction zones in the past. These data can be compared to geophysical observations that provide us with a snapshot of the conditions inside subduction zones today.

Metamorphism takes place in three general regions of subduction zone systems: in the **accretionary wedge** (which sits in the corner between the over-riding plate and subducting plate), within the subducting plate itself, and in the boundary zone between the (coherent) subducting plate and the over-riding plate, a region known as the **subduction channel** (Figure 10.4).

Accretionary wedges are thick, thrust-fault-stacked wedges of intercalated oceanic, continental clastic and volcanogenic sediments formed as the subducting plate moves past the over-riding (or **hanging wall**) plate. Every active

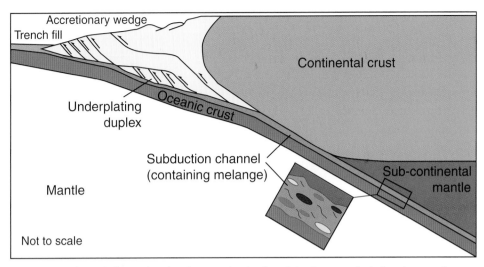

Figure 10.4 Schematic figure showing the upper levels of a subduction zone, including the accretionary-wedge (meta)sediments and the melange forming in the subduction channel. Modified from Cawood *et al.* (2009).

subduction zone today has an accretionary wedge building up above it. Different sizes and styles of wedges reflect a combination of sediment type and thickness, and overall convergence rate. The shape, size and thickness of the wedge influence the progression of metamorphic reactions within it. Metamorphic grade in accretionary wedges is generally low: prehnite–pumpellyite through to greenschist facies with rare blueschist. Many slate belts form in accretionary wedges.

Beneath the accretionary wedge, the oceanic crust and any remaining sedimentary cover subduct into the mantle. Metamorphic changes in oceanic crust may create greenschist-, blueschist- or eclogite-facies assemblages in response to changing P–T conditions and water availability (Chapter 5). Coherent (kilometre-scale) tracts of metamorphosed oceanic metabasalts, metagabbros and ultramafic rocks are relatively common from different localities around the world, see review by Agard *et al.* (2009).

Most of these terranes record maximum pressures of 2.0–2.3 GPa (equating to depths of c. 70 km). One exception is a large and relatively complete section of *HP* oceanic crust exposed in the Western Alps (Angiboust *et al.* 2009). Some of the rocks exposed near the Lago di Cignana in northern Italy contain coesite, demonstrating the attainment of particularly high pressures in some parts. This locality represents the currently only known oceanic crustal *UHP* terrane.

The 'normal' maximum pressure ceiling of 2.0–2.3 GPa for subducted oceanic crust appears to place a general physical limit for its return to the surface (Agard *et al.* 2009). The subduction of oceanic crust is extremely common, and happens along every subduction zone on Earth today. However, we see relatively little evidence for this very common tectonic process in the meta-morphic record. The answer has to do with buoyancy. While plagioclase remains stable, oceanic crust is less dense than the mantle (allowing it to 'float' on the underlying mantle), but as the oceanic crust subducts, primary minerals transform into denser phases. At depths of around 80–100 km (pressures of 3.0 GPa and temperatures of 700 °C), basaltic eclogite becomes denser than the mantle (with a density of 3.35–3.7 g/cm^3 compared to an average harzbur-gite density of 3.3 g/cm^3; Walsh & Hacker 2004). The mechanisms by which such rocks can nevertheless be returned to the surface are discussed in Section 10.3.

Between the subducting plate and the overlying mantle wedge lies a highly-deformed shear zone, essentially a layer of weak rocks that lubricates the interface between the subducting plate and the over-riding plate. This zone, known as the **subduction channel** (Figure 10.4), contains a mixture of subducted oceanic and shelf sediments such as greywackes, shales, sandstones, calcareous rocks and cherts, serpentinised mantle from the over-riding plate, and blocks of subducted oceanic crust. The serpentine is particularly important here as it is both weak and buoyant. It therefore facilitates the lubrication of the channel and the transport (predominantly the exhumation) of stronger blocks of material within it.

The metamorphosed remnants of subduction-channel material are present in a number of metamorphic belts, and are usually spatially and temporally associ-ated with remnants of former accretionary wedges. These terranes consist of a jumbled-up mixture of blocks of contrasting lithology and metamorphic grades in a highly-deformed matrix of lower-grade serpentinites or mudstones, and are known as mélanges (Figure 10.5). The juxtaposition of blocks that have experi-enced different metamorphic histories occurs during exhumation. The unit con-taining blueschists and eclogites on the SE side of the Ile de Groix (Chapter 5, Sections 5.5.1 and 5.5.2) is considered to represent a mélange terrane (Bosse *et al.* 2002).

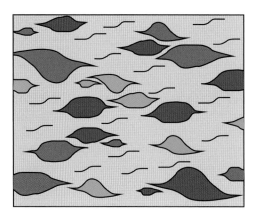

Figure 10.5 Schematic image of a subduction-channel mélange containing pods and boudins of greenschist- (pale green), blueschist- (blue) and eclogite-facies (dark green) metabasites in a low-grade matrix (grey). The scale of this figure could be anything from ~50 cm to ~500 m wide.

10.2.2   Low-Pressure, High-Temperature (*LP–HT*) Metamorphism

LP–HT metamorphic rocks are found in more than one type of terrane. The designation of *LP–HT* is *relative*: metamorphism may occur at relatively high absolute pressures, but the associated temperatures are higher than those expected for a typical 'undisturbed' continental geotherm (Figure 1.6). At the lowest pressures, typical minerals include cordierite and andalusite. At higher pressures, diagnostic features include the relative timing of melting compared to the formation of sillimanite (see Figure 4.18). *P–T–t* of *LP–HT* rocks paths tend to show heating at near-constant pressure.

Tectonic settings in which *LP–HT* metamorphism might be expected include regions of extension and/or regions of volcanic or magmatic activity (Figure 1.6). The Ivrea Zone of the southern European Alps provides an ancient example of *HT–LP* metamorphism caused by crustal extension. This terrane consists of a sequence of mid to lower continental crustal rocks, originally of sedimentary and igneous origin, metamorphosed to mid-amphibolite to granulite facies conditions in the Permian. Migmatites produced in mafic lithologies record *UHT* conditions of up to 950 °C at ~900 MPa in some units (Figure 10.6) (Kunz & White 2019). Detailed study of the timing of different aspects of the *P–T* paths of the different lithologies has shown that the high-grade metamorphism both preceded regional magmatism and outlasted it for at least 20 Ma (e.g. Barboza *et al.* 1999). A high geothermal gradient caused by upwelling of the asthenosphere beneath thinned continental crust likely caused both metamorphism and magmatism in this region.

Areas of magmatic activity can cause *LP–HT* metamorphism throughout the crust. At shallow levels in areas of volcanic activity, geothermal systems develop and cause localised metamorphic and metasomatic changes (Chapter 5, Section 5.6.1). Deeper in the crust, the emplacement of magmas can lead to the development

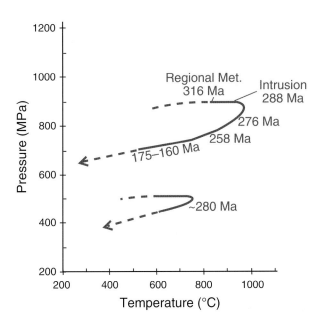

Figure 10.6 *P–T* paths for granulite-facies rocks metamorphosed during rifting in the Ivrea Zone, Alps. Data from Ewing *et al.* (2013, 2015); Kunz *et al.* (2014, 2018); Kunz & White (2019).

of discrete contact metamorphic aureoles (Chapter 4 and Figure 10.7) as well as more regionally-extensive *LP–HT* metamorphism. In the latter case, metamorphism is broadly associated with magmatic activity, including volcanism, rather than localised around individual intrusions. In the Connemara region of Ireland, for example, *HT–LP* metamorphic rocks are exposed in a belt that contains a suite of calc-alkaline intrusions (Yardley *et al.* 1987). The protoliths of the Connemara metasediments correlate with those of the southern Scottish Highlands and have a similar early metamorphic history (Chapter 1, Section 1.2.1): garnet and staurolite grades were reached during crustal thickening along an *MP–MT* metamorphic trajectory. In the Connemara region, however, an arc volcanic centre subsequently developed, which emplaced significant volumes of magma into the thickened pile of metasediments. This led to further heating, and the replacement of earlier *MP–MT* assemblages with higher-temperature assemblages that formed during ongoing exhumation (Figure 10.7). This later *HT–LP* metamorphic overprint correlates with the magmatic intrusions on a large (regional) scale.

At the deepest crustal levels, magmas intruding into the lower crust can crystallise as batholiths tens of kilometres in diameter. The lower crust in these regions likely comprises a mixture of older metasedimentary material as well as igneous rocks. Both may experience metamorphic changes during younger intrusive events. The Fiordland region in New Zealand, for example, exposes one of the world's largest and most complete cross-sections through a (Late-Jurassic and Early-Cretaceous) crustal arc

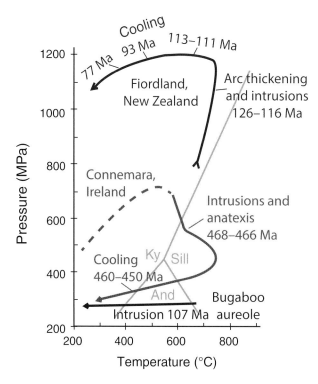

Figure 10.7 Typical *P–T–t* paths of rocks metamorphosed in magmatic arcs and contact aureoles. Note the Fiordland path is 'anti-clockwise' because magmatic intrusion occurred at roughly the same time as crustal thickening; the rocks then remained at depth as they cooled. The Connemara path is similar to paths from the Himalaya (Figure 10.8) but the rocks reached higher temperatures at lower pressures. Fiordland data from Flowers *et al.* (2005) and references therein. Connemara data from Friedrich *et al.* (1999) and Yardley *et al.* (1987); Bugaboo aureole data from Pattison & DeBuhr (2015).

system. The deepest exposed portions of the arc consist of a complex of granulite-facies para- and orthogneiss into which plutonic rocks of gabbroic to granitic compositions intruded. Detailed geochronological investigation has shown that the magmatic activity caused metamorphism in the original crust (Figure 10.7; Flowers *et al.* 2005).

10.2.3 Moderate-Pressure, Moderate-Temperature (*MP–MT*) Metamorphism

MP–MT metamorphism is very common in the geological record and arguably the earliest rocks recognised as being metamorphic formed within this facies series. *MP–MT* metamorphic rocks are found in broad elongate terranes that are commonly bounded by major faults. Typical index minerals include biotite, garnet, staurolite, kyanite and sillimanite in metapelites, and hornblende in metabasites, as described in Chapters 4 and 5. Granulite-facies minerals such as orthopyroxene may also form. Minerals such as cordierite, andalusite and glaucophane are absent. Peak granulite or amphibolite metamorphic mineral assemblages may be overprinted by lower-grade assemblages, but may be evidenced by relics such as inclusions in garnet, zoned minerals or other mineral textures (Chapter 7).

Typical *MP–MT P–T–t* paths show heating throughout burial with extensive exhumation before significant cooling. This pattern is a consequence of low thermal conductivities and tectonic processes that can move rocks more quickly than they can heat or cool. The paths generally show a 'clockwise' sequence (Figure 10.8).

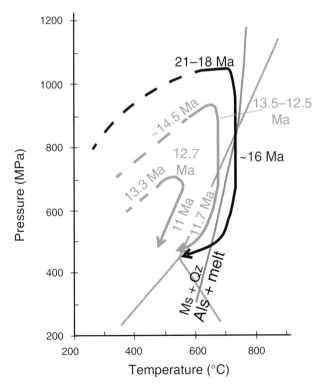

Figure 10.8 *P–T–t* paths for *MP–MT* metamorphic rocks formed in the Himalaya. The pale and dark blue paths relate to different metamorphic units. Data from Mottram *et al.* (2014, 2015).

The moderate geothermal gradients that give rise to *MP–MT* metamorphism are developed in a number of ways. Crustal thickening due to plate interactions is the most common cause of prograde *MP–MT* metamorphism. Furthermore, at any time when colder rocks stop being buried (either during subduction or during collision), they will slowly start heating up as the local heat supply gets more equally distributed. This is a process known as **thermal relaxation** (also shown on Figure 9.3a) and explains why many metamorphic rocks have an overprint that formed at a relatively higher *T/P* ratio than the main assemblage.

10.3   Building Orogens

After careful investigation, most metamorphic terranes reveal details of their *P–T–t* history, which can then be interpreted in terms of the tectonic setting(s) in which they developed. The terranes can then be placed into a tectonic context and contrasted with neighbouring terranes. The most complex metamorphic histories are those of orogenic belts which are an amalgamation of tectonic units that were originally metamorphosed in other tectonic settings. The ages and environments of formation of the different terranes, and the metamorphic, igneous and structural histories they document, together allow the whole history of the orogen to be pieced together. Two geologically-young collisional orogens, the Himalaya and the Alps, provide excellent case studies of continental collision; the formation of accretionary orogens such as the example from south-east Asia (Section 10.1.5) shares similar traits. The *P–T–t* histories of the different terranes that together comprise these orogens provide a good example of how different tectonic snapshots can be combined to decipher the orogenic history.

10.3.1   Tectonic Assembly of the Himalaya

The Himalayan orogen formed as a result of the collision between India and Asia after the closure of the intervening Tethyan ocean basin. The precise timing of collision is still somewhat debated but, on the basis of the last evidence of oceanic sediments, and the first evidence of Asian sediments on the Indian continent, collision occurred at c. 50 Ma ago (Najman *et al.* 2010).

The oldest evidence for Himalayan metamorphism is exposed in two relatively small localities in the western Himalaya, in Kaghan, Pakistan, and Tso Morari, India. In both localities, *UHP* metamorphic terranes of Indian continental crustal origin lie close to, and structurally underneath, the **suture zone** between the Indian and Asian continents (O'Brien *et al.* 2001; Sachan *et al.* 2004). The suture zone is marked by fragments of oceanic crust and upper mantle. Coesite inclusions in garnets in eclogite-facies metabasaltic boudins, which are themselves hosted by felsic gneiss, suggest that the Indian continental margin reached pressures of at least 2.6 GPa (depths of 80 km). Zircons texturally associated with coesite yield ages of between 47 Ma and 45 Ma (Chapter 9, Section 9.4.2), i.e. within 10 Ma of the last marine sediments to be deposited in the seaway that separated the two continental masses.

The main region of metamorphosed rocks in the Himalaya comprises two orogen-parallel belts that formed from Indian continental margin sediments, separated by a major thrust-sense shear zone (Figure 10.9; also see review by Cottle *et al.* 2015). In the central part of the orogen, the structurally higher, greenschist- to migmatite-grade Greater Himalayan Sequence reached peak kyanite-grade metamorphism at around 30 Ma, and then started to melt as pressure dropped during exhumation and it crossed the muscovite-dehydration reaction (see Figure 4.13). The resulting granites yield ages clustering around 20 Ma. The timing of different stages of metamorphism is somewhat diachronous along the strike of the orogen and metamorphism youngs towards the east.

One difference between the regional metamorphism in the Himalayas and that in many other orogens is that the generation of magmas appears to have been a *consequence* of metamorphic heating rather than its *cause*. The magmas formed by mica melting reactions (reactions [4.14]–[4.20] in Chapter 4) are granitic in composition and are particularly rich in aluminium. In contrast, arc migmatites are associated with voluminous basic to intermediate magmatism. The granite plutons exposed at the highest structural levels of the Himalaya (Figure 10.9) formed by the coalescence of multiple generations of anatectic melts formed deeper in the crust. Some contain cordierite, suggesting that they crystallised at high crustal levels. These granites are characteristic of melting in regions where heat was not brought into the system by mantle-derived magmatic activity. High-precision isotopic analyses (Sr, Nd and U–Pb) are required to demonstrate an origin from crustal

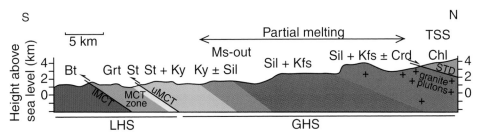

Figure 10.9 Schematic cross-section through the central Himalayan orogen (Nepal) showing changes in metamorphic grade. MCT = Main Central Thrust; l = lower boundary, u = upper boundary; STD = South Tibetan Detachment; LHS = Lesser Himalayan Sequence; GHS = Greater Himalayan Sequence; TSS = Tethyan Sedimentary Sequence. Figure modified from Imayama *et al.* (2010).

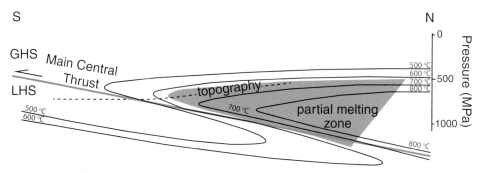

Figure 10.10 Schematic figure, based on thermal modelling, showing inversion of isotherms caused by thrusting of hotter Greater Himalayan Sequence (GHS) rocks on top of colder Lesser Himalayan Sequence (LHS) rocks along the MCT (Main Central Thrust) during the Himalayan orogeny. Modified from Le Fort (1975).

melting alone (Hopkinson *et al.* 2017). The presence of significant quantities of melt in the Himalayan middle crust today (equivalent to the source region for the granite plutons) has been inferred from geophysical investigations across the Himalayas and Tibet (Nelson *et al.* 1996).

The metamorphism in the structurally-lower Lesser Himalayan Sequence is somewhat younger, with the timing of peak metamorphism varying with grade down the section (Mottram *et al.* 2014). This **inverted metamorphic sequence**, in which the highest-grade rocks are exposed at structurally higher levels, is associated with the thrust-sense shear zones that separates the Greater and Lesser Himalayan Sequences (Figure 10.10). Burial of colder rocks along this thrust fault, under hotter metamorphic rocks, led to heat flowing from the hotter hanging wall into the colder footwall.

10.3.2   Tectonic Assembly of the Alps

Geological investigation over many decades has shown that the Western Alps formed during a complex collision between Europe and Africa (or, specifically, a fragment of Africa called Adria), and the closure of a multi-basin ocean that once lay between them; Beltrando *et al.* (2010) provide a clear summary. There are ongoing discussions about the exact details of the original location and role of different units, but modern geochronological techniques have played a major role in resolving some of the greatest debates. Critically, many of the units caught up in the Alpine orogeny record an older Permian metamorphic history (part of the Variscan orogeny) followed by several phases of Alpine overprinting ranging from the Jurassic until recent times. It has taken much detailed work to disentangle which metamorphic assemblages formed at which times in rocks of different bulk composition. This overprinting phenomenon is not unique to the Alps, but untangling earlier Variscan (*LP–HT*) from later Alpine (*HP–LT*) metamorphism has been particularly problematic in the Alps because the Alpine orogeny achieved lower peak temperatures. Furthermore, the Alpine collision resulted in an extremely complex structural history with many sheet-like units (**nappes**) being folded and faulted together. The structural history of the Alps is beyond the scope of this book but is summarised in Beltrando *et al.* (2010).

The oldest Alpine-aged metamorphic history in the Western Alps is recorded by the rocks of the Adriatic continental margin (the Sesia zone, see map in Figure 4.16) which reached peak greenschist- to eclogite-facies conditions at around 65 Ma (Rubatto *et al.* 1999). This metamorphism is interpreted as recording the subduction of the southernmost ocean basin southwards beneath the Adriatic continent at that time.

Geochronological evidence for further, later, deep subduction is provided by the *UHP* metamorphism of a slice of oceanic crust and upper mantle (the Piemonte ocean, Figure 10.11) at 44 Ma (Rubatto *et al.* 1998). Volumetrically, very little of the subducted oceanic crust is now preserved at the surface, although seismic tomography has shown the presence of several slabs beneath the Alps that may represent the still-subducted remnants (Lippitsch *et al.* 2003).

The metamorphism to *UHP* conditions of the Piemonte oceanic crust was followed by initial continental collision, recorded by the *UHP* metamorphism of blocks of crystalline continental crust (including the Dora Maira, Gran Paradiso and Monte Rosa massifs, Figures 10.11 and 4.16) at around 35 Ma (Gebauer *et al.* 1997). The continental shelf sediments of the Briançonnais terrane, which provide most of the metapelite *HP* history (Chapter 4), equilibrated at around 40–35 Ma at a range of pressures from greenschist through to blueschist facies (Berger & Bousquet 2008).

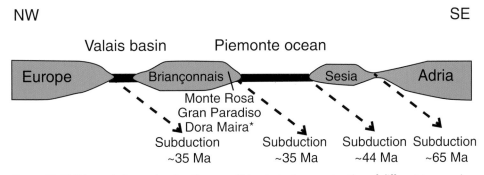

Figure 10.11 Schematic figure showing the pre-collision tectonic reconstruction of different terranes in the Western Alps. The cross-section is marked as B–B′ on the map in Figure 4.16. *Note that the location of the Dora Maira Massif is a northwards projection (as suggested by Michard *et al.* (2004)). In the southern part of the Western Alps there is no evidence for the Valais basin and the Briançonnais is considered part of the European margin. Figure modified from Rubatto *et al.* (1998).

10.4   How Do Tectonic Processes Drive Exhumation?

We saw in Chapter 9 (Section 9.1.3, Figure 9.4), that both surface and tectonic processes operate to bring rocks from depth up to the surface. Different exhumation mechanisms dominate in different settings – for example, erosion is more important in regions of plate collision than extension as higher and steeper topography is created. The tectonic setting *itself* does not drive burial or exhumation, rather it is the combination of different exhumation mechanisms within that setting that facilitates the transport of rocks back to the surface.

The *P–T–t* paths followed by metamorphic rocks provide information about the different mechanisms that likely operated to bury and exhume them at different stages. For example, rocks metamorphosed to great depths under *(U)HP–LT* conditions commonly record rapid exhumation rates in the order of many cm/a, i.e. at similar rates to horizontal plate tectonic rates (see e.g. Figure 9.13). Temperature remains high as pressure drops because there is no time for the rocks to cool. These rates suggest that the processes that return these rocks to the surface involve significant tectonic forces. Meanwhile, rocks that have been metamorphosed at *MP–MT* conditions often record slower exhumation rates, in the order of 1–2 cm/a (e.g. Figure 10.8), and might therefore involve a mixture of both tectonic forces and erosion. Rocks metamorphosed in the deep lower crust may record considerably slower exhumation rates, <1 mm/a (e.g. Figure 10.7), and they may have experienced extended periods of residence in the deep crust before exhuming. These slow rates are more suggestive of erosion-related exhumation.

A combination of gravity and shear traction aid the burial of dense oceanic crust and its overlying sedimentary cover in subduction zones. Traction forces, related to the overall rock strength, keep the sediments attached to the underlying crystalline crust. As the rocks become more deeply buried, upwards-directed buoyancy forces become stronger due to the difference in density between the mantle and the (metamorphosed) crustal rocks. Once buoyancy forces become strong enough, the traction forces are overcome and the buoyant rocks will start to exhume back towards the surface (Warren *et al.* 2008). This balance between buoyancy and rock strength is the reason why different protolith types, such as sediments, oceanic crust or continental crust, are exhumed from different depths in subduction zones, and why many shelf sediments are scraped off at high structural levels to form accretionary wedges. The overall weakness and buoyancy of material in the subduction channel, especially serpentinite and shale, facilitates the exhumation of blocks of crust that have been subducted to different depths by lowering the shear traction.

The *P–T–t* paths of rocks exhumed from *HP* and *UHP* conditions commonly show two stages of exhumation (Figure 9.13): a steep (and fast) decompression stage, and a later slower, cooling stage. The first stage has been attributed to buoyancy forces, and the second stage to upper crustal faulting coupled with erosion once the rocks reach neutral buoyancy at middle to upper crustal depths.

Buoyancy plays a much more minor role in exhuming rocks during collisional tectonics that lead to crustal thickening, because the buoyancy differences between rocks at different crustal levels is smaller than between the crust and the mantle. In collision zones, rocks are buried and uplifted along thrust- and normal-sense faults and shear zones. Folding may also both bury and uplift rocks. Exhumation is subsequently facilitated by erosion of the resulting topography (Figure 9.4).

The focussed exhumation of high-grade rocks in the core of the Himalayan orogen (the Greater Himalayan Sequence that was introduced above in Section 10.3.1) illustrates the importance of the coupling between tectonic and surface processes in such environments. In the Himalayas, the greenschist- to migmatite-grade Greater Himalayan Sequence crops out along the length of the orogen bounded by the normal-sense South Tibetan Detachment at its roof and the Main Central Thrust at its base (Figure 10.9). The terranes on the other sides of those structures reached only much lower peak metamorphic grades. Geochronologic evidence shows that both the bounding faults were in motion at the same time, suggesting that the high-grade metamorphic core was transported southwards ('extruded' towards the foreland) by a combination of thrust-sense movement at its base and normal-sense movement at its roof (reviewed by Godin *et al.* 2006). Geodynamic models and experimental data suggest that this extrusion may be facilitated by weakening of the orogenic middle crust (here represented by the Greater Himalayan Sequence) by partial melting, in combination with erosion focussed at the orogenic front (Beaumont *et al.* 2001; Rosenberg & Handy 2005). In this respect the exhumation of the high-grade metamorphic core of the Himalaya can be considered as analogous to cream being squeezed from between two layers of cake when pressure is applied on top (e.g. Figure 9.4b). In the Himalayan example the pressure is provided by the gravitational force of the 5 km high Tibetan Plateau.

10.5   Changes in Metamorphism Through Geological Time

Metamorphic rocks preserve records of the Earth's thermal structure and tectonic behaviour throughout most of geological time. It has long been recognised that different metamorphic assemblages appear to be more or less common at different times in Earth history. Compilations of global metamorphic occurrences suggest that three different tectonic regimes have existed at different times in Earth history: the 'Archean', the 'Proterozoic' and the 'modern' (Figure 10.12; Brown & Johnson 2019).

The Archean metamorphic record, for example, appears to be dominated by rather moderate *MP–MT* conditions (Figure 10.12). There are currently no known

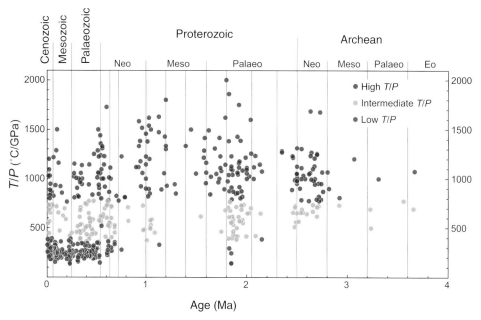

Figure 10.12 Change in thermal gradient as recorded by metamorphic rocks of different ages. Lightly modified from Brown & Johnson (2019).

Archean-aged examples of the more 'extreme' styles of metamorphism (*HP–LT* or *UHT*) that are relatively common later in Earth history. These changes over time may be 'real', due to changes in the composition and style of Earth's plates as the Earth has cooled. However, the changes may also be 'apparent', due to biases in the way the rock record preserves evidence for Earth's evolution billions of years ago. Very little rock record of Archean-age rocks exists on the surface of the Earth today.

Towards the end of the Archean, from the Neoarchean onwards to the Cambrian, metamorphic terranes preserving evidence for *UHT* metamorphism become both more abundant in the geological record, and larger in extent (see the red dots trending towards higher values on Figure 10.12). Many of the Neoarchean examples appear to be spatially and temporally associated with back-arc settings along the edges of the Gondwana supercontinent (Brown 2007).

Eclogites also start to appear in the record from the Neoarchean onwards. The oldest 2.9 Ga examples are suggested by eclogite-facies mineral inclusions in diamonds erupted in kimberlites through the Kaapvaal and Siberian cratons (Gurney *et al.* 2010). The geochemical composition of these diamonds suggests that they formed from crustal (and even surface) carbon. This implies that deep subduction of continental crust must have taken place in the Archean and Proterozoic, even if no other metamorphic record of this process survives. The mechanism responsible for this recycling is still under active debate. These ancient examples, and other Archean and Neoproterozoic whole-rock eclogites exposed in Canada and the Baltic

Shield, record warmer peak temperatures at equivalent pressures than Phanerozoic examples, and are commonly spatially associated in the field with high-pressure granulites (Volodichev *et al.* 2004; Weller & St-Onge 2017). These associations of eclogites with high-pressure granulites continue through the record until the Paleozoic before becoming less common.

Blueschists only rarely date from before the Phanerozoic, although Neoproterozoic (800–700 Ma) examples have been reported from western China, west Africa and India (e.g. Maruyama *et al.* 1996). Eclogites containing lawsonite (recorders of cold subduction zones) and coesite (recording *UHP* conditions) appear only in the Phanerozoic.

SUMMARY

The vast majority of metamorphic rocks on Earth form at or near plate boundaries where tectonic plates are colliding or pulling apart. The main tectonic settings we have discussed in this chapter include mid-ocean ridges, subduction zones, orogenies, regions of high magmatic and/or volcanic activity, and regions of plate extension. Different styles of metamorphism, including those forming the different facies series we introduced in Chapters 4 and 5, are produced in each of these settings. Each setting has a number of characteristic physical features, including the geothermal profile and the relative geometry of the plate interactions. In detail, subtle variations in protoliths, pressure, temperature, stress field and tectonic history make the metamorphic rocks that form in each location unique. Orogens preserve a particularly complex patchwork of different metamorphic terranes, each of which reveals individual *P–T–t* histories that allow the orogenic history to be pieced together.

Plate-tectonic processes facilitate both the burial of rocks and their exhumation under different thermal regimes. The preservation of rocks of certain facies series, and especially the *HP–LT* series forming in subduction zones, is critically dependent on the speed and mechanism of their exhumation. This is because many of the minerals that are stable under high pressures and relatively cold conditions need to remain cold in order for them not to react back to species that are stable under lower pressure and/or warmer conditions.

As the Earth has slowly lost heat since its formation 4.6 billion years ago, the pattern and style of plate tectonics has changed. The fairly sparse record of the oldest metamorphic rocks on Earth suggests that Early Earth was dominated by tectonic processes that facilitated metamorphism under moderate-temperature, moderate-pressure conditions. High- and ultra-high-temperature conditions were first experienced (or at least preserved) from the Neoarchean onwards, and become rarer after the

Cambrian. Their dominance in the Precambrian record appears to be particularly associated with the margins of the Gondwana supercontinent. The record of higher-pressure metamorphism suggests cooling through time: the earliest examples appear to record hotter temperatures than those found at equivalent pressures in the younger Phanerozoic record. Blueschists, and especially those containing lawsonite, are key markers of the 'cold' subduction we see on Earth today.

EXERCISES

1. Explain the difference between a tectonic setting and a metamorphic terrane.
2. Name three index minerals that are typically found in rocks metamorphosed in subduction zones.
3. Name two index minerals that are diagnostic of a rock metamorphosed in a contact aureole at high levels in the crust.
4. Sketch the shape of the P–T path you might expect for:
   (a) an oceanic-crustal basalt subducted to, and exhumed from, 80 km depth within a subduction zone;
   (b) a pelitic sediment buried to 30 km depth in the lower plate of a continental collision zone;
   (c) the pelitic sediment in (b) that receives extra heat input from a nearby magmatic intrusion during exhumation.

   On your sketches note common features of the P–T paths, and those that diverge, and note any assumptions you have made with regards to converting depth to pressure.
5. As mentioned in Chapter 1, Box 1.1, in the 1960s geologist Akiho Miyashiro recognised that HP–LT and HT–LP metamorphic terranes are commonly juxtaposed in the geological record – a phenomenon he called 'paired metamorphic belts'.
   (a) By reference to plate tectonics (and Figure 1.6), explain how these paired metamorphic belts form.
   (b) Where on Earth might paired belts be forming today (even if not yet exposed!)?

FURTHER READING

Brown, M. (2007). Metamorphic conditions in orogenic belts: a record of secular change. *International Geology Review*, **49**, 193–234.

Cawood, P. A., Kröner, A., Collins, W. J. *et al.* (2009). Accretionary orogens through Earth history. *Geological Society, London, Special Publications*, **318**(1), 1–36.

Fowler, C. M. R.(2005). *The Solid Earth: an Introduction to Global Geophysics*, 2nd edition. Cambridge University Press, 685 pages.

Godin, L., Grujic, D., Law, R. D. & Searle, M. P. (2006). Channel flow, ductile extrusion and exhumation in continental collision zones: an introduction. *Geological Society, London, Special Publications*, **268**(1), 1–23.

Jamieson, R. A. & Beaumont, C. (2013). On the origin of orogens. *Bulletin*, **125**(11–12), 1671–702.

Kelsey, D. E. & Hand, M. (2015). On ultra-high-pressure crustal metamorphism: phase equilibria, trace element thermometry, bulk composition, heat sources, timescales and tectonic settings. *Geoscience Frontiers*, **6** (3), 311–56.

Pfiffner, O. A. (2014). *Geology of the Alps*. John Wiley & Sons.

Searle, M. P. (2013). *Colliding Continents: a Geological Exploration of the Himalaya, Karakoram, and Tibet*. Oxford University Press.

Warren, C. J. (2013). Exhumation of (ultra-) high-pressure terranes: concepts and mechanisms. *Solid Earth*, **4**(1), 75–92.

Whitney, D. L., Teyssier, C., Rey, P. & Buck, W. R. (2013). Continental and oceanic core complexes. *GSA Bulletin*, **125**(3–4), 273–98.

APPENDIX 1

Glossary of Mineral Names and Abbreviations Used in the Text

Mineral name	Abbreviation (from Whitney & Evans 2010)	Chemical formula
actinolite	Act	see amphibole
adularia	Adl	see feldspar
aegirine	Aeg	see pyroxene
albite	Ab	see feldspar
alkali feldspar	Afs	see feldspar
allanite	Aln	$(Ce,Ca,Y,La)_2(Al, Fe^{3+})_3(SiO_4)_3(OH)$
almandine	Alm	see garnet
Al-silicate	Als	Al_2SiO_5
amphibole group:	Amp	
clinoamphiboles		
actinolite	Act	$Ca_2(Mg,Fe^{2+})_5Si_8O_{22}(OH)_2$
cummingtonite	Cum	$(Mg,Fe^{2+})_7Si_8O_{22}(OH)_2$
glaucophane	Gln	$Na_2(Mg,Fe^{2+})_3Al_2Si_8O_{22}(OH)$
hornblende	Hbl	$Na_{0-1}Ca_2(Mg,Fe^{2+},Fe^{3+},Al)_5Al_{2-1}Si_{6-7}O_{22}(OH)_2$
riebeckite	Rbk	$Na_2Fe^{2+}_3Fe^{3+}_2Si_8O_{22}(OH)_2$
tremolite	Tr	$Ca_2Mg_5Si_8O_{22}(OH)_2$
orthoamphiboles		
anthophyllite	Ath	$(Mg,Fe^{2+})_7Si_8O_{22}(OH)_2$
gedrite	Ged	$(Mg,Fe^{2+})_5Al_4Si_6O_{22}(OH)_2$
analcite (analcime)	Anl	see zeolite
andalusite	And	Al_2SiO_5
andradite	Adr	see garnet
ankerite	Ank	$Ca(Mg,Fe^{2+},Mn)(CO_3)_2$
anorthite	An	see feldspar
anthophyllite	Ath	see amphibole
antigorite	Atg	see serpentine
apatite	Ap	$CaPO_4$
aragonite	Arg	$CaCO_3$
augite	Aug	see pyroxene
barroisite	Brs	see amphibole
biotite	Bt	see mica
brucite	Brc	$Mg(OH)_2$

(cont.)

Mineral name	Abbreviation (from Whitney & Evans 2010)	Chemical formula
calcite	Cal	$CaCO_3$
carpholite	Car	$(Mn^{2+}Mg^{2+},Fe^{2+})Al_2Si_2O_6(OH)_4$
celadonite	Cel	see clay minerals
chlorite	Chl	$(Mg,Fe^{2+},Al)_{12} (Si,Al)_8O_{20}(OH)_{16}$
chloritoid	Cld	$(Fe^{2+},Mg,Mn)_2Al_4Si_2O_{10}(OH)_4$
chrysotile	Ctl	see serpentine
clay mineral group:		
celadonite	Cel	$K_2Al_2(Mg,Fe^{2+})_2Si_8O_{20}(OH)_4$
illite	Ilt	$K_{1-5}Al_{5-5.5}Si_{7-6}O_{20}(OH)_4$
kaolinite	Kln	$Al_4Si_4O_{10}(OH)_8$
montmorillonite	Mnt	$(0.5Ca,Na)_{0.7}(Al,Mg,Fe^{2+})_4(Si,Al)_8O_{20}(OH)_4.nH_2O$
clinopyroxene	Cpx	see pyroxene
clinozoisite	Czo	see epidote
coesite	Coe	SiO_2
cordierite	Crd	$(Mg,Fe^{2+})_2Al_4Si_5O_{18}$
corundum	Crn	Al_2O_3
cummingtonite	Cum	see amphibole
diamond	Dia	C
diopside	Di	see pyroxene
dolomite	Dol	$Mg(CO_3)_2$
enstatite	En	see pyroxene
epidote group:		
clinozoisite	Czo	$Ca_2Al_3Si_3O_{12}(OH)$
epidote	Ep	$Ca_2Fe^{3+}Al_3Si_3O_{12}(OH)$
fayalite	Fa	see olivine
feldspar group:		
adularia	Adl	$(Na,K)AlSi_3O_8$
albite	Ab	$NaAlSi_3O_8$
alkali feldspar	Afs	$(Na,K)AlSi_3O_8$
anorthite	An	$CaAl_2Si_2O_8$
K-feldspar	Kfs	$KAl Si_3O_8$
orthoclase	Or	$KAl Si_3O_8$
plagioclase	Pl	$(Ca,Na)Al(Al,Si) Si_2O_8$
oligoclase		70–90% Ab
labradorite		30–50% Ab
forsterite	Fo	see olivine
garnet group:	Grt	$(Mg,Ca,Mn, Fe^{2+})_3Al_2Si_3O_{12}$
almandine	Alm	$Fe_3Al_2Si_3O_{12}$
grossular	Grs	$Ca_3Al_2Si_3O_{12}$
pyrope	Prp	$Mg_3Al_2Si_3O_{12}$

(*cont.*)

Mineral name	Abbreviation (from Whitney & Evans 2010)	Chemical formula
spessartine	Sps	$Mn_3Al_2Si_3O_{12}$
gedrite	Ged	see amphibole
glaucophane	Gl	see amphibole
graphite	Gr	C
grossular	Grs	see garnet
hematite	Hem	Fe_2O_3
heulandite	Hul	see zeolite
hornblende	Hbl	see amphibole
hypersthene	Hyp	see pyroxene
illite	Ilt	see clay minerals
ilmenite	Ilm	$FeTiO_3$
jadeite	Jd	see pyroxene
kaolinite	Kln	see clay minerals
K-feldspar	Kfs	see feldspar
kornerupine	Krn	$(Mg,Fe^{2+})_4(Al,Fe^{3+})_6(SiO_4,BO_4)_5(O,OH)_2$
kyanite	Ky	Al_2SiO_5
labradorite		see feldspar
larnite	Lrn	Ca_2SiO_4
laumontite	Lmt	see zeolite
lawsonite	Lws	$CaAl_2Si_2O_7(OH)_2.H_2O$
magnetite	Mag	Fe_3O_4
margarite	Mrg	see mica
mesolite	Mes	see zeolite
mica group:		
biotite	Bt	$K(Mg,Fe)_3(AlSi_3O_{10})(F,OH)_2$
margarite	Mrg	$CaAl_4Si_2O_{10}(OH)_2$
muscovite	Ms	$KAl_2(AlSi_3O_{10})(OH)_2$
paragonite	Pg	$NaAl_3Si_3O_{10}(OH)_2$
phengite	Ph	$K(Mg,Fe^{2+})_{0-1}Al_{3-2}Si_3O_{10}(OH)_2$
phlogopite	Phl	$KMg_3AlSi_3O_{10}(OH)_2$
sericite	Ser	(fine Ph $\pm$ Pa $\pm$ other sheet silicates)
monazite	Mnz	$(Ce, La, Nd, Th)PO_4$
montmorillonite	Mnt	see clay minerals
mordenite	Mor	see zeolite
muscovite	Ms	see mica
oligoclase		see feldspar
olivine group:	Ol	$(Mg,Fe^{2+})SiO_4$
fayalite	Fa	Fe_2SiO_4
forsterite	Fo	Mg_2SiO_4
omphacite	Omp	see pyroxene
orthopyroxene	Opx	see pyroxene

(cont.)

Mineral name	Abbreviation (from Whitney & Evans 2010)	Chemical formula
osumilite	Osm	$(K,Na)(Fe,Mg)_2(Al,Fe)_3(Si,Al)_{12}O_{30}$
paragonite	Pg	see mica
phengite	Ph	see mica
phlogopite	Phl	see mica
plagioclase	Pl	see feldspar
prehnite	Prh	$Ca_2Al_2Si_3O_{10}(OH)_2$
pumpellyite	Pmp	$Ca_4(Mg,Fe^{2+})(Al,Fe^{3+})_5Si_6O_{23}(OH)_3.2H_2O$
pyrite	Py	FeS_2
pyrope	Prp	see garnet
pyrophyllite	Prl	$Al_2Si_4O_{10}(OH)_2$
pyroxene group:		
clinopyroxenes	Cpx	
aegirine	Aeg	$NaFeSi_2O_6$
augite	Aug	$(Na,Ca,Fe,Mg,Al)_2(Al,Si)_2O_6$
diopside	Di	$CaMgSi_2O_6$
jadeite	Jd	$NaAlSi_2O_6$
omphacite	Omp	$(Ca,Na)(Mg,Al)Si_2O_6$
orthopyroxenes	Opx	
enstatite	En	$Mg_2Si_2O_6$
hypersthene	Hyp	$Mg_{1-1.4}Fe^{2+}_{1-0.6}Si_2O_6$
pyrrhotite	Po	$Fe_{(1-x)}S$
quartz	Qz	SiO_2
rutile	Rut	TiO_2
sapphirine	Spr	$(Mg,Al)_8(Al,Si)_6O_{20}$
sericite	Ser	see mica
serpentine group:	Srp	
antigorite	Atg	$(Mg,Fe^{2+})_3Si_2O_5(OH)_4$
chrysotile	Ctl	$Mg_3Si_2O_5(OH)_4$
sillimanite	Sil	Al_2SiO_5
spessartine		see garnet
spinel	Spl	$(Mg,Fe)Al_2O_4$
spurrite	Spu	$Ca_4Si_2O_8.CaCO_3$
staurolite	St	$(Fe^{2+},Mg)_4 Al_{18}Si_{7.5}O_{44}(OH)_4$
stilbite	Stb	see zeolite
stilpnomelane	Stp	$(K,Ca)_{0-1.4}(Fe,Mg,Al)_{5.9-8.2}Si_8O_{20}(OH)_4(O,OH, H_2O)_{3.6-8.5}$
talc	Tlc	$Mg_6Si_8O_{20}(OH)_4$
thomsonite	Thm	see zeolite
titanite	Ttn	$CaTiSiO_5$
tourmaline	Tur	$Na(Fe^{2+},Mg)_3Al_6B_3Si_6O_{27}(OH,F)_4$
tremolite	Tr	see amphibole

(cont.)

Mineral name	Abbreviation (from Whitney & Evans 2010)	Chemical formula
vesuvianite (idocrase)	Ves	$Ca_{10}(Mg,Fe^{2+})_2Al_4Si_9O_{34}(OH,F)_4$
wairakite	Wrk	see zeolite
wollastonite	Wo	$CaSiO_3$
xenotime	Xen	YPO_4
zeolite group:		
analcite	Anl	$Na(AlSi_2O_6) \cdot H_2O$
heulandite	Hul	$(Ca,Na)_2(Al_2Si_7O_{18}).6H_2O$
laumonite	Lmt	$CaAl_2Si_4O_{12}.4H_2O$
mesolite	Mes	$Na_2Ca_2Al_6Si_9O_{30}.8H_2O$
mordenite	Mor	$(Na_2,K_2,Ca)Al_2Si_{10}O_{24}.7H_2O$
stilbite	Stb	$(Ca,Na_2,K_2)Al_2Si_7O_{18}.7H_2O$
thomsonite	Thm	$NaCa_2Al_5Si_5O_{20}.6H_2O$
wairakite	Wrk	$CaAl_2Si_4O_{12}.2H_2O$
zircon	Zrn	$ZrSiO_4$
zoisite	Zo	$Ca_2Al_3Si_3O_{12}(OH)$

APPENDIX 2
Schreinemakers Methods for the Construction of Phase Diagrams

This Appendix describes a geometrical approach to constructing phase diagrams for any chemical system, requiring only a knowledge of phase compositions. From this starting point, it is possible to identify all the possible reactions in the system and find out where they sit relative to one another on a phase diagram. The methods were originally published by F. A. H. Schreinemakers in a series of 29 articles between 1915 and 1925, and were introduced to geologists by the publication of several detailed accounts, in particular that of Zen (1966). While today you are unlikely to need to use the methods to construct a new phase diagram for the first time, the basic rules are surprisingly useful for resolving ambiguous relationships.

Although originally used primarily to predict the relationship between univariant curves on a P–T diagram, the same methods can be applied equally to any phase diagram whose axes are intensive variables which control the Gibbs free energy of the mineral assemblages plotted. Commonly used axes, in addition to P and T, include mol fractions in a fluid phase (as in T–X_{CO_2} diagrams, Chapter 6), chemical potentials of fluid components (e.g. $\mu_{CO_2}, \mu_{H_2O}, \mu_{SiO_2}$) and fugacity (e.g. $\log f_{O_2}$).

Following the terminology of Chapter 2, Sections 2.2 and 2.3, phase diagrams are divided into divariant fields, i.e. segments within which a specific divariant assemblage ($P = C$) is stable, separated by univariant curves ($P = C + 1$) along which the divariant assemblages of both adjacent segments can co-exist in equilibrium. Univariant curves intersect at an invariant point ($P = C + 2$) and the complete set of univariant curves intersecting at an invariant point is often referred to as a **Schreinemakers bundle**.

A2.1 The Basics: How Many Reactions in a System?

Schreinemakers methods are applied to a system defined as a group of phases that can all be made from a set of C components. For a system of $C + 2$ phases, there is one invariant point where all phases can co-exist. This lies at the intersection of a series of univariant curves along each of which $C + 1$ phases can co-exist at equilibrium, i.e. along each univariant curve one phase must be missing, so there will be a univariant curve for each of the $C + 2$ phases. By convention, univariant curves are labelled by the name of the phase that is *absent*, placed in brackets. For complex reactions this is obviously simpler than writing out the reaction in full.

A system containing more than $C + 2$ phases will have more than one univariant point, and there will be a cluster of univariant curves about each. Such systems are known as **multisystems** and will not be considered further here, because they are both more complex and less unambiguous to interpret than systems with only a single invariant point.

A2.1.1 Degeneracy

In some systems of $C + 2$ phases there are less than $C + 2$ possible univariant reactions. For example all the possible reactions between the phases corundum (Al_2O_3), quartz (SiO_2), andalusite (Al_2SiO_5) and sillimanite (Al_2SiO_5) in the binary system $Al_2O_3 - SiO_2$ are:

corundum + quartz = sillimanite (And)
corundum + quartz = andalusite (Sil)
sillimanite = andalusite (Crn, Qz).

(Note the use of the phase-absent convention to identify each reaction.) Only three reactions are possible instead of four because one of them involves only two phases and therefore serves simultaneously as both reaction (Crn) and reaction (Qz). This situation arises because this reaction effectively requires only one chemical component Al_2SiO_5. Such reactions are termed **degenerate**, because they require fewer than C components.

Degeneracy also arises in systems with more than two components, where it is no longer restricted to polymorphic transitions. In the system $MgO-SiO_2-H_2O$, illustrated in Figure 2.5a, several of the possible assemblages of $C + 2$ phases include degenerate reactions. For example forsterite, enstatite and quartz lie on a straight line along the $MgO-SiO_2$ join, and it is therefore possible to write a reaction in which the central phase, enstatite, breaks down to the two outer phases, forsterite and quartz. Again this will be a degenerate reaction since less than $C + 1$ phases are involved and it can be written in terms of only two components. Similarly enstatite, anthophyllite and talc plot on a straight line so that reaction between them is also degenerate.

A2.2 The Fundamental Axiom

The Gibbs free energy of any phase or assemblage of phases varies continuously with change in intensive parameters (e.g. Figure 3.1). As a result, any univariant curve, representing conditions where two assemblages are equally stable, must be continuous and each of the two assemblages must be unambiguously more stable on one side and less stable on the other. This remains true even if a third equivalent assemblage is actually more stable than either of the two under consideration. For example, it is still possible to identify *P–T* conditions within the stability of sillimanite (Figure 2.1) along which kyanite and andalusite are equally stable, and these define a univariant curve of **metastable equilibrium** which is simply an extension of the boundary between the andalusite and kyanite stability fields, into the sillimanite field. Such a curve is known as the **metastable extension** of the stable univariant curve.

A2.3   Morey–Schreinemakers Rule

If we define a system of, say, four components and six phases, A–F, then all phases can co-exist at a single invariant point from which six stable univariant curves radiate. Along each of these curves, five of the six phases can co-exist. These curves divide the diagram up into divariant fields within each of which four phases can occur. Although various assemblages are possible, according to bulk composition, one will be unique to the segment. Consider the segment in which A, B, C, D co-exist. This must be bounded by stable univariant curves along each of which A, B, C, D are still stable phases, but are joined by one other. In other words, they must be curves (F) and (E). On the other side of each of these curves the assemblage A + B + C + D is no longer stable. Morey–Schreinemakers rule states that the angle of intersection of curves (F) and (E), or in general any two univariant curves defining the limits of a single divariant field, where they meet at the invariant point, must not exceed 180°. This rule (also known as the 180° rule) is illustrated in Figure A2.1. Because each curve has a metastable extension continuing as a smooth curve beyond the invariant point, the case in which the divariant field occupies an angle of greater than 180° (Figure A2.1b) is a logical absurdity. Within the field of stability of the assemblage A + B + C + D lie the metastable extensions of both bounding univariant curves (F) and (E). However, on one side of each of these curves A + B + C + D is, by definition, *not* the most stable assemblage.

A2.4   Relative Positioning of Univariant Curves About the Invariant Point

Continuing with the four-component system we have just discussed, it is apparent from Figure A2.1a that, even though reaction (F) is not defined, we know that phase E is stable along curve F because only F is absent. It follows that it must be a stable member of the assemblage in the sector to the right of this curve. By the same

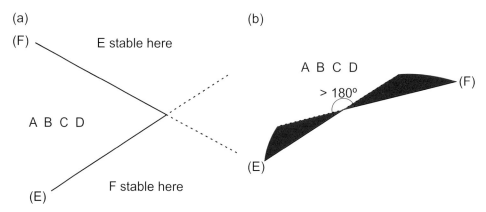

Figure A2.1 Morey–Schreinemakers rule (or 180° rule), illustrated for a four-component system containing the phases A, B, C, D, E, F. (a) Drawn correctly, shows a divariant field within which the assemblage A + B + C + D is stable, bounded by univariant curves (F) and (E) which intersect at an angle less than 180° at the invariant point. (b) Drawn incorrectly, shows the angle between (E) and (F) greater than 180°. As a result the metastable extension of each of these curves (shown dashed) lies within the stability field of A + B + C + D. However, on crossing each metastable extension into the shaded area, an alternative assemblage becomes more stable than A + B + C + D, which is logically inconsistent with being already within the stability field of that assemblage.

argument, phase F must be stable in the divariant field below curve (E). This leads us to a general rule: the stable portion of the univariant curve from which a phase is absent must lie on the *opposite side* of each of the other univariant curves in the system from that on which it appears as a reaction product. Phase E is a product on the upper right side of reaction (F) and the stable portion of reaction (E) lies to the lower left in Figure A2.1a.

In combination with Morey–Schreinemakers rule this allows us to produce only two possible solutions for the sequence in which univariant curves (and their metastable extensions) are encountered around the invariant point. These two solutions are simply mirror images.

As an example, Figure A2.2a shows the composition of five phases; J, K, L, M, N, made up of three components. The five univariant reactions that can be written between them are:

$$K + L = M + N \ (J)$$
$$J + M = N + L \ (K)$$
$$N = J + K + M \ (L)$$
$$N = J + K + L \ (M)$$
$$K + L = J + M \ (N)$$

In Figure A2.2b the univariant curve (J) has been drawn in quite arbitrarily (no axes are indicated), the invariant point in the system has been located at one end of it and the metastable extension is shown beyond the invariant

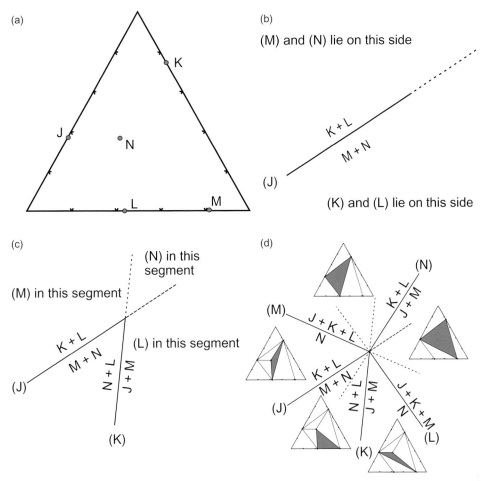

Figure A2.2 Construction of a phase diagram. (a) Compositions of the phases J–N in the three-component system. (b) Step 1: curve (J) and the invariant point have been arbitrarily located. Constraints on the locations of the stable portions of the remaining curves are indicated. (c) Step 2: curve (K) added to curve (J), showing additional constraints generated. (d) The completed diagram, including compositional triangles to show the compatible assemblages in each segment; the unique assemblages are stippled.

point. It can be seen that the stable portions of univariant curves (K) and (L) must lie to one side, while (M) and (N) lie on the other, according to the way in which (J) is labelled. Adding curve (K) (Figure A2.2c) provides further constraints on the location of (L), (M) and (N). The completed diagram is shown in Figure A2.2d. Remember that this solution is only unique in so far as it predicts the sequence in which univariant curves and their metastable extensions are encountered in going around the invariant point, the angles between them and whether they are encountered in a clockwise or counter-clockwise sense have been decided arbitrarily.

A2.4.1   Positioning Degenerate Reactions

The way in which degenerate reactions are positioned follows logically in the same way as for normal reactions, except that for some the univariant curve is stable on both sides of the invariant point. In the example used above, of the system corundum–quartz–andalusite–sillimanite, positioning curves (And) and (Sil) requires curves (Crn) and (Qz) to appear in the same sector. These are of course both the same reaction, which is therefore plotted within that sector in the normal manner (Figure A2.3a).

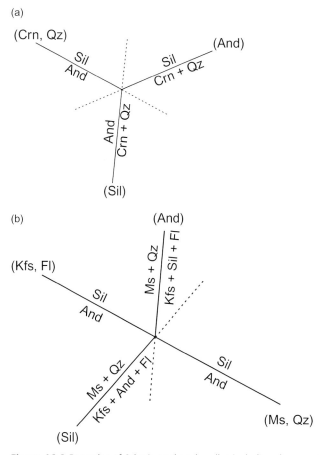

Figure A2.3 Examples of Schreinemakers bundles including degenerate reactions. (a) System corundum–quartz–andalusite–sillimanite. Degenerate reaction (Crn, Qz) is stable on one side of the invariant point only. (b) System muscovite–quartz–K-feldspar–sillimanite–andalusite–fluid. Degenerate reaction (Ms, Qz, Kfs, Fl) is stable on both sides of the invariant point.

In contrast, consider the system K-feldspar–muscovite–quartz–sillimanite–andalusite–fluid, made up of the components, K_2O–SiO_2–Al_2O_3–H_2O. The possible reactions are:

muscovite + quartz = sillimanite + K-feldspar + fluid (And)
muscovite + quartz = andalusite + K-feldspar + fluid (Sil)
andalusite = sillimanite (Ms, Kfs, Qz, Fl).

The first reaction (And) requires the stable portions of reactions (Ms) and (Qz) to lie on one side of it, while (Kfs) and (Fl) lie on the other. However, these four reactions are all one and the same. This means that univariant curve (Ms, Kfs, Qz, Fl) is stable on both sides of the invariant point and we label one end of it (Ms, Qz), the other (Kfs, Fl). This has been done in Figure A2.3b.

A2.5   Orienting the Schreinemakers Bundle with Respect to Intensive Variables

Although the application of Schreinemakers rules is a purely geometric exercise, it is obviously desirable to locate the curves relative to P, T or other parameters. In the case where P and T are the axes then, since assemblages with smaller volumes are stabilised by increased pressure, it is usually possible to decide which is the high-pressure side of any curve. Since the molar volume of water is variable, it is obviously important to choose an appropriate P–T condition. Similarly, higher temperatures stabilise higher-entropy assemblages. As a rule of thumb, devolatilisation reactions normally proceed with increased temperature.

Where one or both of the axes is a parameter such as X_{CO_2}, f_{O_2} or the chemical potential of a particular component, then the fluid species or component concerned is released by reaction as its concentration, fugacity or chemical potential decreases. Thus we see in Chapter 6 that calcite + quartz break down, releasing CO_2, at low values of X_{CO_2}, if pressure and temperature are constant (Figure 6.2).

APPENDIX 3

Application of the Phase Rule to Rocks Undergoing Hydrothermal Metamorphism

Rocks undergoing metamorphism can behave either as **closed systems,** in which case no material is added or removed, or as **open systems** involving loss or gain of some components. Most metamorphic rocks behave as open systems to some extent, because they lose volatiles during progressive heating. However, the amounts involved are relatively small and so there is no appreciable loss of other components in solution. In contrast, in hydrothermal metamorphism where large volumes of fluid may pass through the rock, the concentrations of a number of constituents of the rock may be changed appreciably, giving rise to significant metasomatism.

It was pointed out by D. S. Korzhinskii in the 1930s (see Korzhinskii (1959) for an English account) that the Gibbs phase rule as outlined in Chapter 2 can be modified to be applied to open systems. Korzhinskii distinguished between **inert components,** i.e. those that neither enter nor leave the rock, and **perfectly mobile components,** which are present in the fluid phase and may be added to, or removed from, the rock according to whether their chemical potential in the fluid is greater or less than in the mineral matter of the rock. The important point is that chemical potentials of perfectly mobile components are controlled by external factors; for example, that of a solute component may be controlled by mineral solubility in the source region from which the fluid was derived. If we consider that metamorphic P–T conditions will normally lie within divariant fields on a P–T diagram for the system concerned (i.e. $F = 2$), the normal situation is that the *total* number of phases will equal the number of components during isochemical metamorphism, i.e. $P = C$. Korzhinskii showed that if we designate the number of inert components as C_i and the number of perfectly mobile components as C_m, then the number of *solid* phases $P_s = C_i = C - C_m$. This is often known as the Korzhinskii phase rule.

As an example, consider the mineral assemblage talc + forsterite + tremolite in the aureole of the Bergell tonalite (Chapter 2). The chemical system is MgO–CaO–SiO_2–H_2O, and if a fluid phase is present also, from the Gibbs phase rule $P = C = 4$. On the other hand, we can treat water as a perfectly mobile component because its pressure, and hence chemical potential, is independently controlled by the mechanical strength of the rock which prevents P_{fluid} exceeding $P_{lithostatic}$ (Chapter 1, Section 1.4.2). In this case $P_s = C_i = 3$ from the Korzhinskii phase rule, effectively the same result as from the Gibbs phase rule because only water was considered to be mobile.

Korzhinskii's phase rule is really of use when it comes to hydrothermal processes where large fluxes of fluid through rock volumes permit a range of normally inert components, including most of the major element constituents of rocks, to become mobile. Clearly, as more components become mobile there must be fewer inert components, and hence the number of phases in the rock will be smaller. As a result, *hydrothermally metamorphosed rocks and related veins contain only a small number of co-existing minerals, and may be monomineralic.* It could be argued that some rocks, such as amphibolites, have a very restricted number of phases in any case, even though their metamorphism was isochemical. However, in amphibolite the dominant minerals are amphibole and plagioclase solid solutions and these can display considerable compositional variability between nearby samples. Metasomatic rocks formed in open systems not only have few minerals, those minerals are of very uniform composition even if they have potential for solid solution. The reason is that element ratios in the solid solution are being fixed at a constant value by the composition of the infiltrating fluid.

Keeping to the example of metamorphosed ultrabasic rocks, ultrabasic pods emplaced along shear zones into siliceous rocks are not uncommon in orogenic belts, and have often recrystallised to monomineralic talc rocks with addition of SiO_2 and removal of CaO. In this case only MgO can be considered inert, and so the final rock has only one mineral. Where less water gained access, CaO does not behave as a perfectly mobile component however, and bimineralic talc–tremolite rocks result.

Petrologists sometimes invoke large-scale fluid movements as an explanation for features of the rocks they are studying; the Korzhinskii phase rule provides a simple check on whether such an interpretation is reasonable. Processes that generate mineral assemblages with large numbers of co-existing minerals are unlikely to involve large-scale fluid flow and metasomatism, while mineralogically simple rocks are more likely to be of metasomatic origin.

References

Agard, P., Yamato, P., Jolivet, L. & Burov, E. (2009). Exhumation of oceanic blueschists and eclogites in subduction zones: timing and mechanisms. *Earth-Science Reviews*, **92**(1–2), 53–79.

Ague, J. J. & Carlson, W. D. (2013). Metamorphism as garnet sees it: the kinetics of nucleation and growth, equilibration, and diffusional relaxation. *Elements*, **9**(6), 439–45.

Amato, J. M., Johnson, C. M., Baumgartner, L. P. & Beard, B. L. (1999). Rapid exhumation of the Zermatt-Saas ophiolite deduced from high-precision Sm–Nd and Rb–Sr geochronology. *Earth and Planetary Science Letters*, **171**(3), 425–38.

Anczkiewicz, R., Platt, J. P., Thirlwall, M. F. & Wakabayashi, J. (2004). Franciscan subduction off to a slow start: evidence from high-precision Lu–Hf garnet ages on high grade-blocks. *Earth and Planetary Science Letters*, **225**(1–2), 147–61.

Angiboust, S., Agard, P., Jolivet, L. & Beyssac, O. (2009). The Zermatt-Saas ophiolite: the largest (60-km wide) and deepest (c. 70–80 km) continuous slice of oceanic lithosphere detached from a subduction zone? *Terra Nova*, **21**(3), 171–80.

Árkai, P. (1991). Chlorite crystallinity: an empirical approach and correlation with illite crystallinity, coal rank and mineral facies as exemplified by Palaeozoic and Mesozoic rocks of northeast Hungary. *Journal of Metamorphic Geology*, **9**(6), 723–34.

Austrheim, H. (1987). Eclogitization of lower crustal granulites by fluid migration through shear zones. *Earth and Planetary Science Letters*, **81**(2–3), 221–32.

Austrheim, H. & Griffin, W. L. (1985). Shear deformation and eclogite formation within granulite-facies anorthosites of the Bergen Arcs, western Norway. *Chemical Geology*, **50**(1–3), 267–81.

Bach, W. & Früh-Green, G. L. (2010). Alteration of the oceanic lithosphere and implications for seafloor processes. *Elements*, **6**(3), 173–8.

Balashov, V. N. & Yardley, B. W. D. (1998). Modeling metamorphic fluid flow with reaction-permeability feedbacks and at a range of effective stress. *American Journal of Sciences*, **298**, 441–70.

Baldwin, S. L. & Das, J. P. (2015). Atmospheric Ar and Ne returned from mantle depths to the Earth's surface by forearc recycling. *Proceedings of the National Academy of Sciences*, **112**(46), 14174–9.

Baldwin, S. L., Monteleone, B. D., Webb, L. E. *et al.* (2004). Pliocene eclogite exhumation at plate tectonic rates in eastern Papua New Guinea. *Nature*, **431**(7006), 263.

Barboza, S. A., Bergantz, G. W. & Brown, M. (1999). Regional granulite facies metamorphism in the Ivrea zone: Is the Mafic Complex the smoking gun or a red herring? *Geology*, **27**(5), 447–50.

Baxter, E. F. & Scherer, E. E. (2013). Garnet geochronology: timekeeper of tectonometamorphic processes. *Elements*, **9**(6), 433–8.

Beach, A. (1976). The interrelations of fluid transport, deformation, geochemistry and heat flow in early Proterozoic shear zones in the Lewisian complex. *Philosophical Transactions for the Royal Society of London. Series A*, **280**, 569–604.

Beaumont, C., Jamieson, R. A., Nguyen, M. H. & Lee, B. (2001). Himalayan tectonics explained by extrusion of a low-viscosity crustal channel coupled to focused surface denudation. *Nature*, **414**(6865), 738–42.

Beltrando, M., Compagnoni, R. & Lombardo, B. (2010). (Ultra-) High-pressure metamorphism and orogenesis: an Alpine perspective. *Gondwana Research*, **18**(1), 147–66.

Berger, A. & Bousquet, R. (2008). Subduction-related metamorphism in the Alps: review of isotopic ages based on petrology and their geodynamic consequences. *Geological Society, London, Special Publications*, **298**, 117–44.

Berger, A., Burri, T., Alt-Epping, P. & Engi, M. (2008). Tectonically controlled fluid flow and water-assisted melting in the middle crust: an example from the Central Alps. *Lithos*, **102**(3–4), 598–615.

Beyssac, O., Cox, S. C., Vry, J. & Herman, F. (2016). Peak metamorphic temperature and thermal history of the Southern Alps (New Zealand). *Tectonophysics*, **676**, 229–49.

Beyssac, O., Goffé, B., Petitet, J.-P. *et al.* (2003). On the characterization of disordered and heterogeneous carbonaceous materials by Raman spectroscopy. *Spectrochimica Acta Part A: Molecular and Biomolecular Spectroscopy*, **59**(10), 2267–76.

Boles, J. R. & Coombs, D. S. (1975). Mineral reactions in zeolitic Triassic tuff, Hokonui Hills, New Zealand. *Geological Society of America Bulletin*, **86**(2), 163–73.

Bosse, V., Ballèvre, M. & Vidal, O. (2002). Ductile thrusting recorded by the garnet isograd from blueschist-facies metapelites of the Ile de Groix, Armorican Massif, France. *Journal of Petrology*, **43**(3), 485–510.

Bourdelle, F. & Cathelineau, M. (2015). Low-temperature chlorite geothermometry: a graphical representation based on a $T–R^{2+}–Si$ diagram. *European Journal of Mineralogy*, **27**(5), 617–26.

Bourdelle, F., Parra, T., Beyssac, O., Chopin, C. & Vidal, O. (2013). Clay minerals as geo-thermometer: A comparative study based on high spatial resolution analyses of illite and chlorite in Gulf Coast sandstones (Texas, USA). *American Mineralogist*, **98**(5–6), 914–26.

Bousquet, R., Oberhansli, R., Schmid, S. *et al.* (2012). *Metamorphic Framework of the Alps - Carte metamorphique des Alpes*, CCGM/CGMW.

Bowman, J. R., Valley, J. W. & Kita, N. T. (2009). Mechanisms of oxygen isotopic exchange and isotopic evolution of $^{18}O/^{16}O$-depleted periclase zone marbles in the Alta aureole, Utah: insights from ion microprobe analysis of calcite. *Contributions to Mineralogy and Petrology*, **157**(1), 77–93.

Bradshaw, J. Y. (1990). Geology of crystalline rocks of northern Fiordland: details of the granulite facies Western Fiordland Orthogneiss and associated rock units. *New Zealand Journal of Geology and Geophysics*, **33**(3), 465–84.

Brown, M. (2007). Metamorphic conditions in orogenic belts: a record of secular change. *International Geology Review*, **49**(3), 193–234.

Brown, M. & Johnson, T. (2019). Metamorphism and the evolution of subduction on Earth. *American Mineralogist*, **104**(8), 1065–82.

Brown, W. L. & Parsons, I. (1989). Alkali feldspars: ordering rates, phase transformations and behaviour diagrams for igneous rocks. *Mineralogical Magazine*, **53**(369), 25–42.

Butler, J. P., Jamieson, R. A., Steenkamp, H. M. & Robinson, P. (2013). Discovery of coesite–eclogite from the Nordøyane UHP domain, Western Gneiss Region, Norway: field relations, metamorphic history, and tectonic significance. *Journal of Metamorphic Geology*, **31**(2), 147–63.

Cann, J. R. (1969). Spilites from the Carlsberg Ridge, Indian Ocean. *Journal of Petrology*, **10**, 1–19.

Cann, J. R., McCaig, A. M. & Yardley, B. W. D. (2015). Rapid generation of reaction permeability in the roots of black smoker systems, Troodos ophiolite, Cyprus. *Geofluids*, **15**(1–2), 179–92.

Carlson, W. D. (2006). Rates of Fe, Mg, Mn, and Ca diffusion in garnet. *American Mineralogist*, **91**(1), 1–11.

Carlson, W. D. (2011). Porphyroblast crystallization: linking processes, kinetics, and microstructures. *International Geology Review*, **53**(3–4), 406–45.

Carlson, W. D. & Rosenfeld, J. L. (1981). Optical determination of topotactic aragonite–calcite growth kinetics: metamorphic implications. *The Journal of Geology*, **89**(5), 615–38.

Carmichael, D. M. (1969). On the mechanism of prograde metamorphic reactions in quartz-bearing pelitic rocks. *Contributions to Mineralogy and Petrology*, **20**, 244–67.

Castro, A. E. & Spear, F. S. (2017). Reaction overstepping and re-evaluation of peak P–T conditions of the blueschist unit Sifnos, Greece: implications for the Cyclades subduction zone. *International Geology Review*, **59**(5–6), 548–62.

Cathelineau, M. (1988). Cation site occupancy in chlorites and illites as a function of temperature. *Clay Minerals*, **23**(4), 471–85.

Cawood, P. A., Kröner, A., Collins, W. J. et al. (2009). Accretionary orogens through Earth history. *Geological Society, London, Special Publications*, **318**, 1–36.

Chapman, D. S. & Furlong, K. P. (1992). Thermal state of the lower continental crust. *Continental Lower Crust*, eds. D. M. Fountain, R. Arculus and R.W. Kay, Elsevier, Amsterdam, pp. 179–99.

Chatterjee, N. D. (1972). The upper stability limit of the assemblage paragonite+

quartz and its natural occurrences. *Contributions to Mineralogy and Petrology*, **34**(4), 288–303.

Chatterjee, N. D. & Johannes, W. (1974). Thermal stability and standard thermodynamic properties of synthetic 2M1-muscovite, KAl$_2$ [AlSi$_3$O$_{10}$(OH)$_2$]. *Contributions to Mineralogy and Petrology*, **48**(2), 89–114.

Cherniak, D. J. (2000). Pb diffusion in rutile. *Contributions to Mineralogy and Petrology*, **139**(2), 198–207.

Chopin, C. (1984). Coesite and pure pyrope in high-grade blueschists of the Western Alps: a first record and some consequences. *Contributions to Mineralogy and Petrology*, **86**(2), 107–18.

Connolly, J. A. D. (2005). Computation of phase equilibria by linear programming: a tool for geodynamic modeling and its application to subduction zone decarbonation. *Earth and Planetary Science Letters*, **236**(1–2), 524–41.

Connolly, J. A. D. & Trommsdorff, V. (1991). Petrogenetic grids for metacarbonate rocks: pressure-temperature phase-diagram projection for mixed-volatile systems. *Contributions to Mineralogy and Petrology*, **108**(1–2), 93–105.

Cook, S. J. & Bowman, J. R. (2000). Mineralogical evidence for fluid–rock interaction accompanying prograde contact metamorphism of siliceous dolomites: Alta Stock Aureole, Utah, USA. *Journal of Petrology*, **41**(6), 739–57.

Coombs, D. S. (1954). The nature and alteration of some Triassic sediments from Southland, New Zealand. *Transactions of the Royal Society of New Zealand*, **82**, 65–109.

Coombs, D. S., Ellis, A. J., Fyfe, W. S. & Taylor, A. M. (1959). The zeolite facies, with comments on the interpretation of hydrothermal syntheses. *Geochimica et Cosmochimica Acta*, **17**(1–2), 53–107.

Cottle, J. M., Larson, K. P. & Kellett, D. A. (2015). How does the mid-crust accommodate deformation in large, hot collisional orogens? A review of recent research in the Himalayan orogen. *Journal of Structural Geology*, **78**, 119–33.

Day, H. W. (2012). A revised diamond-graphite transition curve. *American Mineralogist*, **97**(1), 52–62.

de Capitani, C. & Petrakakis, K. (2010). The computation of equilibrium assemblage diagrams with Theriak/Domino software. *American Mineralogist*, **95**(7), 1006–16.

De Paoli, M. C., Clarke, G. L. & Daczko, N. R. (2012). Mineral equilibria modeling of the granulite–eclogite transition: effects of whole-rock composition on metamorphic facies type-assemblages. *Journal of Petrology*, **53**(5), 949–70.

Dempster, T. J. & Tanner, P. W. G. (1997). The biotite isograd, Central Pyrenees: a deformation-controlled reaction. *Journal of Metamorphic Geology*, **15**(4), 531–48.

Denison, C., Carlson, W. D. & Ketcham, R. A. (1997). Three-dimensional quantitative textural analysis of metamorphic rocks using high-resolution computed X-ray tomography: Part I. Methods and techniques. *Journal of Metamorphic Geology*, **15**(1), 29–44.

Diener, J. F. A. & Powell, R. (2012). Revised activity–composition models for clinopyroxene and amphibole. *Journal of Metamorphic Geology*, **30**(2), 131–42.

Dodson, M. (1973). Closure temperature in cooling geochronological and petrological systems. *Contributions to Mineralogy and Petrology*, **40**(3), 259–74.

Dungan, M. A., Vance, J. A. & Blanchard, D. P. (1983). Geochemistry of the Shuksan greenschists and blueschists, North Cascades, Washington: variably fractionated and altered metabasalts of oceanic affinity. *Contributions to Mineralogy and Petrology*, **82**(2–3), 131–46.

Dyck, B., Waters, D. J., St-Onge, M. R. & Searle, M. P. (2020). Muscovite dehydration melting; reaction mechanisms, microstructures, and implications for anatexis. *Journal of Metamorphic Geology*, **38**, 29–52.

El Korh, A., Schmidt, S. T., Vennemann, T. & Ballevre, M. (2013). Trace element and isotopic fingerprints in HP–LT metamorphic rocks as a result of fluid–rock interactions (Ile de Groix, France). *Gondwana Research*, **23**(3), 880–900.

Elmer, F. L., White, R. W. & Powell, R. (2006). Devolatilization of metabasic rocks during greenschist–amphibolite facies metamorphism. *Journal of Metamorphic Geology*, **24**(6), 497–513.

England, P. C. & Katz, R. F. (2010). Melting above the anhydrous solidus controls the location of volcanic arcs. *Nature*, **467**(7316), 700–3.

England, P. C. & Thompson, A. B. (1984). Pressure–temperature–time paths of regional metamorphism I. Heat transfer during the evolution of regions of thickened continental crust. *Journal of Petrology*, **25**(4), 894–928.

Ewing, T. A., Hermann, J. & Rubatto, D. (2013). The robustness of the Zr-in-rutile and Ti-in-zircon thermometers during high-temperature metamorphism (Ivrea-Verbano Zone, northern Italy). *Contributions to Mineralogy and Petrology*, **165**(4), 757–79.

Ewing, T. A., Rubatto, D., Beltrando, M. & Hermann, J. (2015). Constraints on the thermal evolution of the Adriatic margin during Jurassic continental break-up: U–Pb dating of rutile from the Ivrea–Verbano Zone, Italy. *Contributions to Mineralogy and Petrology*, **169**(4), 44.

Farley, K. A. (2000). Helium diffusion from apatite: General behavior as illustrated by Durango fluorapatite. *Journal of Geophysical Research: Solid Earth*, **105**(B2), 2903–14.

Ferry, J. M. (1983). Regional metamorphism of the Vassalboro Formation, south-central Maine, USA: a case study of the role of fluid in metamorphic petrogenesis. *Journal of the Geological Society*, **140**(4), 551–76.

Ferry, J. M. (1994). Overview of the petrologic record of fluid flow during regional metamorphism in northern New England. *American Journal of Science*, **294**(8), 905–88.

Ferry, J. M. & Spear, F. S. (1978). Experimental calibration of the partitioning of Fe and Mg between biotite and garnet. *Contributions to Mineralogy and Petrology*, **66**(2), 113–17.

Fettes, F. & Desmons, J. (Eds.) (2007). *Metamorphic Rocks: A Classification and Glossary of Terms*. Cambridge University Press.

Flowers, R. M., Bowring, S. A., Tulloch, A. J. & Klepeis, K. A. (2005). Tempo of burial and exhumation within the deep roots of a magmatic arc, Fiordland, New Zealand. *Geology*, 33(1), 17–20.

Forshaw, J. B., Waters, D. J., Pattison, D. R. M., Palin, R. M. & Gopon, P. (2019). A comparison of observed and thermodynamically predicted phase equilibria and mineral compositions in mafic granulites. *Journal of Metamorphic Geology*, 37(2), 153–79.

Frey, M., Capitani, C. de & Liou, J. G. (1991). A new petrogenetic grid for low-grade metabasites. *Journal of Metamorphic Geology*, 9(4), 497–509.

Friedrich, A. M., Bowring, S. A., Martin, M. W. & Hodges, K. V. (1999). Short-lived continental magmatic arc at Connemara, western Irish Caledonides: Implications for the age of the Grampian orogeny. *Geology*, 27(1), 27–30.

Gebauer, D., Schertl, H.-P., Brix, M. & Schreyer, W. (1997). 35 Ma old ultrahigh-pressure metamorphism and evidence for very rapid exhumation in the Dora Maira Massif, Western Alps. *Lithos*, 41(1–3), 5–24.

Giuntoli, F., Menegon, L. & Warren, C. J. (2018). Replacement reactions and deformation by dissolution and precipitation processes in amphibolites. *Journal of Metamorphic Geology*, 36(9), 1263–86.

Glazner, A. F., Bartley, J. M., Coleman, D. S., Gray, W. & Taylor, R. Z. (2004). Are plutons assembled over millions of years by amalgamation from small

magma chambers? *GSA Today*, 14(4/5), 4–12.

Godin, L., Grujic, D., Law, R. D. & Searle, M. P. (2006). Channel flow, ductile extrusion and exhumation in continental collision zones: an introduction. *Geological Society, London, Special Publications*, 268(1), 1–23.

Goffe, B. & Chopin, C. (1986). High-pressure metamorphism in the Western Alps: zoneography of metapelites, chronology and consequences. *Schweizerische Mineralogische Und Petrographische Mitteilungen*, 66(1–2), 41–52.

Goldstein, R. H. & Samson, I. (2003). Petrographic analysis of fluid inclusions. *Fluid Inclusions: Analysis and Interpretation*, 32, 9–53.

Greenwood, H. J. (1962). Metamorphic reactions involving two volatile components. *Carnegie Institute Washington Yearbook*, 61, 82–5.

Greenwood, H. J. (1967). Wollastonite: Stability in H_2O–CO_2 mixtures and occurrence in a contact–metamorphic aureole near Salmo, British Columbia, Canada. *American Mineralogist: Journal of Earth and Planetary Materials*, 52(11–12), 1669–80.

Greenwood, H. J. (1975). Buffering of pore fluids by metamorphic reactions. *American Journal of Science*, 275(5), 573–93.

Guenthner, W. R., Reiners, P. W., Ketcham, R. A., Nasdala, L. & Giester, G. (2013). Helium diffusion in natural zircon: Radiation damage, anisotropy, and the interpretation of zircon (U-Th)/He thermochronology. *American Journal of Science*, 313(3), 145–98.

Gurney, J. J., Helmstaedt, H. H., Richardson, S. H. & Shirey, S. B. (2010). Diamonds through time. *Economic Geology*, **105**(3), 689–712.

Haar, L. (1984). *NBS/NRC Steam Tables*. CRC Press.

Haas, J. L. (1976). Thermodynamics properties of the coexisting phases and thermochemical properties of the NaCl component in boiling NaCl solutions. *US Geological Survey Bulletin*, 1421-A.

Hacker, B. R., Abers, G. A. & Peacock, S. M. (2003). Subduction factory 1. Theoretical mineralogy, densities, seismic wave speeds, and H_2O contents. *Journal of Geophysical Research: Solid Earth*, **108**(B1), doi: 10.1029/2001JB001127.

Harker, R. I. & Tuttle, O. F. (1956). Experimental data on the P–T curve for the reaction; calcite–quartz $\leftrightarrow$ wollastonite–carbon dioxide. *American Journal of Science*, **254**(4), 239–56.

Harley, S. L. & Green, D. H. (1982). Garnet–orthopyroxene barometry for granulites and peridotites. *Nature*, **300**(5894), 697–701.

Harrison, T. M., Célérier, J., Aikman, A. B., Hermann, J. & Heizler, M. T. (2009). Diffusion of ^{40}Ar in muscovite. *Geochimica et Cosmochimica Acta*, **73**(4), 1039–51.

Harrison, T. M., Duncan, I. A. N. & McDougall, I. A. N. (1985). Diffusion of ^{40}Ar in biotite: temperature, pressure and compositional effects. *Geochimica et Cosmochimica Acta*, **49**(11), 2461–8.

Hawthorne, F. C., Oberti, R., Harlow, G. E. *et al.* (2012). Nomenclature of the amphibole supergroup. *American Mineralogist*, **97**(11–12), 2031–48.

Heinrich, W. & Gottschalk, M. (1995). Metamorphic reactions between fluid inclusions and mineral hosts. I. Progress of the reaction calcite+ quartz= wollastonite+ CO_2 in natural wollastonite-hosted fluid inclusions. *Contributions to Mineralogy and Petrology*, **122**(1–2), 51–61.

Helgeson, H. C. (1978). Summary and critique of the thermodynamic properties of rock-forming minerals. *American Journal of Science*, **278**, 1–229.

Henry, D. J., Guidotti, C. V & Thomson, J. A. (2005). The Ti-saturation surface for low-to-medium pressure metapelitic biotites: Implications for geothermometry and Ti-substitution mechanisms. *American Mineralogist*, **90**(2–3), 316–28.

Hilchie, L. J. & Jamieson, R. A. (2014). Graphite thermometry in a low-pressure contact aureole, Halifax, Nova Scotia. *Lithos*, **208**, 21–33.

Hoisch, T. D., Wells, M. L. & Grove, M. (2008). Age trends in garnet-hosted monazite inclusions from upper amphibolite facies schist in the northern Grouse Creek Mountains, Utah. *Geochimica et Cosmochimica Acta*, **72**(22), 5505–20.

Holland, T. & Blundy, J. (1994). Non-ideal interactions in calcic amphiboles and their bearing on amphibole-plagioclase thermometry. *Contributions to Mineralogy and Petrology*, **116**(4), 433–47.

Holland, T. J. B. (1980). The reaction albite = jadeite+ quartz determined experimentally in the range 600–1200 °C. *American Mineralogist*, **65**(1–2), 129–34.

Holland, T. J. B. & Powell, R. (2011). An improved and extended internally consistent thermodynamic dataset for phases of petrological interest, involving a new equation of state for solids. *Journal of Metamorphic Geology*, **29**(3), 333–83.

Holmes, A. (1911). The association of lead with uranium in rock-minerals, and its application to the measurement of geological time. *Proceedings of the Royal Society of London. Series A*, **85**(578), 248–56.

Holness, M. (1992). Metamorphism and fluid infiltration of the calc-silicate aureole of the Beinn an Dubhaich granite, Skye. *Journal of Petrology*, **33**(6), 1261–93.

Hopkinson, T. N., Harris, N. B. W., Warren, C. J. *et al.* (2017). The identification and significance of pure sediment-derived granites. *Earth and Planetary Science Letters*, **467**, 57–63.

Huang, W.-L. & Wyllie, P. J. (1973). Melting relations of muscovite-granite to 35 kbar as a model for fusion of metamorphosed subducted oceanic sediments. *Contributions to Mineralogy and Petrology*, **42**(1), 1–14.

Iaccarino, S., Montomoli, C., Carosi, R. *et al.* (2015). Pressure–temperature–time-deformation path of kyanite-bearing migmatitic paragneiss in the Kali Gandaki valley (Central Nepal): investigation of Late Eocene–Early Oligocene melting processes. *Lithos*, **231**, 103–21.

Imayama, T., Takeshita, T. & Arita, K. (2010). Metamorphic *P–T* profile and *P–T* path discontinuity across the far-eastern Nepal Himalaya: investigation of channel flow models. *Journal of Metamorphic Geology*, **28**(5), 527–49.

Indares, A. & Dunning, G. (2001). Partial melting of high-P–T metapelites from the Tshenukutish Terrane (Grenville Province): petrography and U–Pb geochronology. *Journal of Petrology*, **42**(8), 1547–65.

Jaeger, J. C. (1964). Thermal effects of intrusions. *Reviews of Geophysics*, **2**(3), 443–66.

Jamieson, R. A. & Beaumont, C. (2013). On the origin of orogens. *Bulletin*, **125**(11–12), 1671–702.

Jamieson, R. A., Hart, G. G., Chapman, G. G. & Tobey, N. W. (2012). The contact aureole of the South Mountain Batholith in Halifax, Nova Scotia: geology, mineral assemblages, and isograds. *Canadian Journal of Earth Sciences*, **49**(11), 1280–96.

Jamtveit, B., Moulas, E., Andersen, T. B. *et al.* (2018). High pressure metamorphism caused by fluid induced weakening of deep continental crust. *Scientific Reports*, **8**(1), 1–8.

Johannes, W. (1984). Beginning of melting in the granite system Qz-Or-Ab-An-H_2O. *Contributions to Mineralogy and Petrology*, **86**(3), 264–73.

Johannes, W. & Holtz, F. (1992). Melting of plagioclase in granite and related systems: composition of coexisting phases and kinetic observations. *Earth and Environmental Science Transactions of The Royal Society of Edinburgh*, **83**(1–2), 417–22.

Johannes, W. & Puhan, D. (1971). The calcite-aragonite transition, reinvestigated. *Contributions to Mineralogy and Petrology*, **31**(1), 28–38.

Johnson, T. E., Brown, M. & Solar, G. S. (2003). Low–pressure subsolidus and suprasolidus phase equilibria in the MnNCKFMASH system: Constraints on conditions of regional metamorphism in western Maine, northern Appalachians. *American Mineralogist*, **88**(4), 624–38.

Johnson, T. E., Clark, C., Taylor, R. J. M., Santosh, M. & Collins, A. S. (2015). Prograde and retrograde growth of monazite in migmatites: An example from the Nagercoil Block, southern India. *Geoscience Frontiers*, **6**(3), 373–87.

Kaneko, Y. & Miyano, T. (2004). Recalibration of mutually consistent garnet–biotite and garnet–cordierite geothermometers. *Lithos*, **73**(3–4), 255–69.

Kawachi, Y. (1975). Pumpellyite–actinolite and contiguous facies metamorphism in the Upper Wakatipu district, southern New Zealand. *New Zealand Journal of Geology and Geophysics*, **17**, 169–208.

Kelsey, D. E. (2008). On ultrahigh-temperature crustal metamorphism. *Gondwana Research*, **13**(1), 1–29.

Kelsey, D. E. & Hand, M. (2015). On ultrahigh temperature crustal metamorphism: phase equilibria, trace element thermometry, bulk composition, heat sources, timescales and tectonic settings. *Geoscience Frontiers*, **6**(3), 311–56.

Kerrick, D. M. (1968). Experiments on the upper stability limit of pyrophyllite at 1.8 kb and 3.9 kb water pressure. *American Journal of Science*, **266**(3), 204–14.

Kerrick, D. M. (1974). Review of metamorphic mixed-volatile (H_2O–CO_2) equilibria. *American Mineralogist*, **59**(7–8), 729–62.

Kirchner, J. W., Finkel, R. C., Riebe, C. S. *et al.* (2001). Mountain erosion over 10 yr, 10 ky, and 10 My time scales. *Geology*, **29**(7), 591–4.

Kirschner, D. L., Sharp, Z. D. & Masson, H. (1995). Oxygen isotope thermometry of quartz–calcite veins: Unraveling the thermal–tectonic history of the subgreenschist facies Morcles nappe (Swiss Alps). *Geological Society of America Bulletin*, **107**(10), 1145–56.

Kisch, H. J. (1991). Illite crystallinity: recommendations on sample preparation, X-ray diffraction settings, and interlaboratory samples. *Journal of Metamorphic Geology*, **9**(6), 665–70.

Kohn, M. J. & Penniston-Dorland, S. C. (2017). Diffusion: Obstacles and opportunities in petrochronology. *Reviews in Mineralogy and Geochemistry*, **83**(1), 103–52.

Konrad-Schmolke, M., Zack, T., O'Brien, P. J. & Jacob, D. E. (2008). Combined thermodynamic and rare earth element modelling of garnet growth during subduction: examples from ultrahigh-pressure eclogite of the Western Gneiss Region, Norway. *Earth and Planetary Science Letters*, **272**(1–2), 488–98.

Korzhinskii, D. S. (1959). *Physicochemical Basis of the Analysis of the Paragenesis of Minerals: English Translation*. New York, Consultants Bureau. Inc., 142 pages.

Koziol, A. M. & Bohlen, S. R. (1992). Solution properties of almandine-pyrope garnet as determined by phase equilibrium experiments. *American Mineralogist*, **77**(7–8), 765–73.

Kretz, R. (1974). Some models for the rate of crystallization of garnet in metamorphic rocks. *Lithos*, **7**, 123–31.

Kruhl, J. H., Wirth, R. & Morales, L. F. G. (2013). Quartz grain boundaries as fluid pathways in metamorphic rocks. *Journal of Geophysical Research: Solid Earth*, **118**(5), 1957–67.

Kunz, B. E., Johnson, T. E., White, R. W. & Redler, C. (2014). Partial melting of metabasic rocks in Val Strona di Omegna, Ivrea Zone, northern Italy. *Lithos*, **190**, 1–12.

Kunz, B. E., Regis, D. & Engi, M. (2018). Zircon ages in granulite facies rocks: decoupling from geochemistry above 850 °C? *Contributions to Mineralogy and Petrology*, **173**(3), 26.

Kunz, B. E. & White, R. W. (2019). Phase equilibrium modelling of the amphibolite to granulite facies transition in metabasic rocks (Ivrea Zone, NW Italy). *Journal of Metamorphic Geology*, **37**(7), 935–50.

Lacroix, B. & Vennemann, T. (2015). Empirical calibration of the oxygen isotope fractionation between quartz and Fe–Mg-chlorite. *Geochimica et Cosmochimica Acta*, **149**, 21–31.

Landis, C. A. & Coombs, D. S. (1967). Metamorphic belts and orogenesis in southern New Zealand. *Tectonophysics*, **4**(4–6), 501–18.

Le Fort, P. (1975). Himalayas: the collided range. Present knowledge of the continental arc. *American Journal of Science*, **275**(1), 1–44.

Liou, J. G., Hacker, B. R. & Zhang, R. Y. (2000). Into the forbidden zone. *Science*, **287**, 1215–16.

Lippitsch, R., Kissling, E. & Ansorge, J. (2003). Upper mantle structure beneath the Alpine orogen from high-resolution teleseismic tomography. *Journal of Geophysical Research: Solid Earth*, **108** (B8), doi: 10.1029/2002JB002016.

Mancktelow, N. S. & Pennacchioni, G. (2004). The influence of grain boundary fluids on the microstructure of quartz-feldspar mylonites. *Journal of Structural Geology*, **26**(1), 47–69.

Maruyama, S., Liou, J. G. & Terabayashi, M. (1996). Blueschists and eclogites of the world and their exhumation. *International Geology Review*, **38**(6), 485–594.

McDade, P. & Harley, S. L. (2001). A petrogenetic grid for aluminous granulite facies metapelites in the KFMASH system. *Journal of Metamorphic Geology*, **19**(1), 45–59.

Menzies, C. D., Wright, S. L., Craw, D. *et al.* (2018). Carbon dioxide generation and drawdown during active orogenesis of siliciclastic rocks in the Southern Alps, New Zealand. *Earth and Planetary Science Letters*, **481**, 305–15.

Michard, A., Avigad, D., Goffé, B. & Chopin, C. (2004). The high-pressure metamorphic front of the south Western Alps (Ubaye-Maira transect, France, Italy). *Schweizerische Mineralogishe und Petrographische Mitteilungen*, **84**, 215–35.

Mizuochi, H., Satish-Kumar, M., Motoyoshi, Y. & Michibayashi, K. (2010). Exsolution of dolomite and application of calcite–dolomite solvus geothermometry in high-grade marbles: an example from Skallevikshalsen, East Antarctica. *Journal of Metamorphic Geology*, **28**(5), 509–26.

Moore, S. J., Cesare, B. & Carlson, W. D. (2015). Epitaxial nucleation of garnet on biotite in the polymetamorphic metapelites surrounding the Vedrette di Ries intrusion (Italian Eastern Alps). *European Journal of Mineralogy*, **27**(1), 5–18.

Morgan, D. J., Jollands, M. C., Lloyd, G. E. & Banks, D. A. (2014). Using titanium-in-quartz geothermometry and geospeedometry to recover temperatures in the aureole of the Ballachulish Igneous Complex, NW Scotland. *Geological Society, London, Special Publications*, **394**, 145–65.

Mottram, C. M., Parrish, R. R., Regis, D. *et al.* (2015). Using U-Th–Pb petrochronology to determine rates of ductile thrusting: Time windows into the Main Central Thrust, Sikkim Himalaya. *Tectonics*, **34**(7), 1355–74.

Mottram, C. M., Warren, C. J., Regis, D. *et al.* (2014). Developing an inverted Barrovian sequence; insights from monazite petrochronology. *Earth and Planetary Science Letters*, **403**, 418–31.

Muffler, L. J. P. & White, D. E. (1969). Active metamorphism of upper Cenozoic sediments in the Salton Sea geothermal field and the Salton Trough, southeastern California. *Geological Society of America Bulletin*, **80**, 157–82.

Müller, T., Dohmen, R., Becker, H. W., Ter Heege, J. H. & Chakraborty, S. (2013). Fe–Mg interdiffusion rates in clinopyroxene: experimental data and implications for Fe–Mg exchange geothermometers. *Contributions to Mineralogy and Petrology*, **166**(6), 1563–76.

Najman, Y., Appel, E., Boudagher-Fadel, M. *et al.* (2010). Timing of India–Asia collision: Geological, biostratigraphic, and palaeomagnetic constraints. *Journal of Geophysical Research: Solid Earth*, **115**(B12), doi:10.1029/2010JB007673.

Nelson, K. D., Zhao, W., Brown, L. D. *et al.* (1996). Partially molten middle crust beneath southern Tibet: synthesis of project INDEPTH results. *Science*, **274**(5293), 1684–8.

Newton, R. C. & Haselton, H. T. (1981). Thermodynamics of the garnet–plagioclase–Al_2SiO_5–quartz geobarometer. In *Thermodynamics of Minerals and Melts*, eds. R. C. Newton, A. Navrotsky and B. J. Wood, Springer, New York, pp. 131–47.

Nitsch, K.-H. (1971). Stabilitätsbeziehungen von prehnit-und pumpellyit-haltigen Paragenesen. *Contributions to Mineralogy and Petrology*, **30**(3), 240–60.

O'Brien, P. J. & Rötzler, J. (2003). High-pressure granulites: formation, recovery of peak conditions and implications for tectonics. *Journal of Metamorphic Geology*, **21**(1), 3–20.

O'Brien, P. J., Zotov, N., Law, R., Khan, M. A. & Jan, M. Q. (2001). Coesite in Himalayan eclogite and implications for models of India–Asia collision. *Geology*, **29**(5), 435–8.

Orville, P. M. (1972). Plagioclase cation exchange equilibria with aqueous chloride solution; results at 700 $°C$ and 2000 bars in the presence of quartz. *American Journal of Science*, **272**(3), 234–72.

Palin, R. M., White, R. W., Green, E. C. R. *et al.* (2016). High-grade metamorphism and partial melting of basic and intermediate rocks. *Journal of Metamorphic Geology*, **34**(9), 871–92.

Parrish, R. R., Gough, S. J., Searle, M. P. & Waters, D. J. (2006). Plate velocity exhumation of ultrahigh-pressure eclogites in the Pakistan Himalaya. *Geology*, **34**(11), 989–92.

Parsons, I. & Lee, M. R. (2009). Mutual replacement reactions in alkali feldspars I: microtextures and mechanisms. *Contributions to Mineralogy and Petrology*, **157**(5), 641.

Pattison, D. R. M. (1992). Stability of andalusite and sillimanite and the Al_2SiO_5 triple point: constraints from the Ballachulish aureole, Scotland. *The Journal of Geology*, **100**(4), 423–46.

Pattison, D. R. M. (2003). Petrogenetic significance of orthopyroxene-free garnet+ clinopyroxene+ plagioclase±quartz-bearing metabasites with respect to the amphibolite and granulite facies. *Journal of Metamorphic Geology*, **21**(1), 21–34.

Pattison, D. R. M. & DeBuhr, C. L. (2015). Petrology of metapelites in the Bugaboo aureole, British Columbia, Canada. *Journal of Metamorphic Geology*, **33**(5), 437–62.

Penniston-Dorland, S. C. & Ferry, J. M. (2006). Development of spatial variations in reaction progress during regional metamorphism of micaceous carbonate rocks, northern New England. *American Journal of Science*, **306**(7), 475–524.

Perrillat, J. P., Daniel, I., Lardeaux, J. M. & Cardon, H. (2003). Kinetics of the coesite–quartz transition: application to the exhumation of ultrahigh-pressure rocks. *Journal of Petrology*, **44**(4), 773–88.

Poli, S. & Schmidt, M. W. (2002). Petrology of subducted slabs. *Annual Review of Earth and Planetary Sciences*, **30**(1), 207–35.

Pollington, A. D. & Baxter, E. F. (2010). High resolution Sm–Nd garnet geochronology reveals the uneven pace of tectonometamorphic processes. *Earth and Planetary Science Letters*, **293**(1–2), 63–71.

Powell, R. & Holland, T. J. B. (2008). On thermobarometry. *Journal of Metamorphic Geology*, **26**(2), 155–79.

Powell, R., Holland, T. & Worley, B. (1998). Calculating phase diagrams involving solid solutions via non-linear equations, with examples using THERMOCALC. *Journal of Metamorphic Geology*, **16**(4), 577–88.

Pownall, J. M., Hall, R., Armstrong, R. A. & Forster, M. A. (2014). Earth's youngest known ultrahigh-temperature granulites discovered on Seram, eastern Indonesia. *Geology*, **42**(4), 279–82.

Priestley, K., Jackson, J. & McKenzie, D. (2008). Lithospheric structure and deep earthquakes beneath India, the Himalaya and southern Tibet. *Geophysical Journal International*, **172**(1), 345–62.

Ravna, E. K. (2000). Distribution of Fe^{2+} and Mg between coexisting garnet and hornblende in synthetic and natural systems: an empirical calibration of the garnet–hornblende Fe–Mg geothermometer. *Lithos*, **53**(3–4), 265–77.

Regis, D., Warren, C. J. , Young, D. & Roberts, N. M. W. (2014). Tectono-metamorphic evolution of the Jomolhari massif: Variations in timing of syn-collisional metamorphism across western Bhutan. *Lithos*, **190**, 449–66.

Regis, D., Warren, C. J., Mottram, C. M. & Roberts, N. M. W. (2016). Using monazite and zircon petrochronology to constrain the P–T–t evolution of the middle crust in the Bhutan Himalaya. *Journal of Metamorphic Geology*, **34**(6), 617–39.

Rice, J. M. (1977). Progressive metamorphism of impure dolomitic limestone in the Marysville aureole, Montana. *American Journal of Science*, **277**(1), 1–24.

Richards, J. P. (2011). Magmatic to hydrothermal metal fluxes in convergent and collided margins. *Ore Geology Reviews*, **40**(1), 1–26.

Rieder, M., Cavazzini, G., D'yakonov, Y. S. *et al.* (1998). Nomenclature of the Micas. *Clays and Clay Minerals*, **46**(5), 586–95.

Rimstidt, J. D. (1997). Quartz solubility at low temperatures. *Geochimica et Cosmochimica Acta*, **61**(13), 2553–8.

Roselle, G. T., Baumgartner, L. P. & Chapman, J. A. (1997). Nucleation-dominated crystallization of forsterite in the Ubehebe Peak contact aureole, California. *Geology*, **25**(9), 823–6.

Rosenberg, C. L. & Handy, M. R. (2005). Experimental deformation of partially melted granite revisited: implications for the continental crust. *Journal of Metamorphic Geology*, **23**(1), 19–28.

Rubatto, D., Gebauer, D. & Compagnoni, R. (1999). Dating of eclogite-facies zircons: the age of Alpine metamorphism in the Sesia–Lanzo Zone (Western Alps). *Earth and Planetary Science Letters*, **167**(3–4), 141–58.

Rubatto, D., Gebauer, D. & Fanning, M. (1998). Jurassic formation and Eocene subduction of the Zermatt–Saas-Fee ophiolites: implications for the geodynamic evolution of the Central and Western Alps. *Contributions to Mineralogy and Petrology*, **132**(3), 269–87.

Rubatto, D. & Hermann, J. (2001). Exhumation as fast as subduction? *Geology*, **29**(1), 3–6.

Rubatto, D. & Hermann, J. (2007). Zircon behaviour in deeply subducted rocks. *Elements*, **3**(1), 31–5.

Rubatto, D., Williams, I. S. & Buick, I. S. (2001). Zircon and monazite response to prograde metamorphism in the Reynolds Range, central Australia. *Contributions to Mineralogy and Petrology*, **140**(4), 458–68.

Rubie, D. C. (1983). Reaction-enhanced ductility: The role of solid-solid univariant reactions in deformation of the crust and mantle. *Tectonophysics*, **96**(3–4), 331–52.

Ruiz-Agudo, E., Putnis, C. V. & Putnis, A. (2014). Coupled dissolution and precipitation at mineral–fluid interfaces. *Chemical Geology*, **383**, 132–46.

Sachan, H. K., Mukherjee, B. K., Ogasawara, Y. *et al.* (2004). Discovery of coesite from Indus Suture Zone (ISZ), Ladakh, India: evidence for deep subduction. *European Journal of Mineralogy*, **16**(2), 235–40.

Schiffman, P. & Liou, J. G. (1983). Synthesis of Fe-pumpellyite and its stability relations with epidote. *Journal of Metamorphic Geology*, **1**(2), 91–101.

Schliestedt, M. & Matthews, A. (1987). Transformation of blueschist to greenschist facies rocks as a consequence of fluid infiltration, Sifnos (Cyclades), Greece. *Contributions to Mineralogy and Petrology*, **97**(2), 237–50.

Schmidt, M. W. & Poli, S. (1994). The stability of lawsonite and zoisite at high pressures: Experiments in CASH to 92 kbar and implications for the presence of hydrous phases in subducted lithosphere. *Earth and Planetary Science Letters*, **124**(1–4), 105–18.

Schulz, B., Triboulet, C., Audren, C., Pfeifer, H.-R. & Gilg, A. (2001). Two-stage prograde and retrograde Variscan metamorphism of glaucophane-eclogites, blueschists and greenschists from Ile de Groix (Brittany, France). *International Journal of Earth Sciences*, **90**(4), 871–89.

Schumacher, J. C. (2007). Metamorphic amphiboles: composition and coexistence. *Reviews in Mineralogy and Geochemistry*, **67**, 359–416.

Scoates, J. S. & Friedman, R. M. (2008). Precise age of the platiniferous Merensky Reef, Bushveld Complex, South Africa, by the U–Pb zircon chemical abrasion ID-TIMS technique. *Economic Geology*, **103**(3), 465–71.

Selverstone, J., Morteani, G. & Staude, J. (1991). Fluid channelling during ductile shearing: transformation of granodiorite into aluminous schist in the Tauern Window, Eastern Alps. *Journal of Metamorphic Geology*, **9**(4), 419–31.

Sherlock, S. C., Kelley, S. P., Zalasiewicz, J. A. *et al.* (2003). Precise dating of low-temperature deformation: Strain-fringe analysis by $^{40}Ar–^{39}Ar$ laser microprobe. *Geology*, **31**(3), 219–22.

Silver, E. A. & Beutner, E. C. (1980). Melanges. *Geology*, **8**(1), 32–4.

Skippen, G. (1974). An experimental model for low pressure metamorphism of siliceous dolomitic marble. *American Journal of Science*, **274**(5), 487–509.

Slaughter, J., Kerrick, D. M. & Wall, V. J. (1975). Experimental and thermodynamic study of equilibria in the system $CaO–MgO–SiO_2–H_2O–CO_2$. *American Journal of Science*, **275**(2), 143–62.

Smit, M. A., Scherer, E. E. & Mezger, K. (2013). Lu–Hf and Sm–Nd garnet geochronology: Chronometric closure and implications for dating petrological processes. *Earth and Planetary Science Letters*, **381**, 222–33.

Smith, D. C. (1984). Coesite in clinopyroxene in the Caledonides and its implications for geodynamics. *Nature*, **310**(5979), 641–4.

Sobolev, N. V. & Shatsky, V. S. (1990). Diamond inclusions in garnets from metamorphic rocks: a new environment for diamond formation. *Nature*, **343**(6260), 742–6.

Spear, F. S. (1981). An experimental study of hornblende stability and compositional variability in amphibolite. *American Journal of Science*, **281**(6), 697–734.

Spear, F. S. (1993). Metamorphic phase equilibria and pressure–temperature–time paths. *MineralogicalSociety of America Monographs*, **1**, 799.

Spear, F. S. (2017). Garnet growth after overstepping. *Chemical Geology*, **466**, 491–9.

Spear, F. S. & Cheney, J. T. (1989). A petrogenetic grid for pelitic schists in the system $SiO_2–Al_2O_3–FeO–MgO–K_2O–H_2O$. *Contributions to Mineralogy and Petrology*, **101**(2), 149–64.

Spear, F. S. & Pattison, D. R. M. (2017). The implications of overstepping for metamorphic assemblage diagrams (MADs). *Chemical Geology*, **457**, 38–46.

Spear, F. S., Selverstone, J., Hickmott, D., Crowley, P. & Hodges, K. V. (1984). PT paths from garnet zoning: A new technique for deciphering tectonic processes in crystalline terranes. *Geology*, **12**(2), 87–90.

St-Onge, M. R. (1987). Zoned poikiloblastic garnets: PT paths and syn-metamorphic uplift through 30 km of structural depth, Wopmay Orogen, Canada. *Journal of Petrology*, **28**(1), 1–21.

Staudigel, H., Plank, T., White, B. & Schmincke, H. (1996). Geochemical fluxes during seafloor alteration of the basaltic upper oceanic crust: DSDP Sites 417 and 418. *Subduction: Top to Bottom*, eds. G. E. Bebout, D. W. Scholl, S. H. Kirby and J. P. Platt, AGU Geophysical Monograph Series **96**, 19–38.

Tagami, T. & O'Sullivan, P. B. (2005). Fundamentals of fission-track thermochronology. *Reviews in Mineralogy and Geochemistry*, **58**(1), 19–47.

Taylor, R. J. M., Clark, C., Fitzsimons, I. C. W. *et al.* (2014). Post-peak, fluid-mediated modification of granulite facies zircon and monazite in the Trivandrum Block, southern India. *Contributions to Mineralogy and Petrology*, **168**(2), 1044.

Taylor, R. J. M., Kirkland, C. L. & Clark, C. (2016). Accessories after the facts: Constraining the timing, duration and conditions of high-temperature metamorphic processes. *Lithos*, **264**, 239–57.

Thompson, A. B. (1982). Dehydration melting of pelitic rocks and the generation of H_2O-undersaturated granitic liquids. *American Journal of Science*, **282**(10), 1567–95.

Tilley, C. E. (1925). Metamorphic zones in the southern Highlands of Scotland. *Quarterly Journal of the Geological Society*, **81**, 100–12.

Todd, C. S. & Engi, M. (1997). Metamorphic field gradients in the Central Alps. *Journal of Metamorphic Geology*, **15** (4), 513–30.

Tómasson, J. & Kristmannsdóttir, H. (1972). High temperature alteration minerals and thermal brines, Reykjanes, Iceland. *Contributions to Mineralogy and Petrology*, **36**(2), 123–34.

Touret, J. L. R. R. (2001). Fluids in metamorphic rocks. *Lithos*, **55**(1–4), 1–25.

Trommsdorff, V. (1966). Progressive metamorphose kiesliger Karbonatgesteine in den Zentralalpen zwischen Bernina und Simplon.

Schweizerische Mineralogische und Petrographische Mitteilungen, **46**, 431–60.

Trommsdorff, V. (1972). Change in TX during metamorphism of siliceous dolomitic rocks of the Central Alps. *Schweizerische Mineralogische und Petrographische Mitteilungen*, **52**(3), 567–71.

Trommsdorff, V. & Connolly, J. A. D. (1996). The ultramafic contact aureole about the Bregaglia (Bergell) tonalite: isograds and a thermal model. *Schweizerische Mineralogische Und Petrographische Mitteilungen*, **76**, 537–47.

Trommsdorff, V. & Evans, B. W. (1972). Progressive metamorphism of antigorite schist in the Bergell tonalite aureole (Italy). *American Journal of Science*, **272**(5), 423–37.

Trommsdorff, V., Skippen, G. & Ulmer, P. (1985). Halite and sylvite as solid inclusions in high-grade metamorphic rocks. *Contributions to Mineralogy and Petrology*, **89**(1), 24–9.

Tsujimori, T. & Ernst, W. G. (2014). Lawsonite blueschists and lawsonite eclogites as proxies for palaeo-subduction zone processes: A review. *Journal of Metamorphic Geology*, **32**(5), 437–54.

van Keken, P. E., Currie, C., King, S. D., *et al.* (2008). A community benchmark for subduction zone modeling. *Physics of the Earth and Planetary Interiors*, **171**(1), 187–97.

Vidal, O., Parra, T. & Trotet, F. (2001). A thermodynamic model for Fe-Mg aluminous chlorite using data from phase equilibrium experiments and natural pelitic assemblages in the 100 to 600 °C, 1 to 25 kb range. *American Journal of Science*, **301**(6), 557–92.

Villa, I. M. (2016). Diffusion in mineral geochronometers: Present and absent. *Chemical Geology*, **420**, 1–10.

Volodichev, O. I., Slabunov, A. I., Bibikova, E. V, Konilov, A. N. & Kuzenko, T. I. (2004). Archean eclogites in the Belomorian mobile belt, Baltic Shield. *Petrology*, **12**(6), 540–60.

Wakabayashi, J., Moores, E. M., Sloan, D. & Stout, D. L. (1999). Subduction and the rock record: Concepts developed in the Franciscan Complex, California. *Geological Society of America Special Paper*, **338**, 123–33.

Walsh, E. O. & Hacker, B. R. (2004). The fate of subducted continental margins: Two-stage exhumation of the high-pressure to ultrahigh-pressure Western Gneiss Region, Norway. *Journal of Metamorphic Geology*, **22**(7), 671–87.

Wang, C. Y, Chen, W.-P. P. & Wang, L.-P. P. (2013). Temperature beneath Tibet. *Earth and Planetary Science Letters*, **375**, 326–37.

Wark, D. A. & Watson, E. B. (2006). TitaniQ: a titanium-in-quartz geothermometer. *Contributions to Mineralogy and Petrology*, **152**(6), 743–54.

Warr, L. N. (2018). A new collection of clay mineral 'Crystallinity' Index Standards and revised guidelines for the calibration of Kübler and Árkai indices. *Clay Minerals*, **53**(3), 339–50.

Warr, L. N. & Cox, S. C. (2016). Correlating illite (Kübler) and chlorite (Árkai) "crystallinity" indices with metamorphic mineral zones of the South Island, New Zealand. *Applied Clay Science*, **134**, 164–74.

Warren, C. J., Beaumont, C. & Jamieson, R. A. (2008). Deep subduction and rapid exhumation: Role of crustal strength and strain weakening in continental subduction and ultrahigh-pressure rock exhumation. *Tectonics*, **27**(6), doi:10.1029/2008TC002292.

Wassmann, S. & Stöckhert, B. (2013). Rheology of the plate interface – dissolution precipitation creep in high pressure metamorphic rocks. *Tectonophysics*, **608**, 1–29.

Waters, D. J. & Lovegrove, D. P. (2002). Assessing the extent of disequilibrium and overstepping of prograde metamorphic reactions in metapelites from the Bushveld Complex aureole, South Africa. *Journal of Metamorphic Geology*, **20**(1), 135–49.

Watson, E. B. & Baxter, E. F. (2007). Diffusion in solid-Earth systems. *Earth and Planetary Science Letters*, **253**(3–4), 307–27.

Wei, C. J. & Clarke, G. L. (2011). Calculated phase equilibria for MORB compositions: a reappraisal of the metamorphic evolution of lawsonite eclogite. *Journal of Metamorphic Geology*, **29**(9), 939–52.

Wei, C. J., Powell, R. & Clarke, G. L. (2004). Calculated phase equilibria for low- and medium-pressure metapelites in the KFMASH and KMnFMASH systems. *Journal of Metamorphic Geology*, **22**(5), 495–508.

Wei, C. & Powell, R. (2003). Phase relations in high-pressure metapelites in the system KFMASH (K_2O–FeO–MgO–Al_2O_3–SiO_2–H_2O) with application to natural rocks. *Contributions to Mineralogy and Petrology*, **145**(3), 301–15.

Wei, C. & Powell, R. (2006). Calculated phase relations in the system NCKFMASH (Na_2O–CaO–K_2O–FeO–MgO–Al_2O_3–SiO_2–H_2O) for high-pressure metapelites. *Journal of Petrology*, **47**(2), 385–408.

Weinberg, R. F. & Hasalová, P. (2015). Water-fluxed melting of the continental crust: A review. *Lithos*, **212**, 158–88.

Weller, O. M. & St-Onge, M. R. (2017). Record of modern-style plate tectonics in the Palaeoproterozoic Trans-Hudson orogen. *Nature Geoscience*, **10**(4), 305–11.

Weller, O. M., St-Onge, M. R., Waters, D. J. et al. (2013). Quantifying Barrovian metamorphism in the Danba structural culmination of eastern Tibet. *Journal of Metamorphic Geology*, **31**(9), 909–35.

Wheeler, J. (1992). Importance of pressure solution and Coble creep in the deformation of polymineralic rocks. *Journal of Geophysical Research: Solid Earth*, **97**(B4), 4579–86.

Whitney, D. L. & Evans, B. W. (2010). Abbreviations for names of rock-forming minerals. *American Mineralogist*, **95**, 185–7.

Whitney, D. L., Teyssier, C., Rey, P. & Buck, W. R. (2013). Continental and oceanic core complexes. *Geological Society of America Bulletin*, **125**(3–4), 273–98.

Wintsch, R. P., Christoffersen, R. & Kronenberg, A. K. (1995). Fluid-rock reaction weakening of fault zones. *Journal of Geophysical Research: Solid Earth*, **100**(B7), 13021–32.

Wirth, R. (2009). Focused Ion Beam (FIB) combined with SEM and TEM: Advanced analytical tools for studies of chemical composition, microstructure and crystal structure in geomaterials on a nanometre scale. *Chemical Geology*, **261**(3–4), 217–29.

Wood, B. J. & Banno, S. (1973). Garnet-orthopyroxene and orthopyroxene-clinopyroxene relationships in simple and complex systems. *Contributions to Mineralogy and Petrology*, **42**(2), 109–24.

Wright, T. O. & Platt, L. B. (1982). Pressure dissolution and cleavage in the Martinsburg Shale. *American Journal of Science*, **282**(2), 122–35.

Wu, C.-M. & Chen, H.-X. (2015). Revised Ti-in-biotite geothermometer for ilmenite- or rutile-bearing crustal metapelites. *Science Bulletin*, **60**(1), 116–21.

Wu, C.-M. & Cheng, B.-H. (2006). Valid garnet–biotite (GB) geothermometry and garnet–aluminum silicate–plagioclase–quartz (GASP) geobarometry in metapelitic rocks. *Lithos*, **89**(1–2), 1–23.

Wu, C.-M. & Zhao, G. (2006). Recalibration of the garnet–muscovite (GM) geothermometer and the garnet–muscovite–plagioclase–quartz (GMPQ) geobarometer for metapelitic assemblages. *Journal of Petrology*, **47**(12), 2357–68.

Wu, C.-M. (2015). Revised empirical garnet-biotite-muscovite-plagioclase geobarometer in metapelites. *Journal of Metamorphic Geology*, **33**(2), 167–76.

Yardley, B. W. D. (2009). The role of water in the evolution of the continental crust. *Journal of the Geological Society*, **166**(4), 585–600.

Yardley, B. W. D. & Barber, J. P. (1991). Melting reactions in the Connemara Schists: The role of water infiltration in the formation of amphibolite facies migmatites. *American Mineralogist*, **76**(5–6), 848–56.

Yardley, B. W. D., Barber, J. P. & Gray, J. R. (1987). The metamorphism of the Dalradian rocks of western Ireland and its relation to tectonic setting. *Philosophical Transactions of the Royal Society of London. Series A*, **321**, 243–70.

Yardley, B. W. D., Leake, B. E. & Farrow, C. M. (1980). The metamorphism of Fe-rich pelites from Connemara, Ireland. *Journal of Petrology*, **21**(2), 365–99.

Yardley, B. W. D. & Lloyd, G. E. (1989). An application of cathodoluminescence microscopy to the study of textures and reactions in high-grade marbles from Connemara, Ireland. *Geological Magazine*, **126**(4), 333–7.

Yardley, B. W. D. & Valley, J. W. (1997). The petrologic case for a dry lower crust. *Journal of Geophysical Research*, **102**(B6), 12173–85.

Zen, E.-A. (1966). Construction of Pressure-Temperature Diagrams for Multicomponent Systems After the Method of Schreinemakers: A Geometric Approach, *US Geological Survey Bulletin*, 1225.

Index

Page numbers in italics indicate figures.